THE CIVIL CONTINGENCIES ACT 2004

Risk, Resilience, and the Law in the United Kingdom

GW00546538

THE CIVIL CONTINGENCIES ACT 2004

Risk, Resilience, and the Law in the United Kingdom

CLIVE WALKER & JIM BRODERICK

OXFORD

UNIVERSITY PRESS

This book has been printed digitally and produced in a standard specification
in order to ensure its continuing availability

OXFORD
UNIVERSITY PRESS

Great Clarendon Street, Oxford OX2 6DP

Oxford University Press is a department of the University of Oxford.
It furthers the University's objective of excellence in research, scholarship,
and education by publishing worldwide in

Oxford New York

Auckland Cape Town Dar es Salaam Hong Kong Karachi
Kuala Lumpur Madrid Melbourne Mexico City Nairobi
New Delhi Shanghai Taipei Toronto
With offices in
Argentina Austria Brazil Chile Czech Republic France Greece
Guatemala Hungary Italy Japan South Korea Poland Portugal
Singapore Switzerland Thailand Turkey Ukraine Vietnam

Oxford is a registered trade mark of Oxford University Press
in the UK and in certain other countries

Published in the United States
by Oxford University Press Inc., New York

© Clive Walker and James Broderick 2006

ISBN 978-0-19-929626-2

CONTENTS—SUMMARY

CONTENTS

PREFACE

Salus populi suprema est lex
'The safety of the people is the highest law'
attr. Marcus Tullius Cicero—106 to 43 BC

Fiat justitia, ruat caelum
'Though the heavens fall, let justice be done'
Lord Mansfield in *R v Wilkes* (1770) 98 ER 327 at 347

The contrasting attitudes represented above lie at the heart of many debates on how societies should protect themselves against risk, threat, and attack. The United Kingdom Government's conclusion to its own deliberations on the matter was announced in the Queen's Speech on the opening of Parliament on 26 November 2003:

> The threat of international terrorism and a changing climate have led to a series of emergencies and heightened concerns for the future. My Government will introduce a Bill creating a long-term foundation for civil contingencies capable of meeting these challenges at national and local level.[1]

That proposal had been driven by various pressures and concerns—disasters[2] such as floods[3] and droughts, Foot and Mouth disease, and other potential health calamities, as well as manufactured crises such as climate change, widespread civil unrest (including the fuel protests in 2000), and terrorism on the cataclysmic scale witnessed on September 11, 2001. An important stage in policy formation was marked by the *Consultation Paper on a Draft Civil Contingencies Bill* issued by the Cabinet Office.[4] That paper was debated by several parliamentary committees, primarily by a Joint Committee of both Houses. Eventually, there emerged the Civil Contingencies Act 2004, which marks an important recognition of the increasingly important discourses of risk and resilience within society (outlined in Chapter 1).

---

[1] HC Debs vol 415 col 5.

[2] From 1995 to 2004, 5,989 reported disasters killed 901,177 people and affected over 2.5 billion people; from 1985 to 1994, there were 643,418 reported killed and 1.74 billion affected: Centre for Research on the Epidemiology of Disasters, http://www.cred.be/. See also International Federation of Red Cross and Red Crescent Societies, World Disaster Report 2004 (http://www.ifrc.org/publicat/wdr/, 2005).

[3] These have increased in Europe over the period 1985 to 2004: http://www.em-dat.net/documents/working_paper/FloodDisastersinEurope85–04.pdf.

[4] Cm 5843, London 2003.

The Act does not, however, represent the genesis of the regulation of contingencies and emergencies. The United Kingdom has long entertained laws dealing with certain threats in certain situations (the background to the Act is provided in Chapter 2). Most prominent and controversial in recent decades has been legislation about terrorism, though other statutes have addressed debilitating industrial strikes, power and water supply shortages, and food chain dangers. But, especially after the events of 2000 and 2001, a more comprehensive approach to risk and resilience was warranted. Thus, the Act brings about a much more widespread and active requirement to engage in risk and resilience planning.

Legislation in this field can be expected to attract great attention and interest from various sectors—local authorities, police and security services, health and emergency services, civil liberties groups, as well as lawyers. In responding to those audiences, this book will offer a survey and analysis of the Civil Contingencies Act 2004, including the full text of the legislation (in the Appendix). After exploring the meanings of 'emergency' adopted by the statute (Chapter 3), the book will concentrate upon the contents of the Civil Contingencies Act 2004, its planning aspects in Part I (Chapter 4) and the reactive and potentially draconian emergency powers in Part II (Chapter 5). As well as describing the Act, there will be a critical stance, based not only on the political and organizational theories about risk and resilience but also on the tenets of constitutionalism, consideration of which will be evident in assessing the constitutional and human rights impacts of the Act (Chapters 6 and 7).

There will follow explorations in Part III of the practical implications of the legislation, including management structures and resources, for the wide range of agencies and bodies which must take heed of it (Chapter 8).

Finally, the amount of counterpart legislation and organizational structures in foreign jurisdictions is vast. This book does not purport to offer a full comparative survey, but some examples are related in the final chapter (Chapter 9), followed by our final reflections as to the worth of the legislative scheme.

The Government's handling of risks and emergencies in recent years has failed to inspire public confidence. In a range of crises, from the Foot and Mouth outbreak through to the grounds for war in Iraq, official predictions or capabilities have been found wanting. The Civil Contingencies Act 2004 tenders reassurance by the promise of systemic planning and activity in civil resilience, though defence lies beyond its scope. The wide-ranging powers in the Act have the capability of delivering on the promise. But, as shall be revealed in this book, efforts will be hampered because the legislation is hesitant and uneven.

The authors thank a number of sources and correspondents. Chief amongst these are Ian Mackley and Andrew Kennon, clerks to the Joint Select Committee on which we both served in 2003–4, as well as Baron Lewis Moonie, its Chair, and its specialist, Jennifer Smookler. We thank in the Cabinet Office, Dan Greaves, Roger Hargreaves, and Bruce Mann, who provided comments on the text. As ever, we thank our academic colleagues for their support (including Adam Crawford and David Wall at Leeds). Our sponsors, the editors of Oxford University Press, Jane Kavanagh and Annabel Moss, have sustained us with commendable tolerance and efficiency.

Jim took prime responsibility for Chapter 1 and the section on practical guides for local planners in Chapter 8, Clive was primarily responsible for the remainder. However, the book nevertheless represents a collaborative exploration of civil contingencies, and both authors endorse the text as a whole. We have stated the law and materials available as at 1 October 2005.

Professor Clive Walker
Centre for Criminal Justice Studies
School of Law
University of Leeds
Leeds LS2 9JT
law6cw@leeds.ac.uk

Dr Jim Broderick
Institute of Lifelong Learning
University of Leicester
Leicester LE1 7RH
jpb11@le.ac.uk

TABLE OF CASES

EUROPEAN COURT OF JUSTICE

MALAYSIA

SOUTH AFRICA

UNITED KINGDOM

UNITED STATES OF AMERICA

TABLE OF LEGISLATION

RISK, REGULATION, AND RESILIENCE IN EMERGENCY

1

RISK AND RISK MANAGEMENT THEORY

A. Introduction

This part of the book is designed to elicit the broad themes by which to **1.01** understand the nature of emergencies and risk as well as responsive strategies and operational plans. The emergence of the Civil Contingencies Act 2004 should not be divorced from its context in terms of wider societal developments. The planning function in Part I of the Civil Contingencies Act and the emergency powers in Part II are influenced by a number of theoretical approaches and concerns associated with contemporary social scientific analyses of the concept of risk, in particular, the management of diffuse and complex risks that shape or even threaten the notion of 'modernity' itself.[1] This chapter will first survey a number of theoretical discourses as well as meanings of risk and risk management which, we suggest, can provide a meaningful framework by which to comprehend and assess the objectives, structure, and content of the Civil Contingencies Act. Equally of interest here is the related notion of a 'regulatory state' in which the role of government is increasingly that of risk regulator, rather than guarantor and provider of security and welfare.

This survey is undertaken with some caveats in mind. First, the study of risk is **1.02** not new. The concept of risk-taking, it can be argued, is an inherent component of what it means to be human. The belief in 'Acts of God' (an external circumstance triggered by ungovernable natural causes and not an expression of

---

[1] See Beck, U, *Risk Society* (Sage, London, 1992), Beck, U, Giddens, A, and Lash, S, *Reflexive Modernization* (Polity Press, Cambridge, 1994).

socially constructed vulnerabilities or uncertainties)[2] represent pre-modern social attempts to understand catastrophes and hazards. However, 'risk' and 'risk management' in their modern guises have been most closely associated with scientific or quantitative techniques of defining and measuring external events. Exemplars would be the empirical science framework that shapes the study of engineering in all its forms and the actuarial techniques which underpin insurance industry practices. Such assumptions underline the definition of 'risk' adopted by the BSE Inquiry:

> A risk is not the same as a hazard. A hazard is an intrinsic propensity to cause harm. Natural phenomena, physical substances, human activities will be hazardous if they have an intrinsic propensity to cause harm. A risk is the likelihood that a hazard will result in harm. A risk can usually be evaluated once the nature of the hazard and the degree of exposure to it are identified. Risk evaluation involves considering both the likelihood that a hazard will cause harm and the severity of the harm that is threatened.[3]

When formulating the Civil Contingencies Act, the Government adopted a broadly similar perspective on the definitions of risk and risk management:

> Risk refers to uncertainty of outcome, whether positive opportunity or negative threat, of actions and events. . . . Risk management covers all the processes involved in identifying, assessing and judging risks, taking actions to mitigate or anticipate them, and monitoring and reviewing progress. Effective risk management requires processes in place to monitor risks; access to reliable, up-to-date information about risk; the right balance of control in place to deal with those risks; and decision-making processes supported by a framework of risk analysis and evaluation.[4]

1.03 Within the primacy of such rational and quantitative approaches, the study of risk encompasses an enormous range of (often competing) definitions and perspectives. This complexity in part derives from the compartmentalization of hazards into specific, technical spheres each with its own experts,[5] language, and analytical methods. It is also influenced by the upsurge in interest in the management of risk as a systems problem from the 1950s onwards and in the development of wider social scientific debate since the 1980s, in particular those social and psychological studies of risk that emphasize cultural and cognitive factors as key sources of causation in the perception, analysis, and management of risk. Such studies have sought to undermine the 'objectivity' of risk analysis and evaluation by exploring the underlying value-judgements that affect risk discourse. Some authors even conclude that that there is no such thing as real or 'objective' risk, merely differing cultural or psychological approaches and value-systems that determine how the

---

[2] See Gilbert, C, 'Studying Disaster' in Quarantelli, EL (ed), *What Is a Disaster?* (Routledge, London, 1998) 11. For non-sociological approaches, see Alexander, D, *Natural Disasters* (UCL Press, London, 1993) 13–14.

[3] BSE Inquiry, *Report* (1999–00 HC 887) vol 1 ch 2, para 162.

[4] Cabinet Office, *The Draft Civil Contingencies Bill*, para 2.4.

[5] Some argue that the growing importance of risk is being driven through the efforts of new risk elites (inspectors and auditors): Power, M., *The Audit Society* (Oxford University Press, Oxford, 1997).

concept is constructed and understood.[6] This enhanced interest in risk and risk management over recent decades has produced a diverse and highly differentiated field of study which is characterized by fierce debate, disagreement, or outright disjuncture of approach within and across the social and physical sciences. Indeed, for Hood and Jones, the 'field' of risk should rather be considered an 'archipelago' of distinct, even disconnected, perspectives.[7]

Nevertheless, such analyses, particularly those that relate to cultural and organi- **1.04** zational factors, are especially apposite to the Civil Contingencies Act because they address two important problems for the role of the state in facing and managing emergencies. First, how should governments respond to serious adverse events which, by definition, are diffuse, novel, and unforeseen? Secondly, in dealing with the unknown, how can governments best prepare themselves and the wider public? That both concerns are fundamental to the Act can be seen from the fact that the legislation itself is divided into two discrete sections which directly correspond to these queries.

The multidisciplinary and fragmented nature of academic approaches to risk **1.05** creates difficulties for those wishing to navigate the treacherous waters around the archipelago. However, this plurality of perspectives also yields some interesting and useful insights that inform this study of civil contingencies legislation. In the first instance, the work of organizational and cultural theorists reveals that risk analysis and management is not a 'politics-free zone'. Further, the search for a 'grand social narrative' of risk which is a key element (and for many a criticism) of the work of Ulrich Beck and Anthony Giddens, both problematises and locates risk discourse at the centre of wider debate relating to the impact of late modernity on social organization. Their core ideas will now be summarized.

B. The Risk Society

(a) Overview

Beck and Giddens explore what they depict as a new phase of modernity, one **1.06** which is defined by a profound social transformation from industrial to risk society, or from 'simple' to 'reflexive' modernity. It is the very advancement of science and technology itself which is generating and multiplying new and unforeseen threats. Pre-industrial catastrophes and hazards, no matter their scale

---

[6] See Slovic, P, 'Perception of Risk: Reflections on the Psychometric Paradigm', in Krimsky, S and Golding, D (eds), *Social Theories of Risk* (Praeger, New York, 1992); Douglas, M and Wildavsky, A, *Risk and Culture: An Essay on the Selection of Technical and Environmental Dangers* (University of California Press, Berkeley, 1982); Thompson, M, Ellis, R, and Wildavsky, A, *Cultural Theory* (Westview Press, Boulder Colo, 1990).

[7] Hood, C and Jones, DKC (eds), *Accident and Design* (UCL Press, London, 1996).

and devastation, were essentially 'external' in nature. They happened to the human race and were attributed either to the forces of nature or 'acts of God'. But industrial 'risk' in its 'modern' sense has its origin in 'techno-economic' decision-making. This realm is not the domain of any external forces nor even individual human agency, rather it is the concern of entire organizations and political groupings. Therefore, for Beck, 'it is not the number of dead and wounded, but rather a social feature, their industrial self-generation, which makes the hazards of mega-technology a political issue.'[8]

1.07 Three arguments underpin the theory of risk society. First, risk society begins where nature ends. For Giddens and Beck, this means that the focus of our anxiety has switched away from what nature can do to us towards what we can do to nature. Certainly, for Beck such anxiety is a deeply politicizing aspect of the risk society debate not only for environmentalist but also for global capital. Key here is a central paradox of the risk society: 'these internal risks are generated by the processes of modernization which try to control them.'[9] Secondly, risk society represents the end of traditional certainties: 'To put it crudely, simple modernization runs within the framework of categories and principles of industrial society... however we are concerned with a phase of social transformation in which, by dint of its own dynamics modernization changes its shape within industrial society.'[10] The dynamics of modernization, therefore, are viewed as undermining modernization itself. In particular, the calculus of risk, which (seemingly) offers a mechanism for rendering the unforeseen foreseeable, is undermined. Thirdly, the theory of risk society connects with the process of individualization mapped out by Durkheim and Weber.[11] However, in Beck's view, human beings today are not being released *into* industrial society but *from* it into world risk society where we must all contend with myriad, contradictory, incalculable risks that permeate all levels, from the global to the personal.[12]

1.08 The transformed situation challenges or even cancels established safety bureaucracies and destabilizes the social pillars—of limitation, accountability, compensation, and precautionary aftercare—that underpin the very notion of the 'provident' or 'providing' state. The idea that the welfare state can take care of its citizens 'from cradle to grave'[13] and in increasing aspects of their

---

[8] Beck, U, 'From Industrial Society to Risk Society: Questions of Survival, Social Structure and Ecological Enlightenment' (1992) 9 *Theory Culture and Society* 98.

[9] Beck, U, 'Politics of Risk Society' in Franklin, J, (ed), *The Politics of Risk Society* (Polity Press, Cambridge, 1998) 10.

[10] Beck, U, 'From Industrial Society to Risk Society: Questions of Survival, Social Structure and Ecological Enlightenment' (1992) 9 *Theory Culture and Society* 98 at 121.

[11] Beck, U and Beck-Gernsheim, E, *Individualization* (Sage, London, 2001).

[12] Beck, U, 'Risk Society and the Provident State' in Lash, S, Szersynski, B, and Wynne, B, (eds), *Risk, Environment and Modernity* (Sage, London, 1996) 29.

[13] See *Social Insurance and Allied Services* (Cmd 6404, HMSO, London, 1942); Beveridge, W, *Full Employment in a Free Society* (George Allen & Unwin, London, 1944).

lives[14] is no longer tenable, giving rise to a profound debate regarding the appropriate function or role of the state in relation to the individual in times of uncertainty. The 'risk society' is therefore 'an epoch in which the dark sides of progress increasingly come to dominate social debate'.[15] New risks result in novel understandings of risk and danger, for example, 'reflexive criminology' and 'actuarial justice'.[16] They also require new responses; as Beck argues, 'risks presume industrial, that is, techno-economic decisions and considerations of utility.'[17]

Beck makes it clear that his analysis takes as a point of departure the work of **1.09** Francois Ewold regarding the development of the 'provident' state.[18] From this perspective, pre-industrial hazards or threats that are characterized as being essentially incalculable became transformed, through a process of rationalization, into calculable risk in a variety of areas. As a result, the state has responded by promoting multiple and diverse forms of insurance against such calculable risks.[19] With the evolution of the modern industrial state, those forms of calculation have been increasingly extended into different spheres of activity and social life to the extent that society as a whole becomes viewed as a risk group with the state acting as the ultimate insurer against risk. This perspective thereby allows an exploration of the social, political, and institutional development of the modern state, as evidenced by the massive expansion of its bureaucracy and areas of 'legitimate' state intervention. In this sense, the concept of the 'provident' state underpins the various models of what we can term the 'welfare' state in Europe, the United States, and other parts of the developed world. This conjunction can be clearly seen in relation to the United Kingdom. According to Giddens:

> The welfare state became the left's project in the post-1945 period—it became seen above all as a means of achieving social justice and income redistribution. By and large, however, it did not originate as such. It developed as a security state, a way of protecting against risk, where collective rather than private insurance was necessary.[20]

Of course the evolution of the welfare state pre-dates the post-1945 period.[21] Yet, **1.10** Beck's main focus is less on the rise of the welfare state than with its continuing

---

[14] See Luhmann, N, *Political Theory in the Welfare State* (Walter de Gruyter, Berlin, 1990) 21–3.

[15] Beck, U, *Ecological Enlightenment* (Humanities Press, New Jersey, 1995) 2.

[16] Nelken, D, 'Reflexive Criminology?', and Feeley, M and Simon, J, 'Actuarial Justice: The Emerging New Criminal Law', in Nelken, D, (ed) *The Futures of Criminology* (Sage, London, 1994).   [17] Beck, U, *Risk Society* (Sage, London, 1992) 98.

[18] Ewald, F, *L'État Providence* (Editions Grasset, Paris, 1986).

[19] For a recent example, see Reinsurance (Acts of Terrorism) Act 1993.

[20] Giddens, A, 'Risk Society: The Context of British Politics' in Franklin, J, (ed), *The Politics of Risk Society* (Polity Press, Cambridge, 1998) 27.

[21] Early avocations include Thomas Paine's *Rights of Man* (1790) and John Ruskin's *Unto This Last* (1862), and the welfare model was legislated by the Old Age Pension Act 1908 and the National Insurance Act 1911.

operation at the point where the processes that have underpinned the growth of the modern industrial state themselves become problematized: 'The entry into risk society occurs at the moment when the hazards which are now decided and consequently produced by society undermine and/or cancel the established safety systems of the provident state's existing risk calculations.'[22]

1.11 For a number of analysts, Beck's emphasis on a very particular phase of evolution of the modern industrial state is a source of criticism. Beck is accused of projecting a culturally Eurocentric myth about nature and modernity that is born of a specific time and place.[23] It fails to engage with the position of less developed countries which are battling with immediate and major problems of extreme poverty, famine, war, and epidemic, afflictions that, it could be argued, correspond to conditions in Europe during the seventeenth and eighteenth centuries but from which it has largely emerged through the 'benefits' of industrial modernity. Certainly, Beck's later work has seen a shift in emphasis towards the problems of the politics of development and globalization, and Beck himself notes that his earlier works did presuppose a welfare state model much like that of Germany of the 1970s and 1980s.[24]

1.12 But to discount such arguments on the basis of the 'exceptionalism' of a late twentieth century German dispensation ignores the wider consensus—at least in the developed world—that the state's legitimate function has become merely the arbiter of collective provision against risk. Nevertheless, the provision of 'welfare' has not entirely died as a state concern. For example, in a speech in 2005 to the Institute of Public Policy Research (IPPR), Prime Minister Tony Blair reaffirmed a fundamental commitment not only to 'providentially' regulate against shared risk but also to provide shared services:

> Health and safety legislation is necessary to protect people at work. Food standards are necessary to protect people from harm. Protections are necessary to protect children from predatory adults. These are things against which, historically, the state has underwritten the risk. *The pooling of such risks is still the fundamental basis of our case for publicly funded services.*[25]

(b) The changing 'architecture' of risk distribution

1.13 According to Beck, as hazards become reflexive in the process of modernization and defy the attempts of governments and elites to control them, the response of

---

[22] Beck, U, 'Risk Society and the Provident State' in Lash, S, Szersynski, B, and Wynne, B, (eds), *Risk, Environment and Modernity* (Sage, London, 1996) 31.

[23] Adams, J, *Risk* (Routledge, London, 2001).

[24] Beck, U and Willms, J, *Conversations with Ulrich Beck* (Polity Press, Cambridge Press, 2004) 14.

[25] 'Common sense culture not compensation culture' (http://www.number-10.gov.uk/output/Page7562.asp., 26 May 2005 emphasis added).

bureaucratic institutions, rooted in essentially nineteenth century assumptions about the role of the state, has too often been one of denial.[26] According to Giddens: 'The crisis of the welfare state is not purely fiscal, it is a crisis of risk management in a society dominated by a new type of risk.'[27] The risk society thesis therefore seeks to interpret how these 'interconnected processes . . . are altering the epistemological and cultural status of science and the constitution of politics'[28] in a phase of modernity that is imbued with 'manufactured uncertainty'.[29]

In this new risk paradigm, the problem is not primarily one of risk avoidance or abolition,[30] rather it is how risks are distributed and managed. For Beck, 'it is about the architecture of risk definition in the face of growing competition between overlapping discourses of risk (such as nuclear power versus ozone hole).'[31] However, for authors such as Frank Furedi, persons are 'at risk' from dangers which are best avoided since there is autonomy from human attempts to allay them.[32] The future battle grounds for states will therefore not just be about repressive police powers but also about insurance, private security, and the preparedness to embrace risk through health and infrastructure agencies.[33] But this focus is for all, and so the politics of risk society transcends traditional state-centric forms of political decision-making and therefore requires a new political discourse and new organizational forms 'for science and business, science and the public sphere, science and politics, technology and law, and so forth.'[34] The challenge for established governments and legislatures is to retain relevance by giving some institutional form to the dialogue of world risk society that departs from the assumptions of simple modernity and moves to incorporate social groups and interests that currently fall outside the mainstream political—bureaucratic and institutional—domain.[35]

1.14

---

[26] Beck, U, 'Politics of Risk Society' in Franklin, J, (ed), *The Politics of Risk Society* (Polity Press, Cambridge, 1998) 18

[27] Giddens, A, 'Risk Society: The Context of British Politics' in Franklin, J, (ed), *The Politics of Risk Society* (Polity Press, Cambridge, 1998) 33.

[28] Beck, U, 'Politics of Risk Society' in Franklin, J, (ed), *The Politics of Risk Society* (Polity Press, Cambridge, 1998) 10.

[29] Ibid, 12. See also Giddens, A, *Beyond Left and Right* (Polity Press, Cambridge Press, 1994).

[30] But there is still a belief that risk may be reduced: Civil Contingencies Secretariat, *Full Regulatory* Impact, para 22.

[31] Beck, U, 'Risk Society and the Provident State' in Lash, S, Szersynski, B, and Wynne, B, (eds), *Risk, Environment and Modernity* (Sage, London, 1996) 36.

[32] Furedi, F, *Culture of Fear: Risk-taking and the Morality of Low Expectation* (rev ed, Continuum, London, 2002) 19.

[33] See Abraham, K, *Distributing Risk* (Yale University Press, New Haven, 1986); Baker, T and Simon, J, (eds), *Embracing Risk* (University of Chicago Press, Chicago, 2002).

[34] Beck, U, 'From Industrial Society to Risk Society: Questions of Survival, Social Structure and Ecological Enlightenment' (1992) 9 *Theory Culture and Society* 98 at 119. See also Beck, U, *The Reinvention of Politics* (Polity Press, Cambridge Press, 1997).

[35] Giddens, A, 'Risk Society: The Context of British Politics' in Franklin, J, (ed), *The Politics of Risk Society* (Polity Press, Cambridge, 1998) 33–4.

1.15 While it is not the intent here to provide a comprehensive (and inevitably speculative) survey of the contours of these 'new' social boundaries, some general observations can be made in relation to two issues that are exceptionally relevant for the Civil Contingencies Act—the impact of scientific knowledge and the lay understanding of it and the implications for the state as prime regulator of risk.

(c) Scientific rationality and lay knowledge

1.16 In Ulrich Beck's view, the advent of world risk society implies that the social basis of rationality—the assumptions which govern science, the law, economics, the private sector (particularly insurance), and politics—are undermined since society is 'placed under permanent pressure to negotiate foundations without a foundation.'[36] The process of modernization has affected the ability to make the 'unpredictable predictable' or the 'incalculable calculable'.[37] For Beck, the point is highlighted by contemporary scientific 'progress': 'In contrast to early industrial risks, nuclear, chemical, ecological and genetic engineering risks (a) can be limited in terms of neither time nor place, (b) are not accountable according to the established rules of causality, blame and liability, and (c) cannot be compensated or insured against.'[38] These developments therefore introduce a second risk phase, one which is imbued with 'manufactured uncertainty'[39] where the production of risks is the consequence of scientific or political attempts to control them.

1.17 The pace of scientific knowledge and technological development opens up new opportunities and may provide immediate solutions but simultaneously introduces yet more complexity and renders the world more unknowable and uncertain.[40] In this regard: 'It is the successes of science which sow the doubts'[41] since scientific knowledge is revealed to be inherently sceptical and mutable.[42] Beck argues that politicians can no longer rely on

---

[36] Beck, U, 'From Industrial Society to Risk Society: Questions of Survival, Social Structure and Ecological Enlightenment' (1992) 9 *Theory Culture and Society* 98 at 114.

[37] McGuigan, J, *Modernity and Postmodern Culture* (Open University, Cambridge, 1999) 124–6.

[38] Beck, U, 'Risk Society and the Provident State' in Lash, S, Szersynski, B, and Wynne, B, (eds), *Risk, Environment and Modernity* (Sage, London, 1996) 31. See also Ewald, F, 'Insurance and Risk' in Burchell, G, Gordon, C, and Miller, P (ed), *The Foucault Effect: Studies in Governmentality* (Harvester-Wheatsheaf, London, 1991). This is not literally true, even for nuclear installations: Nuclear Installations Act 1965.

[39] See Giddens, A, *Beyond Left and Right* (Polity Press, Cambridge Press, 1994).

[40] For Beck's purposes, uncertainty can be contrasted with risk—what is not objectively measurable risk can be counted as uncertainty: O'Malley, P, *Risk, Uncertainty and Government* (Glasshouse, London, 2004) 13.

[41] Beck, U, *Risk Society* (Sage, London, 1992) 106. See also Luhmann, N, *Risk* (Walter de Gruyter, New York, 1993).

[42] Giddens, A, 'Risk Society: The Context of British Politics' in Franklin, J, (ed), *The Politics of Risk Society* (Polity Press, Cambridge, 1998) 23–4. Of course, falsification has long been established as a prime scientific method.

a 'monopoly' of scientific expertise in the diagnosis of hazards because: 'first
...there are always competing claims and viewpoints from a variety of
actors...Secondly experts can only supply more or less uncertain factual information about probabilities...Thirdly, if politicians just implement scientific
advice they become caught in the mistakes...and uncertainties of scientific
knowledge.'[43] Lastly, for Beck, the distinction between scientific research, theory,
and technological enterprise has become eroded. In other words, the logic of
scientific discovery, which presupposes the testing of hypotheses before putting
them into practice, has been undermined. By way of illustration, Beck suggests
that theories of nuclear reactor safety are testable only after they have been built,
and, in similar vein, Beck points out that genetically modified crops must be
grown first before their behaviour and interaction with other living organisms can
be studied. This has serious implications for Beck as it reveals that the boundary
between the 'laboratory' and 'society' has been lost, and along with it, the con-
trollability of the laboratory environment. Moreover, given inherent uncertainty,
scientists themselves cannot predict what will happen before their research is
applied to (and thereby tested in) the wider social environment, and this short-
coming is now more fully appreciated by society.[44] For Anthony Giddens, here is
the 'defining characteristic of what Ulrich Beck calls risk society. A risk society is a
society where we increasingly live on a high technological frontier which absolutely no one completely understands and which generates a diversity of possible
futures.'[45] In those circumstances, it is difficult to discern which risks are worth
taking and those in error might have to pay in damages for legal negligence.[46]

Amongst the commentators on Beck's treatment of scientific knowledge, **1.18**
Szerszynski, Lash, and Wynne observe that the risk society thesis can be criticized 'for an overly realist account of the generation—via the growth of real risks
which are now universal and unmanageable—of a new cultural consciousness
which introduces modernity and its institutions to pervasive public scepticism,
or "self refutation".'[47] This difficulty has led to an emphasis being placed on
institutional process and, particularly, on scientific or 'expert' knowledge that
neglects (at least in Beck's earlier work) vital aspects of 'transformation processes' such as the 'sub-political', but nevertheless crucial, sphere of 'lay' or
'public' knowledge. Wynne argues that this focus goes further than mere

---

[43] Beck, U, 'Politics of Risk Society' in Franklin, J, (ed), *The Politics of Risk Society* (Polity Press, Cambridge, 1998) 14–15.

[44] See Beck, U, *Ecological Politics in an Age of Risk* (Polity Press, Cambridge Press, 1995); Douglas, M and Wildavsky, A, *Risk and Culture* (University of California Press, Berkeley, 1983).

[45] Giddens, A, 'Risk Society: The Context of British Politics' in Franklin, J, (ed), *The Politics of Risk Society* (Polity Press, Cambridge, 1998) 25.

[46] See Adams, J, *Risk* (UCL Press, London, 1995).

[47] Wynne, B, 'May the Sheep Safely Graze? A Reflexive View of the Expert-Lay Knowledge Divide', in Lash, S, Szersynski, B, and Wynne, B, (eds), *Risk, Environment and Modernity* (Sage, London, 1996) 44.

omission 'because it implicitly reproduces just those fundamental dichotomies which are key parts of the problem of modernity: natural knowledge versus "social" knowledge, nature versus society, expert versus lay knowledge.'[48] On these grounds, although Beck and Giddens identify an erosion of public trust in institutions as a symptom of reflexive modernity, they still explore this process in terms of a 'basic ethos [which] is too rooted in rational-choice neo-classical economistic models of human behaviour and response.'[49]

1.19 Wynne's criticism of the risk society thesis lies in challenging the strict division that is drawn between 'expert' and 'lay' knowledge on the grounds that it places too much emphasis on formal distinctions between propositional, formulaic, and hermeneutic truth claims and assumes that 'science only deals in propositional claims and commitments'.[50] For Szersynski *et al*, Wynne presents a sustained critique of risk society thesis. However, they also note that:

> ... rationalists such as Beck and Giddens who privilege (social) scientific over lay knowledge are not the only ones guilty of this flaw. So are more 'post'- or 'anti'-modernist writers ... who would seem to privilege the hermeneutic truths of lay actors over the propositional truths of the scientists in the understanding of the natural environment. Again the same dichotomy is invoked; the hierarchy is merely stood on its head.[51]

For Wynne, the objective is not to privilege any particular domain of knowledge. Rather, the desire is that: 'In seeking the basis of more legitimate, less alienating forms of public knowledge, and stable authority out of present conditions of incoherence and disorientation, new constitutional norms of valid knowledge may be articulated.'[52]

1.20 But, Meera Nanda has identified a potentially critical political problem arising from these very conditions of incoherence and disorientation in public knowledge. In a study of Hindu nationalism in Indian politics, Meera notes that contemporary Hindu nationalism is sustained by a particular instrumental rationality. This 'vedic science' opportunistically blends key assumptions of scientific rationality, technologic and industrial mechanisms with non-scientific culturally generated myths (arising from the traditions and narratives embodied in sacred texts) to underpin narrow state-centric political aims. Such instrumental fusion of science and myth is characterized by Nanda as a 'reactionary modernity' which is used to justify the pursuit of a nationalist agenda that seeks to privilege some and exclude others from the political process.[53] It might be argued that, by

---

[48] Ibid, 45.   [49] Ibid, 56.   [50] Ibid, 76.

[51] Szersynski, B, Lash, S, and Wynne, B, 'Introduction: Ecology, Realism and the Social Sciences', in Lash, S, Szersynski, B, and Wynne, B, ibid, 7.

[52] Wynne, B, 'May the Sheep Safely Graze? A Reflexive View of the Expert-Lay Knowledge Divide', in ibid, 78.

[53] Nanda, M, *Prophets Facing Backward: Postmodern Critiques of Science and Hindu Nationalism in India* (Rutgers University Press, New Jersey, 2003).

implication, other forms of instrumental rationality or reactionary modernity could be used to underpin the politicization of 'Christian' (or 'Islamic') fundamentalism or form a basis of ideological justification for the promotion of 'anti-Westernism' or a 'war on terror' in post-Cold War politics. Such interaction of scientific knowledge, lay truth claims, and non-rational mythologizing at a societal level could create a dangerously high potential for the generation of politically divisive and destabilizing instrumental rationalities.

The recognition of these discourses, and how best to respond, is clearly occupying decision-makers at the highest levels of government, perhaps chastened by the political horrors of dealing with BSE.[54] This is again illustrated by the Prime Minister's speech to the IPPR which makes explicit two key problems posed by scientific knowledge that have clear relevance for the Civil Contingencies Act, especially for the duty to assess, plan, and advise on contingency planning that is specified in Part I of the legislation. First, for the Prime Minister: ' "Science" is often taken to be a synonym for "certainty" . . . In fact, in the scientific world ambiguity, uncertainty, the wisdom that comes with failing and changing your mind, are all essential to progress.'[55] Hence, although seemingly still attached to the idea of rationalist 'progress' through scientific and technological development, there is also acknowledgement that scientific 'certainties' are often indeterminate or contingent. Secondly, the Prime Minister observed that 'the pace of change can be bewildering and breeds insecurity'. The indeterminacy and pace of scientific development thus present 'a major challenge both to politicians and the media. The structure of political combat tends to invite certainty, or at least a show of certainty, when that idiom is entirely inappropriate for discussing fine-grained risks and the balance of probability.'[56] **1.21**

Such comments accord with Sheila Jasanoff's contention that the interaction of science, biotechnology, politics, and policy in Europe and the United States is fuelling two key sources of destabilizing change. The first of these is cognitive and refers to the migration of individuals and groups away from 'realist' towards 'constructivist' world-views that emphasize the inherent subjectivity and socially constructed sources of all knowledge bases and value-systems. The second is political and relates to the fracturing of the authority of the nation-state under **1.22**

---

[54] According to Alan Irwin (*Citizen Science: A Study of People and Sustainable Development* (Routledge, London, 1995) 23): 'the BSE issue became the focus for a whole series of criticisms and concerns—about food industry practices, about the independence and competence of the government ministry, about the limits to scientific understanding in such a complex and under-researched area. Despite this broad critique, the typical "official" response was to present the issue as a challenge to the "facts".' See further Ch 4.

[55] 'Common sense culture not compensation culture' (http://www.number-10.gov.uk/output/Page7562.asp., 26 May 2005). [56] Ibid.

the onslaught of both locally sourced and supranational pressures (arising from the environment and the forces of globalization). Jasanoff argues that:

> As a result, the 'old' politics of modernity—with its core values of rationality, objectivity, universalism, centralization, and efficiency—is confronting, and possibly yielding to, a 'new' politics of pluralism, localism, irreducible ambiguity, and aestheticism in matters of lifestyle and taste [that have] attenuated the connections between states and citizens, calling into question the capacity of national governments to discern and meet their citizens' needs.[57]

Indeed, for Jasanoff, the very distribution and substance of the 'needs' of the citizenry in these polities is being shaped by the rise of ever more knowledgeable populaces, demanding greater and more meaningful control over the technological changes and processes that are seen to affect their welfare and prosperity. Jasanoff suggests that the new epoch will potentially be the 'proving ground' for new political orders 'whose success depends on living wisely with power to manipulate living things and equally growing uncertainty about the consequences of doing so.'[58]

1.23 In part, the concerns expressed are reflected (albeit rather weakly) in changes in outlook on the part of the scientific establishment itself. In its landmark 1992 study, the Royal Society sought to draw together a variety of perspectives on risk.[59] But, for Hood and Jones, two of its key contributors, reaction to the report was mixed:

> Four chapters good, two chapters bad (with apologies to George Orwell) appears to have been the orthodoxy's response...The 'good' four chapters were those written by distinguished engineers, statisticians and natural scientists, which reflected a view of public risk management as properly the domain of science and engineering rather than of politics and economics.[60]

The two 'bads' referred to those chapters that surveyed the more 'social scientific' concerns associated with risk perception and risk management. Hood and Jones also note: 'A few heretics, of course, responded to the document with the opposite mantra "two chapters good, four chapters bad"—and some social scientists...even criticized the offending social science chapters for not going far enough to embrace a politics-centred approach to risk.'[61] However, that at least some of the social scientific critique has been assimilated into the 'hard' scientific approach can be inferred from the Royal Society report in 2004, *Making the UK Safer: Detecting and Decontaminating Chemical and Biological Agents*. This report into the country's readiness to respond to such contaminants does contain explicit reference to the problems inherent in the perception,

---

[57] Jasanoff, S, *Designs on Nature: Science and Democracy in Europe and the United States* (Princeton University Press, Princeton, 2005) 14. See also the example of GM food: Hinchcliffe, S, 'Living Risky Lives—Eating Outer Nature' in Hinchcliffe, S and Woodward, K (ed), *The Natural and the Social: Uncertainty, Risk, Change* (Open University, London, 2000).
[58] Ibid.
[59] Royal Society, *Risk: Analysis, Perception, Management* (Royal Society, London, 1992).
[60] Hood, C and Jones, DKC, (eds), *Accident and Design* (UCL Press, London, 1996) xi.
[61] Ibid, xii.

communication, and management of scientific uncertainty that are directly relevant to the risk communication elements of the civil protection function. Indeed, the report notes:

> In the face of the inevitable scientific uncertainty that will exist, the perception of risk, communication of risk and informing the public about and unexpected chemical or biological incident will be complicated issues to deal with. Collaboration between experts in national and local Government, natural sciences, psychology and social science is required, in addition to public consultation.[62]

While perhaps still not embracing the politicization of risk to the extent that some 'heretics' might like, the report's recognition of the need to synthesize natural and 'social' scientific concerns in the social handling of risk does reveal a degree of sensitivity to the criticisms levelled at the response to the 1992 survey. In this sense, the report represents a call for 'better' science in respect to public discourse of risk and does perhaps reflect a degree of acceptance of the necessity for the development of a more reflexive, or reflective, approach to the social impact of scientific development.

(d) Risk and the regulatory state

Hood *et al* note that: **1.24**

> As well as a 'risk society' we are also said to live in a 'regulatory state' . . . The idea of the 'regulatory state' is that a new institutional and policy style has emerged, in which government's role as regulator advances while its role as a direct employer or property-owner may decline through privatisation and bureaucratic downsizing.[63]

Implications of this regulatory perspective are that the state functions merely as **1.25** one interest group among many and that governments can seek to regulate, but cannot control, the distribution of risk across social groups. According to O'Malley, this interface between the 'risk society' and the 'regulatory state' can itself be contextualized in that it is possible to 'link changes in risk techniques with changes in broader political rationalities such as the emergence of "neo-liberalism"'.[64] Indeed, O'Malley's view is that risk and uncertainty have been central themes in the 'genealogy of liberalism' since the 'freedom' to make rational choices depends fundamentally on an assumption that the future is indeterminate, not random. This approach governs the exhortations of nineteenth century 'classical' liberalism to thrift, prudence, and independence on the part of the populace, supported by the establishing of actuarial and, later, social insurance mechanisms.[65] It has also given rise to a neo-liberal model of the 'enterprising

---

[62] Royal Society, *Making the UK Safer: Detecting and Decontaminating Chemical and Biological Agents* (Royal Society, London, 2004) 29.

[63] Hood, C, Rothstein, H, and Baldwin, R, *The Government of Risk: Understanding Risk Regulation Regimes* (Oxford University Press, Oxford, 2001) 4.

[64] O'Malley, P, *Risk, Uncertainty and Government* (Glasshouse, London, 2004) 11.

[65] Ibid, 172.

society' that has 're-centred uncertainty—and more precisely the uncertainties associated with the free market and the importance of individual self-reliance—in ways that would also have been thought surprising 50 years ago.'[66]

1.26 Despite these changes, uncertainty and risk should be viewed as being neither inherently determinist nor necessarily a cause for pessimism. Instead, from a neo-liberal perspective, uncertainty begets 'freedom' because indeterminacy creates 'space' for the exercise of rationally based calculation and choice on the part of individuals in areas that are beyond the state's capacity to intervene or control. While O'Malley remains unsure as to whether risk is becoming the dominant organizing principle of government, he is of the opinion that neo-liberalism pushes a 'creative uncertainty' to the foreground of politics and thereby emphasizes individual responsibility:

> ... these responsibilising processes seemingly *democratise* government through the mobilising of risk and uncertainty. Individuals and communities are made free to choose how they will govern themselves in relation to a host of insecurities.[67]

1.27 The style of risk regulation thereby engendered is often referred to as 'governance' and is a response not only to the pragmatism of dealing effectively with risk in late modern society, but is also adopted as a political agenda in an era when government is readily depicted 'not as solution to societal problems but instead as the very root and cause of these problems'.[68] In this context, governance is concerned with a complex pattern of interrelationships between social institutions and individuals[69] rather than simply the formal and hierarchical processes of government. According to Rhodes, 'governance . . . is about regulating relationships in complex systems.'[70] A further element of the concept is explained by Hirst and Thompson whereby 'governance . . . is a function that can be performed by a wide variety of public and private, state and non-state, national and international, institutions and practices.'[71] The discourse is associated with the diminishing role of the nation-state and state power in late modernity[72] and 'governance' can therefore be seen as a 'function that

---

[66] Ibid, 174.

[67] Ibid, 174–5. Consequently, there will be winners and losers in disasters: Quarantelli, EL, (ed), *What Is a Disaster?* (Routledge, London, 1998) 265.

[68] Pierre, J and Peters, BG, *Governance, Politics and the State* (Macmillan, Basingstoke, 2000) 3.

[69] See generally Osborne, D and Gaebler, T, *Reinventing Government* (Addison Wesley, Mass, 1992); Ayres, I and Braithwaite, J, *Responsive Regulation* (Oxford University Press, Oxford, 1992); Jessop, B, 'The Regulation Approach, Governance and Post-Fordism: Alternative Perspectives on Economic and Political Change?' (1995) 24 (3) *Economy and Society* 307; Moran, M, *The British Regulatory State* (Oxford University Press, Oxford, 2003).

[70] Rhodes, RAW, 'The Hollowing Out of the State: The Changing Nature of the Public Services in Britain' (1994) 65 *Political Quarterly* 138 at 151. See further Moran, M, 'The Rise of the Regulatory State in Britain' (2001) 54 *Parliamentary Affairs* 19.

[71] Hirst, P and Thompson, G, 'Globalization and the Future of the Nation State' (1995) 24 (3) *Economy and Society* 408 at 422.

[72] See Giddens, A, *The Consequences of Modernity* (Polity Press, Cambridge, 1990); Rosenau, JN and Czempiel, E-O, *Governance Without Government: Order and Change in World Politics*

can be performed by a wide variety of public and private, state and non-state, national and international, institutions and practices'.[73] Processes such as globalization[74] and devolution also come into play as well in that they may act as forces which generate pressure for further decentralization of responsibility between the public and private sectors. Consequently, Rhodes posits 'the hollowing of the state',[75] though that process does not result in the total enfeeblement of the nation-state, the exclusion of state actors, or a total 'discontinuity' from tradition.[76] Indeed, while forms of economic and political organization have certainly been affected,[77] the idea of 'collective security'—as evidenced by the Civil Contingencies Act—does remain one abiding concern of the late modern state.[78]

Governance networks maintain coherence through a common goal[79] which **1.28** guards against the inherent fragmentary tendencies of the multi-tiered and plural actors. A degree of autonomy for the actors in the network is also a common feature—relationships become horizontal (or heterarchical), decentred, and networked rather than hierarchical command and control relationships,[80] with an emphasis upon shared knowledge and consensus.[81]

For Beck, the self-organization of non-state actors, in the forms of multitudes of **1.29** special interest groupings and alignments will occur as a reflexive mode of dealing with risk, but the predicted 'entrepreneurial' response of individuals and communities to limited state capability and intervention must be tempered by the contention that increasing risk aversion is just as salient a social response to uncertainty. For example, despite unprecedented levels of health, life-expectancy, and material living standards: 'the American public has become more—rather than less—concerned about risk. We have come to perceive ourselves as

---

(Cambridge University Press, Cambridge, 1992); Held, D, *Democracy and The Global Order: From The Modern State to Cosmopolitan Governance* (Polity Press, Cambridge, 1995); Rhodes, RAW, *Understanding Governance: Policy Networks, Governance, Reflexivity And Accountability* (Open University Press, Buckingham, 1997).

[73] Hirst, P and Thompson, G, 'Globalization and the Future of the Nation State' (1995) 24 (3) *Economy and Society* 408 at 422.

[74] See Spybey, T, *Globalization and World Society* (Polity Press, Cambridge, 1996).

[75] Rhodes, RAW, *Understanding Governance: Policy Networks, Governance, Reflexivity and Accountability* (Open University Press, Buckingham, 1997) 17.

[76] Giddens, A, *The Consequences of Modernity* (Polity Press, Cambridge, 1990) 6.

[77] Stewart, A, *Theories of Power and Domination: The Politics of Empowerment in Late Modernity* (Sage, London, 2001); Hutton, W, *The State to Come* (Vintage, London, 1997).

[78] See Giddens, A, *The Third Way and Its Critics* (Polity Press, Cambridge, 2000) 57.

[79] Rosenau, JN and Czempiel, E-O, *Governance Without Government: Order and Change in World Politics* (Cambridge University Press, Cambridge, 1992) 3.

[80] Rhodes, RAW, *Understanding Governance: Policy Networks, Governance, Reflexivity and Accountability* (Open University Press, Buckingham, 1997) 53; Black, J, 'Decentring Regulation' (2001) 53 CLP 103.

[81] See Kooiman, J, 'Social-political Governance: Introduction' in Kooiman, J (ed), *Modern Governance* (Sage, London, 1993) 4; Rosenau, JN and Czempiel, E-O, *Governance Without Government: Order and Change in World Politics* (Cambridge University Press, Cambridge, 1993) 4.

increasingly vulnerable to life's hazards and to believe that our land, air and water are more contaminated by toxic substances than ever before.'[82] The aversion may not be entirely the fault of a credulous or superstitious public but may be fostered more by political elites who rely for their powers of containment and legitimacy not so much on the ability to ward off mutually assured destruction by superpowers but by being seen to take seriously, and be assuring about, identified and imagined risks.[83] For several authors, risk aversion increases because societal estimations of risk and vulnerability to harm are mediated through a 'social amplification framework' rather than a process of rational calculation. According to the Royal Society:

> These researchers adopt a metaphor loosely based upon communications theory, to explain why certain hazards (for example, public transport safety of trains or sea-going ferries) are a particular focus of concern in society, while others (such as motor vehicle accidents) receive comparatively little attention. They suggest that hazards and their objective characteristics (e.g. deaths, injuries, damage and social disruption) interact with a wide range of psychological, social or cultural processes in ways that intensify or attenuate perceptions of risk.[84]

1.30 That the magnitude of the most 'intensified' of these socially perceived risks or uncertainties (particularly environmental hazards), and the potential costs of corrective action, reinforce risk aversion or the avoidance of risk-taking behaviour is illustrated by the increasing centrality of the 'precautionary principle' as a regulatory regime resource.[85] Perceptions of high societal risk do translate into political pressure for governments to engage in preventative intervention—even when a lack of evidence renders the likelihood or severity of a given hazard open to significant doubt.[86] Certainly for Hood *et al*: 'The two ideas of "risk society" and "regulatory state" could, indeed, be linked in so far as risk and safety is often

---

[82] Slovic, P, *The Perception of Risk* (Earthscan, London, 2000) 316. The same point is made by Furedi, F, *Culture of Fear: Risk-taking and the Morality of Low Expectation* (rev ed, Continuum, London, 2002) vii.

[83] The point was made in the BBC documentary, *The Power of Nightmares* (http://news.bbc.co.uk/1/hi/programmes/3755686.stm, 2005).

[84] Royal Society, *Risk: Analysis, Perception, Management* (Royal Society, London, 1992) 114. See also Kasperson, RE, Renn, O, Slovic, P, Brown, HS, Emel, J, Goble, R, Kasperson, JX, and Ratick, S, 'The Social Amplification of Risk: A Conceptual Framework' (1988) 8 *Risk Analysis* 177; Kasperson, RE, Kasperson, JX, and Renn, O, 'The Social Amplification of Risk: Progress in Developing an Integrative Framework' in Krimsky, S, and Golding, D, (eds), *Social Theories of Risk* (Praeger, New York, 1992); Gregory, J and Miller, S, *Science in Public: Communication, Culture and Credibility* (Perseus Publishing, Cambridge, Mass, 2000); Pidgeon, N, Kasperson, RE, and Slovic, P, *The Social Amplification of Risk* (Cambridge University Press, Cambridge, 2003).

[85] Given the incalculability of some of these risks, it may be better to speak in terms of uncertainties, which have been categorized as 'prudential uncertainties' (based on the foresight of harm but no specific prediction of it), 'precautionary uncertainty' (the avoidance or minimization of feasible harm), and 'diagnostic uncertainty' (based on imprecise expert advice): O'Malley, P, *Risk, Uncertainty and Government* (Glasshouse, London, 2004) 25.

[86] For further discussion of the precautionary principle, as well as Ch 8, see Jordan, A, and O'Riordan, T, 'The Precautionary Principle in UK Environmental Law and Policy' in Gray, T, (ed),

held to be one of the major drivers of contemporary regulatory growth, for example in the development of EU regulations.'[87] Similarly, in discussing the concept of 'acceptable' versus 'tolerable' societal risk, the Royal Society noted that: 'Responsibility for environmental safety is a traditional part of government's "police power"... which has become increasingly salient in recent decades with massive increases in bureaucracy and regulation.'[88]

In relation to the Civil Contingencies Act itself, some analysts have argued that **1.31** the impetus to overhaul existing emergency legislation is as much 'subjectively' as 'objectively' driven and that the need for a new Civil Contingencies Act resides in 'diminishing individual self-reliance, the culture of self-indulgence at the expense of community welfare, risk aversion and an expectation that "others" will neutralise all uncertainties'.[89] Concerns about the effects of a risk-averse culture in a variety of spheres of activity also underpinned the Prime Minister's speech to the IPPR. For example, the Prime Minister observed:

> ...there needs to be a proper and proportionate way of assessing risk and the response to it. Government cannot eliminate all risk. A risk-averse scientific community is no scientific community at all. A risk-averse business culture is no business culture at all. A risk-averse public sector will stifle creativity and deny to many the opportunities to be creative while supplying a few with compensation payments... We cannot respond to every accident by trying to guarantee ever more tiny margins of safety. We cannot eliminate risk. We have to live with it, manage it.[90]

Whether the Civil Contingencies Act and its associated regulations should be regarded as a regulatory response to 'diminishing self-reliance', increasing risk aversion, and a shifting of responsibility onto 'others' is a contentious claim. Clearly, the regime being established by the Act and its regulations does contain 'precautionary' elements, particularly in relation to the planning function envisaged in Part I of the Act. However, these do not really conform to the classic definition of the 'precautionary principle' which 'has as its focus prevention by taking no chances'.[91] Instead, the following observation by Pat O'Malley provides a seemingly better fit:

> ...the precautionary principle has become a high profile regulatory resource, especially in the EC. However, in most instances... the principle is put in place as a temporary measure until the risk can be determined (presumably by scientists). For

---

UK Environmental Policy in the 1990s (Macmillan, London, 1995); Furedi, F, *Culture of Fear: Risk-taking and the Morality of Low Expectation* (rev ed, Continuum, London, 2002).

[87] Hood, C, Rothstein, H, and Baldwin, R, *The Government of Risk: Understanding Risk Regulation Regimes* (Oxford University Press, Oxford, 2001) 4.

[88] Royal Society, *Risk: Analysis, Perception, Management* (Royal Society, London, 1992) 92.

[89] Joint Committee on the Draft Civil Contingencies Bill, Appendix 7.

[90] 'Common sense culture not compensation culture' (http://www.number-10.gov.uk/output/Page7562.asp., 26 May 2005).

[91] O'Malley, P, *Risk, Uncertainty and Government* (Glasshouse, London, 2004) 178.

the most part, the appearance of major and potentially catastrophic risks has not given rise to new reflections about the limits of science. At least as often, it has given rise to the intensification of a search for security based on risk and foreseeability.[92]

(e) Risk and the operation of regulatory regimes

1.32 For authors such as Christopher Hood and David Jones, the general discourse on risk is fraught with problems in that, 'instead of converging on agreed definitions of basic terms, the contemporary literature abounds with contra-dictory statements about what the words "hazard" and "risk" mean.'[93] For these analysts, the field of risk and risk management consists of many distinct specialisms and sub-disciplines. As a result, 'risk regulation regimes vary sub-stantially across policy domains in a way that the generalist tone of risk society-type analysis obscures and cannot explain.'[94] From this perspective, variations in the manner in which risk discourse is constructed, how 'acceptable risk' is understood, and, crucially, how variability in risk regulation regimes across policy domains can be explained, requires that we 'go beyond generalizing perspectives like "risk society" to a more disaggregated analysis.'[95]

1.33 This 'meso-level analysis' falls between big-picture macro-political or sociological studies and the use of individual case studies or single-issue policy debates of risk. Of course, 'regime theory' as a conceptual tool to explain policy variation is not a new concept in terms of political theorizing,[96] but Hood *et al* seek more speci-fically to define and apply the term as a: 'complex of institutional geography, rules, practice and animating ideas that are associated with a particular risk or hazard.'[97] By delimiting the idea of a risk regulation regime, they then seek to compare and contrast differing regimes across policy domains in the attempt to explain regulatory variance. Essentially, they adopt a cybernetic view that risk regulation regimes are bounded control systems comprised of three main ele-ments: information gathering, standard setting, and behaviour modification. However, in terms of their content, regimes are differentiated according to size, structure, and decision-making 'style' which then leads to variations policy out-comes. Moreover, regimes operate within specific contexts which also influence the policy process. For Hood *et al* these contextual elements are specifically

---

[92] Ibid, 179–80. O'Malley draws here on the work of Fisher, E, 'Precaution, Precaution Everywhere: Developing a "Common Understanding" of the Precautionary Principle in the European Community' (2002) 9 *Maastricht Journal of European and Comparative Law* 7.

[93] Hood, C and Jones, DKC, (eds), *Accident and Design* (UCL Press, London, 1996) 2.

[94] Hood, C, Rothstein, H, and Baldwin, R, *The Government of Risk: Understanding Risk Regulation Regimes* (Oxford University Press, Oxford, 2001) 171.

[95] Hood, C and Jones, DKC, (eds), *Accident and Design* (UCL Press, London, 1996) 8.

[96] See Krasner, S, (ed), *International Regimes* (Cornell University Press, Ithaca, 1983).

[97] Hood, C, Rothstein, H, and Baldwin, R, *The Government of Risk: Understanding Risk Regulation Regimes* (Oxford University Press, Oxford, 2001) 9.

defined as: (i) the type of risks involved (function), (ii) public attitudes and preferences, and (iii) the structure of organized interests relating to the policy domain. What results from this perspective is a particular conceptual model, or regime matrix, that can be used to evaluate differing risk regulation regimes.[98] By examining differing regimes through the contextual elements, Hood *et al* suggest a means of evaluating difference in regime development and operation that cuts across policy domains and accounts for variations in policy outcomes that the 'grand narratives' of the risk society do not address. From this point of view, the problems associated with risk and risk management are not the inevitable and unmanageable result of problems inherent in the evolution of modernity itself. As such, Hood *et al* do not seek to provide an overarching account of the sources of causation of risk in late modernity. They do, however, attempt to understand how institutional and organizational structure and regulatory context profoundly affect how risks are managed.

Notwithstanding the avowed advantages of this meso-level analysis for **1.34** explaining policy variation across risk domains, Hood *et al* do concede that 'prediction of variety in regime content is an inexact science'.[99] Specifically, in examining nine different risk regulation regimes in the United Kingdom, Hood *et al* found a number of regime variants that were not adequately predicted by the conceptual model they have developed because they had not accounted for the ' "inner life" elements of regulatory regimes'.[100]

Such a meso-level analytic focus sits alongside what Powell and DiMaggio term the **1.35** 'new institutional school of organizational analysis'[101] which, as Vaughan notes, 'emphasizes the way in which non-local environments—industries, professions and the like—penetrate organizations, creating a frame of reference, or worldview, that individuals bring to decision making and action.'[102] Such analysis explores the organizational and cognitive restraints on decision-making that profoundly affect policy outcomes. The focus in these works is on bureaucratic procedure and uncertainty avoidance, organizational learning mechanisms, and the operation of limited, or bounded, rationality in the policy process. However, Vaughan's study of the 1986 Space Shuttle Challenger disaster attempts to further develop this perspective by examining 'bottom up' decision-making in a complex organizational hierarchy. For Vaughan, 'Sensemaking, in this view, is about contextual rationality, so the task is to expose the constraints, both hidden and explicit, both formal and informal, that act on decision makers . . . It underscores the importance of exploring both decision context and the interpretive work of the people making choices.'[103]

---

[98] Ibid, 22.   [99] Ibid, 144.   [100] Ibid.
[101] See Powell, W and DiMaggio, P, (eds), *The New Institutionalism Organizational Analysis* (University of Chicago Press, Chicago, 1991).
[102] Vaughan, D, *The Challenger Launch Decision: Risky Technology, Culture and Deviance at NASA* (University of Chicago Press, Chicago, 1996) 404.   [103] Ibid, 403.

In relation to the Challenger incident, Vaughan concludes that: 'The cause of disaster was a mistake embedded in the banality of organisational life and facilitated by an environment of scarcity and competition, an unprecedented, uncertain technology, incrementalism, patterns of information, routinization, organizational and interrorganizational structures, and a complex culture.'[104]

1.36   But for Charles Perrow, such an account—with its accentuation of the role of organizational culture—simply downplays the influence of power and interest. His interpretation of the launch decision is of a 'damaged' organization whose managers overrode safety concerns as a result of particular production pressures. The corruption of the safety culture in this instance, 'was not the normalisation of deviance or the banality of bureaucratic procedures and hierarchy or the product of an engineering "culture"; it was the exercise of organizational power. We miss a great deal when we substitute culture for power.'[105] Although Perrow's study of high risk technologies is explicitly an exploration of 'Normal Accidents Theory' (NAT) and complex systems failure, its roots are also firmly located in organizational theory.[106] This analysis of NAT and High Reliability Systems Theory (HRT) has been further developed by Scott Sagan in his study of the safety systems put in place to prevent the accidental use of US nuclear weapons.[107] For Sagan, a key influence on safety culture in this system is how the organizations involved tended to proceed on principles of limited or bounded rationality, and he points in particular to Cohen, March, and Olsen's 'garbage can' model of organizational decision-making as being a valuable conceptual tool in understanding how such organizations operate and, crucially, how they learn. According to Sagan, this is essentially 'a more political vision in which 'solutions' are actively looking for problems to attach themselves to, 'problems' are ill-defined and often unrecognized, and 'participants' have limited attention, shifting allegiances, and uncertain intentions.'[108] Sagan also notes the role that power and interest plays in systemic interaction and suggests that interest groupings and power relations 'can exert a strong influence on the frequency of catastrophic accidents, on their interpretation and therefore who receives the blame for failures and, finally, on the degree to which the organizational structures that make normal accidents inevitable are modified or abandoned.'[109] Perrow's work on NAT demonstrates how tightly coupled and

---

[104]   Ibid, xiv.
[105]   Perrow, C, *Normal Accidents: Living with High Risk Technologies* (Princeton University Press, Princeton, 1999) 380.   [106]   Ibid, 368.
[107]   Sagan, S, *The Limits of Safety: Organizations, Accidents, and Nuclear Weapons* (Princeton University Press, Princeton, 1993).
[108]   Ibid, 30. See also March, J and Olsen, J, 'Garbage Can Models of Decision Making' in March, J and Weissinger-Baylon, R, (eds), *Ambiguity and Command: Organizational Perspectives on Military Decision Making* (Pitman, Mass, 1986).
[109]   Sagan, S, *The Limits of Safety: Organizations, Accidents, and Nuclear Weapons* (Princeton University Press, Princeton, 1993) 32.

complex systems are inherently susceptible to catastrophic failure. But such systems are also formal organizations and, as Sagan has rather chillingly demonstrated in relation to the risk of accidental use of US nuclear weapons during the Cold War, system failure is also critically dependent on organizational factors beyond mere technical complexity. For Perrow and Sagan, these factors are characterized as power-political, for Vaughan, the explanation lies in specific cultural factors while, for others, cognitive and perceptual restraints on decision-making are crucial.

It was noted earlier that Hood *et al* take issue with central tenets of the 'risk **1.37** society' thesis, even coming 'close to the conclusion that there is no such thing as the risk society, only different risk regulation regimes.'[110] Yet, they also suggest that variation across risk regulation regimes can only be fully understood by taking into account 'inner life' factors within specific regimes and organizations. In other words, they appear to accept that any 'meso-level' analysis has to rely on exploration of some highly problematic concepts: especially the definition, scope, and relationships between culture, perception, and power. Moreover, the definition and understanding of these critical concepts (which lies at the heart of sociological and political studies) is an ongoing and indeterminate debate regardless of the level of analysis adopted.

C. Towards a Civil Protection Society?

The Civil Contingencies Act represents an attempt to enable the creation of **1.38** a better defined, more comprehensive, and more general (or generic) risk management regime than delivered in previous legislation. Its objectives include communal risk reduction and also, by conscripting a range of non-state bodies, risk spreading.[111] This shift in policy emphasis directly arises from experience of a number of disparate circumstances and political developments since the early 1990s. Perhaps understandably, 'sense-making' in relation to these disconnected events has proved to be a slow process for policy-makers. However, the passing of the Civil Contingencies Act and its associated regulations does reflect recognition, on the part of policy-makers, that revision of the assumptions and theories regarding the nature of emergent and unforeseen hazards with potentially high impact on the United Kingdom is necessary and desirable.

---

[110] Hood, C, Rothstein, H, and Baldwin, R, *The Government of Risk: Understanding Risk Regulation Regimes* (Oxford University Press, Oxford, 2001) 171.

[111] 'Risk' has been categorized as 'insurance risk' (risk spreading), 'clinical risk' (minimizing risk of selected individuals), and 'epidemiological risk' (minimizing the risk of a group): O'Malley, P, *Risk, Uncertainty and Government* (Glasshouse, London, 2004) 21.

1.39 Yet, there remains a debate about how to establish most effectively the nascent civil protection function envisaged by the Act and thereby improve 'resilience'. For the Cabinet Office, 'resilience' is the core organizing principle of the Government's approach to civil contingency planning and is defined as:

> ...the ability to handle disruptive challenges that can lead to or result in crisis...Resilience is built around several key activities. Firstly, risks of disruptive challenge must where possible be identified, either by considering internal weaknesses or scanning the horizon for external threats. Anticipation allows choices to be made. In some circumstances it is possible to prevent disruptive challenges occurring by taking action at an early stage. In other cases, planning has to take place to prepare to deal with a disruptive challenge. If the disruption does occur it becomes necessary to respond, and once the situation is brought under control the focus becomes recovery. This cycle—anticipation, prevention, preparation, response, recovery—is at the heart of resilience.[112]

It follows that resilience can be achieved through integrated action at several stages. Vulnerability can be reduced through anticipation and prevention. Survival and the adaptive maintenance of strength can be secured through preparation and effective response. An elastic restoration of lost facilities or processes can be quickened through attention to recovery processes. However, unless the wide array of stakeholders involved manage to establish a clear and consistent understanding of definitions and relationships between key terms such as 'threat', 'risk', and 'vulnerability',[113] for example, there will always be scope for confusion and conflict in the implementation of the resilience agenda.

1.40 Clearly therefore, in the early stages of development of this new 'civil protection' sphere of activity, it is essential that important questions of institutional function and organization are examined closely. In particular, how should existing organizations respond not only to the immediate duties of the Act, but how should the conceptual context and, indeed, the content of the legislative framework be understood and internalized?

1.41 As a general concept, 'civil protection' (or civil contingency planning) is defined as:

> ...the application of knowledge, measures and practices to anticipate, guard against, prevent, reduce or overcome any hazard, harm or loss that may be associated with natural, technological or man-made crises and disasters in peacetime.[114]

For the Cabinet Office, civil protection is about protecting the public from the effects of emergencies regardless of cause, though not all may fall within the Act because of its local focus. Thus, in relation to terrorism, the actions of the security services are not considered to be part of the civil protection function. However,

---

[112] Cabinet Office, *The Draft Civil Contingencies Bill*, paras 2.1, 2.3. See further Cabinet Office, *Dealing with Disaster* (rev 3rd ed, Cabinet Office, London, 2003) para 1.1.

[113] As well as their relation to the definition of 'emergency' in ss 1 and 19: see Ch 3.

[114] Cabinet Office, *Dealing with Disaster* (rev 3rd ed, Cabinet Office, London, 2003) para 1.2.

dealing with the aftermath of a terrorist attack is part of the function.[115] The Cabinet Office also seeks to differentiate 'civil' from 'public' protection in that:

> Public protection relates to a range of hazards and threats to public safety (with a lower threshold of seriousness or range of impact than is associated with civil protection), covering issues such as child protection, health and safety at work, community safety, protection from crime as well as the effects of emergencies. Some of these are concerned with the protection of individuals. Civil protection is concerned with events or situations that are likely to have an impact on numerous individuals and that generally require a timely and immediate response to limit the harm to the public.[116]

This distinction is not entirely consistent—for example, protection from crime **1.42** has collective as well as individual aspects. Nevertheless, one might distil at least three concepts here, under the heading of societal protection:[117]

- 'civil protection' is about planning against, response to, or mitigating the effects on society (the public as a whole or a section of the public) of disasters and emergencies;
- 'public protection' is based around actions to protect individuals from harm, such as child protection, and health and safety at work;
- 'community safety' relates to collective resilience and protection against non-exceptional harms such as crime, disorder, or public health problems.

These distinctions hopefully enlighten the particular focus of civil protection which is the core of the Civil Contingencies Act.

Within its domain, the Civil Contingencies Act and attendant regulations **1.43** represent a significant departure from previous approaches to the use of emergency powers in peacetime. Indeed, the Civil Contingencies Act provides a 'much broader definition of emergency than has previously existed in United Kingdom legislation. It is intended to cover the full spectrum of current and future events and situations, while at the same time establishing a clear minimum threshold for civil protection planning.'[118] In this sense, the framework outlined fulfils the Government's publicly stated commitment to legislative change, 'to reflect the move from Cold War civil defence to modern civil protection.'[119] This objective is reflected in the expansive definitions of 'emergency' set out in ss 1 and 19, going far beyond the previous preoccupation with disruptive industrial action.

With the abandonment of the civil defence and strike-breaking models of **1.44** emergency powers, it might be argued that the Civil Contingencies Act to a degree 'demilitarizes' this sphere of governmental activity. However, the diffuse and complex array of challenges which are shaping the 'resilience agenda' also tend to blur the traditional boundaries between 'war' and 'peacetime' in policy

---

[115] Cabinet Office, *The Draft Civil Contingencies Bill*, para 2.8.   [116] Ibid, para 2.9.
[117] Ibid, paras 2.8, 2.9.   [118] Ibid, para 2.7.   [119] Ibid, para 1.2.

terms, though not to the extent of declaring a 'war on terror' or to the extent of submerging many forms of societal protection under the heading of 'Homeland Security'.[120] Thus, it is instructive to compare the current fate of the (US) Federal Emergency Management Agency (FEMA) with the (UK) Health Protection Agency (HPA). In 2002, the Chief Medical Officer published a report, *Getting Ahead of the Curve*, which focused both on the natural occurrence of infection and the potential for the deliberate release of infectious agents.[121] The principal recommendation of the report was that an integrated health protection infrastructure be set up with the new HPA at its centre. As well as responsibility for infectious diseases, the remit of the HPA was to be extended to incorporate chemical and radiological issues in a unitary chemical, biological, radiological, or nuclear (CBRN) portfolio. The new framework would provide a more effective and unified health protection function than was the case with comparatively fragmented pre-9/11 health protection arrangements which were spread across four government establishments: the Public Health Laboratory Service; the Centre for Applied Microbiology and Research; the National Focus for Chemical Incidents; and the National Radiological Protection Board. The changes were achieved in part by administrative restructuring, establishing the HPA as a special health authority (SpHA) in 2003 and then, by the Health Protection Agency Act 2004, which founded a non-departmental public body. The HPA is therefore governed by an independent Board of Directors with the Secretary of State for Health providing a guidance framework that informs the work of the agency. The largely economic and bureaucratic rationales for the creation of the HPA, plus the development of its independent status, provide a measure of reassurance that the nascent civil contingencies framework in the United Kingdom has not been subjected to the same degree of military–security encroachment into the civil emergency management realm as is currently evident in the United States.[122] But, given that civil–military relationships are currently evolving as part of the function and that political pressure for change can suddenly be manifest in the wake of unforeseen events, it is imperative that the emerging 'civil protection' framework be subjected to close and ongoing scrutiny to ensure that any such encroachment is minimized. However, who is to take responsibility for such oversight remains a key and, as yet, unanswered question.

1.45 The Prime Minister's speech to the IPPR in May 2005 also acknowledged that an extensive debate about how Government Departments should conceptualize and assimilate 'risk' into their operations is occurring at central government

---

[120] See further Ch 9.

[121] Department of Health, *Getting Ahead of the Curve* (http://www.dh.gov.uk/assetRoot/04/06/03/38/04060338.pdf, 2002). See Walker, C, 'Biological Attack, Terrorism and the Law' (2004) 17 *Journal of Terrorism and Political Violence* 175. See further Ch 4.

[122] See House of Commons Defence Committee, *Defence and Security in the UK* (2001–02) HC 518–1) 71.

level. The thinking encapsulated in the Cabinet Office Strategy Unit's 2002 report, *Risk—Improving Government's Capability to Handle Risk and Uncertainty*, affirms that governments remain 'at root, guarantors of the security of their citizens' and also adopts the position that the pace and scope of scientific and technological change has generated 'manufactured risks' as well as the recognition that governments have to operate in a world of greater 'connectedness'.[123] The report then goes on to outline the overall regime for a resilience framework that defines the government's major roles and responsibilities and provides the contours of a 'comprehensive programme of change' deemed necessary to discharge these functions.[124]

A more detailed analysis of this legislative and organizational framework **1.46** occupies subsequent discussion in this book, but some key points regarding the development of this regime can be made here. Both the Cabinet Office Report and the Prime Minister's speech make explicit reference to the centrality of the Treasury's work on risk and regulatory frameworks. Key documents which outline the Treasury's view include the December 2004 Hampton Review and the Department's own report of October 2004.[125] As the title of the Hampton Review suggests, the thrust of the independently commissioned study is to examine how and where government can significantly lower the 'burden' of regulation on the private sector. In this regard, the views of commentators concerning the 'dangers' of excessive regulation, or regulatory interventions, do appear to have been taken on board by the Treasury—at least in respect of private sector enterprise. Yet, such an emphasis is not consistently reproduced in the Civil Contingencies Act nor in its attendant regulations. Indeed, the Civil Contingencies Act creates new statutory duties and instigates a new sphere of risk management activity located specifically at the local government level.

Whatever regime is being pursued, does the legislative framework create a **1.47** coherent and consistent approach to the distribution of risk across the entire civil protection function? According to the Cabinet Office:

> The Civil Contingencies Bill is part of the Government's wider resilience agenda. Resilience is the ability to handle disruptive challenges that can lead to or result in crisis. The Bill builds resilience by focusing on managing risks associated with events or situations that can lead to emergencies through effective civil protection. This in turn links with the practical civil protection measures the Government has already put in place to build capabilities.[126]

---

[123] Cabinet Office Strategy Unit, *Risk: Improving Government's Capability to Handle Risk and Uncertainty* (Cabinet Office, London, 2002) 4–5.
[124] Ibid, 17.
[125] Hampton, P, *Reducing Administrative Burdens: Effective Inspection and Enforcement* (HMSO, London, 2004); HM Treasury, *The Orange Book, Management of Risk—Principles and Concepts* (HMSO, London, 2004). See further Ch 8.
[126] Cabinet Office, *The Draft Civil Contingencies Bill*, para 2.1.

As is clear from above, at the heart of civil protection (and therefore under-pinning the Civil Contingencies Act and the ensuing civil contingencies regime in the United Kingdom) is the concept of 'resilience'. But, in terms of regu-latory scope and 'burden', this concept remains elusive and ill-defined. In the Consultation Document that accompanied the publishing of the draft Civil Contingencies Bill, the Cabinet Office observes:

> The Government is committed to enhancing the resilience of the United King-dom to disruptive challenge. In recent years, the range of challenges that society faces has broadened as networks have become more complex. We can no longer work on the assumption that disasters in the UK or elsewhere can be localised or occur in isolation. Multiple events can occur at once and—irrespective of malicious intent—can be repeated or cause knock-on effects that demand far greater coordination and integration of activities.[127]

Yet, despite the claim that the civil contingencies legislation 'links' with other 'practical measures' that are 'already in place', the Cabinet Office does acknowledge that: 'In practice, resilience means different things to different organisations because of variations in their size, purpose and interconnectedness. Disruptive challenges can take many forms, as can the responses.'[128] Here then is a primary obstacle facing policy-makers; Government is said to be 'awash with initiatives to promote risk management',[129] but how should different definitions and responses to the resilience agenda be reconciled and coordinated?

1.48 There is next a problem with the nature and scope of the civil contingencies network. The network does incorporate many important private players, as described in Chapter 4, but does not reach out to all. Nor does it veer far towards 'smart' regulation which empowers participants to develop as surrogate reg-ulators.[130]

1.49 Another example of the problems that are being encountered in 'rolling out' the resilience agenda can be seen in relation to the Ministry of Defence's (MOD) reorganization of its readiness reporting processes. According to a National Audit Office (NAO) report, the MOD has developed a, 'sophisticated system for defining, measuring and reporting the readiness of the Armed Forces'.[131] Moreover, the very process of NAO review and the ready usage of terminology such as 'business objectives' in relation to MOD function can itself be char-acterized as an illustration of an 'audit culture' that represents another form of response to the management of collective risk.[132] Yet it should also be noted that

---

[127] Ibid, para 2.2.   [128] Ibid, para 2.5.

[129] Black, J, 'The Emergence of Risk Based Regulation and the New Public Risk Management in the United Kingdom' [2005] PL 512.

[130] See Gunningham, N and Grabosky, P, *Smarter Regulation* (Clarendon Press, Oxford, 1998); Baldwin, R, 'Is Better Regulation Smarter Regulation?' [2005] PL 485.

[131] National Audit Office, *Ministry of Defence: Assessing and Reporting Military Readiness* (2005–06 HC 72) 2.

[132] See Power, M, *The Audit Society* (Oxford University Press, Oxford, 1997).

the audit activities of the NAO and the Public Accounts Committee (PAC) have themselves been criticized for introducing a 'risk averse', even 'blamist', culture into decision-making as a result of their audit-based approach. The Cabinet Office Strategy Unit acknowledges an ingrained:

> ...perception that the PAC and NAO remain a significant deterrent to risk taking. While the PAC and NAO have both taken steps to support the development of well-managed risk taking, concerns were raised about the high profile given to failure in their reports and the need for greater recognition of the context in which decisions under scrutiny were taken.[133]

From this evidence, increasing levels of audit and inspection appear to have engendered unintended and counter-productive consequences within organizations and across spheres of activity.

Finally, the NAO's audit of MOD readiness reporting processes highlights **1.50** another difficulty of auditing resilience. The report notes that, 'The readiness reporting system is continuously evolving and has proven itself over time. Military commanders...have expressed confidence in it.'[134] But, despite such 'confidence', it appears that the MOD are troubled by another succinct question. As the NAO observe: 'Measuring how ready forces are in reality for contingent operations is intrinsically more challenging, not least in answering the question "ready for what?".'[135] In this sense, notwithstanding the importation of the language of risk management, or techniques and processes of 'auditing' risk, or even that a common framework of risk assessment is outlined in Chapter 4 of *Emergency Preparedness*, it is clear from this question that planning assumptions are varyingly defined and assimilated across different spheres of government activity. The problem at an operational level is how the institutions of state—given that they all act as 'interest groupings' in their own right—reconcile the policy concerns being expressed at the highest levels of government with a regulatory inconsistency and complexity that is also imbued with rational choice, economic 'audit' pressures, to create a coherent and consistently understood civil protection regime.

D. Conclusions

Faced with an ever-widening and ever-deepening range of risks and uncertainties, **1.51** governments must engage in their management. But in a risk society, they cannot promise delivery. In terms of response, and bearing in mind reflexive modernization, considered approaches tend to involve either risk evaluation or the

---

[133] Cabinet Office Strategy Unit, *Risk: Improving Government's Capability to Handle Risk and Uncertainty* (Cabinet Office, London, 2002) 17.

[134] National Audit Office, *Ministry of Defence: Assessing and Reporting Military Readiness* (2005–06 HC 72) 1. [135] Ibid.

application of the precautionary principle against uncertainty:

> The detailed response requirement may be established by comparing the probable disruptive challenges (based on historical precedent, identified weaknesses, declared intent of enemies and blue sky thinking) with existing response capability to expose the degree of vulnerability and neutralising response requirement.[136]

Of course, these calculations are far from precise, and account should also be taken (as it is in Chapter 2) that other standards, relating to constitutionalism and human rights, should act as side-constraints to the foregoing calculus.

1.52 For its part, the Civil Contingencies Act 2004 can be seen as part of the response to certain forms of risk and uncertainty by nodes within a newly constituted governance framework. The limitations within its design are notable, but the social conditions of the risk society suggest that it will be neither the exclusive nor the final legislative statement on the subject.

---

[136] Joint Committee on the Draft Civil Contingencies Bill, Appendix 7.

Part II

THE LEGISLATIVE DETAILS

2

BACKGROUND AND OUTLINE

In Part II, we shall provide a comprehensive and critical guide to the legislation. **2.01** The exposition will draw upon the theoretical perspectives adduced in Part I and will inform operational and practical impacts as well as assessments (Part III). This opening Chapter 2 will consider the background history to the Act, an outline of its overall scheme, and the standards by which it should be adjudged.

A. Forerunners to the Act

The most prominent progenitors to the Civil Contingencies Act are (for Part I) **2.02** the Civil Defence Act 1948 and (for Part II) the Emergency Powers Act 1920. Their perceived inadequacies will be briefly explained, as will the range of other legislation which deals with emergencies, including residual prerogative powers.

(a) Forerunners to Part I of the 2004 Act

Civil defence became a vital element of wartime survival in 1939. The Civil **2.03** Defence Act of that year[1] imposed duties upon local authorities (and also public

---

[1] See also Civil Defence Act (Northern Ireland) 1939; Quekett, A, 'Emergency Legislation in Northern Ireland' (1939) 3 *Northern Ireland Legal Quarterly* 170.

utilities) to provide public shelters and granted them powers to undertake civil defence works, while the central government had duties to offer guidance to occupiers and employers.[2] Though these measures were suspended in 1945,[3] the Cold War soon revived their spirit in the shape of the Civil Defence Act 1948[4] and the Civil Defence Act (Northern Ireland) 1950.[5] These in turn fell into decline, beginning in the 1960s and ending in terminal desuetude in the mid-1980s. The 1948 Act envisaged hostile attack by another state which would be countered by public authorities. Accordingly, the Acts were confined to 'civil defence', defined in s 9(1) as:

> . . . any measures not amounting to actual combat for affording defence against any form of hostile attack by a foreign power or for depriving any form of hostile attack by a foreign power of the whole or part of its effect, whether the measures are taken before, at or after the time of the attack . . .

It follows that the legislation became outdated and unable to respond to threats from sub-state sources, such as terrorism or indeed non-state sources, such as disease or meteorological disaster.

2.04 The 1948 Act was essentially an enabling provision. Under s 1, it became 'part of the functions of the designated Minister to take such steps as appear to him from time to time to be necessary or expedient for civil defence purposes'. Further duties under s 2 were imposed upon police forces, fire brigades, and employees of local or police authorities. Training duties were imposed under s 5 upon constables, firemen, and members of civil defence forces and services. The Act also allowed regulations to be issued to utility providers, requiring them to make active civil defence preparations. Amongst the many regulations which resulted were the Civil Defence (General) Regulations 1949[6] and the Civil Defence (Public Protection) Regulations 1949,[7] which made county and county borough councils responsible for collecting and distributing information about possible attack, controlling and coordinating counteraction, including evacuation and emergency care and housing and rescue, protecting against 'the toxic effects of atomic, biological and chemical warfare', and advising the public. The Civil Defence (Planning) Regulations of 1974 laid down a duty to plan for the continuance of essential services in wartime,[8] while the Civil Defence (General Local Authority Functions) Regulations 1983[9] provided duties not only to devise plans but also to revise them, as well as dealing with equipment, control centres, training, and exercises.

---

[2] See O'Brien, TH, *Civil Defence* (HMSO, London, 1955) 187. See further Ch 9.
[3] Civil Defence (Suspension of Powers) Act 1945.
[4] For further information, see http://www.subbrit.org.uk/.
[5] The local authorities were in practice not involved: Topping, I, 'Emergency Planning Law and Practice in Northern Ireland' (1988) 39 *Northern Ireland Legal Quarterly* 336 at 339. The Civil Protection in Peacetime Act 1986 (see below) was not applied to Northern Ireland.
[6] SI No 1342.   [7] SI No 2121.   [8] SI No 70.   [9] SI No 1643.

The county councils and county borough councils were also required to orga-   **2.05**
nize a new Civil Defence Corps, a civilian body of volunteers administered by
the local authorities, distinct from police or military organizations.[10] The Corps
was in some ways a continuance of the Air Raid Precautions wardens, a system
begun in 1935, formalized by the Air-Raid Precautions Act 1937 and Air-Raid
Precautions (General Schemes) Regulations[11] but disbanded in 1946.[12] An
Industrial Civil Defence Service was added in 1951 to organize civil defence
activities at substantial industrial premises. Three civil defence schools were set
up in 1956 (including at Easingwold), and there was a Civil Defence Corps
Staff College at Sunningdale. Recruitment proved problematic and only ever
reached about 25 per cent of establishment (of 800,000). There was 'an air of
vacillation about the whole business',[13] and the Corps was abolished after a
defence spending review in 1965.[14] Nonetheless, the attraction of unpaid
volunteers remained potent, and a Home Office Circular of 1983 encouraged
local authorities to recruit them.[15]

There was also public guidance from the Home Office. The most widely dis-   **2.06**
seminated document was *Protect and Survive*, issued in 1976,[16] followed by
advice on *Domestic Nuclear Shelters* in 1981, and a leaflet, *Civil Defence: Why We
Need It* which responded to criticism of the original document.[17]

Detractors of civil defence increasingly reproved the futile expenditure on   **2.07**
protection against the unprotectable.[18] The mobilization of private sectors
could not be easily secured, though there remained powers under the Civil
Defence Act 1939 to force public utility undertakers.[19] Sometimes, even local
authorities refused to cooperate[20] or made woefully inadequate responses.[21] By
contrast, some commended the assurance given to the public and argued that

---

[10] Civil Defence Corps Regulations 1949 SI No 1433 (repealed by Civil Defence Corps
(Revocation) Regulations 1968 SI No 541). See further Civil Defence (Armed Forces) Act 1954.
[11] 1938 SR & O No 251.
[12] See O'Brien, TH, *Civil Defence* (HMSO, London, 1955) Chs 3, 4.
[13] Laurie, P, *Beneath the City Streets* (Granada, London, 1983) 105.
[14] Hennessy, P, *The Secret State* (Allen Lane, London, 2002) 139.
[15] See Home Office, *Emergency Planning Guidance to Local Authorities* (London, 1983);
Hilliard, L, 'Local Government, Civil Defence and Emergency Planning (1986) 49 MLR 476.
See also Home Office, *Emergency Planning Guidance to Local Authorities* (ISBN 0-86252-196-3,
1985).   [16] See http://www.cybertrn.demon.co.uk/atomic/.
[17] Criticism continued, most notably Raymond Briggs' cartoon book and film, *When the
Wind Blows* (1986)
[18] See Campbell, D, *War Plan UK* (Burnett Books Ltd, London, 1982); Laurie, P, *Beneath
the City Streets* (Granada, London, 1983).
[19] Sections 36–39. Note also the power to requisition land (s 62).
[20] Coventry City Council refused to undertake duties in 1954, and many refused to take part
in the training exercise Hard Rock '82: Campbell, D, *War Plan UK* (Burnett Books Ltd, London,
1982) Ch 1.
[21] HC Debs Standing Committee F col 48 27 January 2004, Nigel Evans. See further
http://www.subbrit.org.uk/rsg/sites/s/swansea/index.html.

a future war would more likely involve conventional and chemical attack rather than nuclear holocaust.[22]

2.08 Attempts were made by the Civil Protection in Peacetime Act 1986[23] to reinvigorate civil defence. Section 2 provided for the use of civil defence resources by local authorities and even encouraged (but did not enjoin) planning for an emergency or disaster involving destruction of, or danger to, life or property.

2.09 The Civil Defence (General Local Authority Functions) Regulations 1993[24] replaced the 1949 regulations and sought to reiterate in contemporary form the duty of county councils to make and keep under review and revise plans for their area, to carry out exercises based on such plans, and to arrange for training, and it was the duty of district councils to assist them.

2.10 Despite these changes, civil defence became an increasingly marginal preoccupation, especially after the end of the Cold War. As a result, 'Overall, it has been generally agreed for some years that a more robust and resilient emergency response culture will not be achieved until there is a new statutory duty covering all the main local responder organisations.'[25]

(b) Forerunners to Part II of the 2004 Act

2.11 The chief precursor to Part II of the 2004 Act was the Emergency Powers Act 1920[26] (plus the Emergency Powers Act (Northern Ireland) 1926[27]). This skeleton legislation enabled Her Majesty to proclaim that a state of emergency existed and thereupon to make regulations to deal with that emergency. Its own legal forebears may in turn be unearthed in the chief World War's Defence of the Realm Act 1914, which had spawned a huge range of muscular regulations and had been extended by the Termination of the Present War (Definition) Act 1918 and the War Emergency Laws (Continuance) Act 1920. It was then

---

[22] Dewar, M, *Defence of the Nation* (Arms and Armour Press, London, 1989).

[23] See Hilliard, L, 'Local Government, Civil Defence and Emergency Planning' (1986) 49 MLR 476; Lewis, J, 'Risk, Vulnerability and Survival' (1987) 13(4) *Local Government Studies* 75.

[24] SI No 1812. See also the Civil Defence (General/Local Authority Functions) (Scotland) Regulations 1993 (SI 1993 No 1774). and Local Government (Transitional and Consequential Provisions and Revocations) (Scotland) Order 1996 SI No 739.

[25] Cabinet Office, *The Draft Civil Contingencies Bill*, para 34.

[26] See Kidd, R, *British Liberty in Danger* (Lawrence & Wishart, London, 1940); Bonner, D, *Emergency Powers in Peacetime* (Sweet & Maxwell, London, 1985) Ch 5; Morris, GS, *Strikes in Essential Services* (Mansell, London, 1986) Ch 3.

[27] The Act was passed on one day. It omitted limits in the 1920 Act relating to conscription and to the expiration at seven days: Topping, I, 'Emergency Planning Law and Practice in Northern Ireland' (1988) 39 *Northern Ireland Legal Quarterly* 336. In the Republic of Ireland, the counterpart is the Protection of the Community (Special Powers) Act 1926 (No 16) which is very similar to the 1920 Act and survives in that jurisdiction. No proclamation has ever been made.

replaced in Ireland by the Regulation of Order in Ireland Act 1920 which allowed for the intensification of military and policing powers during the ongoing conflict.[28] As for British crises, the Emergency Powers Act 1920 (and later, a Northern Ireland counterpart) was passed between 25 and 29 October to deal with major industrial disruption (in particular, a strike by the miners was averted that very week), civil disorders, and aspiring revolutionaries. On the one hand, the Government sought to be emollient about their intentions which were not to attack trade unions.[29] On the other hand, its effect was said by critics at the time to 'clothe [the Government] with exactly the same powers that were used during the war under the Defence of the Realm Act.'[30] Somewhere between the two positions, the Prime Minister, Lloyd George, astutely observed that the Act represented a contingent determination: 'if it was intended by direct action to take the government out of the hands of the constitutional authorities of this country, we shall have to meet it with all the resources of the state...'.[31] The 2004 Act reflects the same resolve, though the circumstances for its use may be as much against the vagaries of nature as human malefaction.

The body of the Emergency Powers Act was a gaunt affair, confined to two **2.12** substantive sections. It was triggered by a Royal Proclamation under s 1:

> 1.—(1) If at any time it appears to His Majesty that any action has been taken or is immediately threatened by any persons or body of persons of such a nature and on so extensive a scale as to be calculated, by interfering with the supply and distribution of food, water, fuel, or light, or with the means of locomotion, to deprive the community, or any substantial portion of the community, of the essentials of life, His Majesty may, by proclamation (hereinafter referred to as a proclamation of emergency), declare that a state of emergency exists....

There followed a regulation-making power in s 2:

> (1) Where a proclamation of emergency has been made and so long as the proclamation is in force, it shall be lawful for His Majesty in Council, by Order, to make regulations for securing the essentials of life to the community, and those regulations may confer or impose on a Secretary of State or other Government department, or any other persons in His Majesty's service or acting on His Majesty's behalf, such powers and duties as His Majesty may deem necessary for the preservation of the peace, for securing and regulating the supply and distribution of food, water, fuel, light, and other necessities, for maintaining the means of transit or locomotion, and for any other purposes essential to the public safety and the life of the community,

---

[28] The regulations are at 1920 SR & O No 1530. See Townshend, C, *The British Campaign in Ireland, 1919–1921* (Oxford University Press, Oxford, 1975); Campbell, C, *Emergency Law in Ireland, 1918–1925* (Clarendon Press, Oxford, 1994).

[29] HC Debs vol 133 col 1401 25 October 1920, Bonar Law.

[30] HL Debs vol 42 col 111 28 October 1920, Lord Buckmaster.

[31] HC Debs vol 133 col 1405 25 October 1920.

and may make such provisions incidental to the powers aforesaid as may appear to His Majesty to be required for making the exercise of those powers effective:

Provided that nothing in this Act shall be construed to authorise the making of any regulations imposing any form of compulsory military service or industrial conscription:

Provided also that no such regulation shall make it an offence for any person or persons to take part in a strike, or peacefully to persuade any other person or persons to take part in a strike. . . .

(3) The regulations may provide for the trial, by courts, of summary jurisdiction, of persons guilty of offences against the regulations; so, however, that the maximum penalty which may be inflicted for any offence against such regulations shall be imprisonment with or without hard labour for a term of three months, or a fine of one hundred pounds, or both such imprisonment and fine, together with the for-feiture of any goods or money in respect of which the offence has been committed: Provided that no such regulations shall alter any existing procedure in criminal cases, or confer any right to punish by fine or imprisonment without trial.

It will be seen that s 2 has many resonances with the 2004 Act, including its open-ended nature, as well as some of its specific limitations.

2.13 Some of the organizational structure associated with the legislation has also found echoes in the current regime, especially at regional level. Earlier examples include the Civil Emergency Organisation formed soon after 1918[32] and brought into operation during the 1926 General Strike, when England and Wales were divided into 11 areas each under a Civil Commissioner who was given special powers under the Emergency Powers Act 1920.[33] A regional system of government was set up in London in September 1938 (extended in early 1939 to the remaining 11 regions and later sanctioned by the Regional Commissioners Act 1939).[34] The Regional Commissioners Act 1939 coordinated and tested the capabilities of emergency services and the Air Raid Precautions (ARP) Services. There were also Regional Councils (akin to the Regional Resilience Forums, described in Chapter 4) allowing for discussion between Commissioners and local stakeholders.[35] The Regional Commissioners were controlled through the Ministry of Home Security War Room, but, in the event of the central government in London being unable to function, the Regional Commissioners could, at their discretion, assume full powers of civil government in their region and were furnished with Regional War Rooms. At that point, the Regional Commissioner moved from 'a dormant commission' of coordination and persuasion sufficient for 'normal' war condition into executive

---

[32] See O'Brien, TH, *Civil Defence* (HMSO, London, 1955) 117.
[33] See ibid, 29; Jeffery, K and Hennessy, P, *States of Emergency: British Governments and Strikebreaking since 1919* (Routledge & Kegan Paul, London, 1983) Ch 8.
[34] Ibid, 118, 157. For details, see 179–180, Chs 7 and 14. See also Laurie, P, *Beneath the City Streets* (Granada, London, 1983) Ch 7. For the corresponding Defence Areas, see 1940 SR & O No 1503.
[35] Ibid, 186.

overlords.[36] The system lingered after 1945.[37] For example, during the 1973–74 miners' strike, it was revealed that, 'Across the country a network of regional commissioners was ready to maintain basic services, as in a nuclear alert.'[38]

Legislative history repeated itself in 1964, when the 1920s Acts were amended **2.14** in order to conserve wartime powers by the Emergency Powers Act 1964[39] and the Emergency Powers (Amendment) Act (Northern Ireland) 1964.[40] The Emergency Powers (Defence) Act 1939, which was passed upon the declaration of war in 1939, allowed for defence regulations to be issued without parliamentary approval. The Act was continued annually after 1945 until repealed by the Emergency Laws (Repeal) Act 1959, subject to the continuance of some regulations. The Emergency Powers Act 1964 sought to preserve some of these surviving regulations by amending the 1920 Act in two ways. First, it widened the causes of 'emergency' to include events of such a nature as to disrupt the life of the community. Secondly, it made permanent a provision (reg 6) from the Defence (Armed Forces) Regulations 1939 to allow without any need to declare a state of emergency the use of the armed forces in direct employment in 'agricultural work or in other work, being urgent work of national importance'. This power still exists and will be described in Chapter 9.

The 1920 Act was invoked on 12 occasions in Britain, all related to industrial **2.15** dislocation, and emergency regulations were issued on ten.[41] The first occasion was on 31 March 1921, when the 'Triple Alliance' (of miners, railwaymen, and transport workers) briefly supported the locked-out miners.[42] There was a

---

[36] Ibid, 175, 185. They could even be empowered to detain without trial: 1940 SR & O No 1135. For further details, see Chs 7 and 14. The commissioners resigned in the first half of 1945; none had ever assumed the full default powers available to them.

[37] In the Cold War era, 12 nuclear bunkers were built, linked to the Central Government War HQ at RAF Rudloe Manor, Corsham, near Bath: Campbell, D, *War Plan UK* (Burnett Books Ltd, London, 1982) Chs 4, 5.

[38] Campbell, J, *Edward Heath: A Biography* (Cape, London, 1993) 571.

[39] See further Whelan, C, 'Military Intervention on Industrial Disputes' (1979) 18 IRJ 222; Peak, S, *Troops in Strikes* (Cobden Trust, London, 1984) Ch 2; Rowe, PJ, and Whelan, CJ (eds), *Military Intervention in Democratic Societies* (Croom Helm, London, 1985); Morris, GS, *Strikes in Essential Services*, (Mansell, London, 1986) Ch 4.

[40] See Creighton, WB, 'Emergency Legislation and Industrial Disputes in Northern Ireland' in Wood, JC, *Encyclopaedia of Northern Ireland Labour Law and Practice* (Labour Relations Agency, Belfast, 1983).

[41] In 1924 and 1948, the dispute was settled before regulations appeared: Morris, GS, 'The Emergency Powers Act 1920' [1979] PL 317 at 318. For further details, see ibid at 320; Bonner, D, *Emergency Powers in Peacetime* (Sweet & Maxwell, London, 1985) 244–54; Ewing, K and Gearty, C, *The Struggle For Civil Liberties: Political Freedom and the Rule of Law in Britain, 1914–1945* (Clarendon, Oxford, 2000) Ch 4. In Northern Ireland, the 1926 Act was proclaimed in 1970, 1972, 1973, 1974, and 1979: Topping, I, 'Emergency Planning Law and Practice in Northern Ireland' (1988) 39 *Northern Ireland Legal Quarterly* 336. Prior to 1973, there could be reliance upon regulations pursuant to the Civil Authorities (Special Powers) Acts (Northern Ireland) 1922–43.

[42] 1921 SR & O Nos 439, 440, 739, 740, 903, 904, 1128, 1129. The Orders came in pairs, one for the Proclamation and one for the regulations. See also *Inkpin v Roll* (1922) 82 JP 61 (further discussed in Ch 5).

proclamation in April 1924 in response to a tram and bus strike, though the action was settled just as the proclamation was being issued so no regulations were produced.[43] The next invocation was the General Strike of 1926, sparked by Trade Union Congress support for the mine workers and lasting from 3 May until 12 May, though the miners did not wholly return to work until December. The initial proclamation was on 30 April 1926, and there followed seven renewals until December.[44] Then there were proclamations relating to dock strikes of June 1948[45] and July 1949,[46] a rail strike in May 1955,[47] and the May 1966 seamen's strike.[48] The period from 1970 to 1974 saw the most active period of usage. There were two invocations in 1970–a dock strike[49] and an electricity power workers' strike.[50] Two further exercises occurred in 1972–a miners' strike[51] and then a dock strike.[52] The final declarations occurred in November 1973[53] in response to the renewed miners' strike and lasted until March 1974.

2.16   Two later uses transpired in Northern Ireland. One was in May 1974,[54] when the Ulster Workers Council called a general strike in opposition to the Northern Ireland Assembly support for the Executive's policy of support for the Sunningdale Agreement[55] as set out in the Northern Ireland Constitution Act 1973.[56] There was also a proclamation in January 1979 relating to the petrol tanker drivers' strike.[57]

2.17   Several trends tended to marginalize the Emergency Powers Act 1920 after 1974. First, more tailored provisions and less confrontational approaches became the fashion. The prime example concerns the Fuel and Electricity (Control) Act 1973, in response to the industrial action by mine workers and

---

[43] The draft is at PRO, CAB 23/47, Cabinet 23(24) 27 March 1924 Appendix III. The circumstances are set out in HC Debs vol 171 col 1682 27 March 1924, J Ramsey MacDonald.

[44] 1926 SR & O Nos 450, 451, 555, 556, 776, 777, 912, 913, 1061, 1062, 1129, 1130, 1333, 1334, 1461, 1499.

[45] Proclamations ceased to be published as secondary legislation on and after this time, presumably since they could not qualify as 'statutory instruments'. On this occasion, the regulations were also not published because the strike action was settled: HC Debs vol 452 col 1385 1 July 1948.   [46] 1949 SI No 1300.

[47] 1955 SI No 791. The relevant Proclamations were reproduced in the volume of statutory instruments at vol II, 3178, 3179.

[48] 1966 SI Nos 600, 740. For the Proclamations see vol II, 2684, 2691, 2692.

[49] 1970 SI No 1042.   [50] 1970 SI No 1864.   [51] 1972 SI No 157.

[52] 1972 SI No 1164.

[53] 1973 SI No 1881; 1973 SI No 2089; 1974 SI No 33; 1974 SI No 175; 1974 SI No 350.

[54] See Emergency Regulations (Northern Ireland) 1974 SI (NI) Nos 16, 27, 28, 29, 46, 47, 88, 89, 247.

[55] Northern Ireland Office, *Northern Ireland Constitutional Proposals* (Cmnd 5259, London, 1973).

[56] See Fisk, R, (1975) *The Point of No Return: The Strike Which Broke the British in Ulster;* Boyle, L (1978) 'The Ulster Workers' Council Strike: May 1974' in, Darby, J and Williamson, A, (eds), *Violence and the Social Services in Northern Ireland* (Heinemann, London, 1978); Rees, M, *Northern Ireland: A Personal Perspective* (Methuen, London, 1985) 65–90.

[57] Emergency Regulations (Northern Ireland) 1979 SI (NI) No 1.

others, a threat exacerbated by disruption to fuel supplies by OPEC countries. The Emergency Powers Act 1920 was at first promulgated, and regulations were issued to restrict the use and supply of fuel.[58] After it was passed in December, the 1973 Act began to replace the 1920 Act regulations. It embodied a number of advantages for the Executive over the Emergency Powers Act. There was no need for a proclamation, directions were not subject to Parliamentary scrutiny, the penalties for contravention were stiffer, and it could persist for one year at a time. It was used extensively in late 1973 and 1974, when there were directions to fix the maximum price of petrol and to restrict electricity supply[59]—the 'three-day week' being declared on 13 December 1973 until March 1974.[60] The 1973 Act was extended even after the industrial action ended and in fact persisted until 30 November 1976,[61] when the Energy Act 1976 (and, since it is still in force, described in Chapter 4) replaced it.

Other measures responding to emergencies proved more controversial, and **2.18** most prominent of all was the Prevention of Terrorism (Temporary Provisions) Act 1974. Current versions are set out in the Terrorism Act 2000, the Anti-Terrorism, Crime and Security Act 2001, and the Prevention of Terrorism Act 2005, with further changes contained in the Terrorism Bill 2005–06. However, these laws against terrorism address the prevention, investigation, detection, and punishment of terrorists. They do not cover planning or recovery.[62]

The second trend acting against the Emergency Powers Act concerned the **2.19** abilities of the military for intervention which have diminished owing to the growing technical sophistication of essential services.[63] In addition, the numerical strength of the armed forces has shrunk while their commitments are arguably more logistically demanding than in the 1970s, when many served in Northern Ireland or Germany rather than in Iraq or Afghanistan. Given the ever-larger numbers of the 'police family', the use of police assets in emergency situations could provide an alternative, though restraint has prevailed in view of the potential conflict with their law enforcement and peace-keeping functions.[64]

---

[58] See Electricity (Heating) (Restriction) Orders 1973 SI Nos 1900, 1913; Electricity (Advertising, Display etc) (Restriction) Order 1973 SI No 1901; Motor Fuel (Restriction of Supplies) Order 1973 SI No 1943.

[59] Fuel Control (Modification of Enactments) (Speed Limits) Order 1973 SI No 2051; Fuel and Electricity (Heating) (Control) Orders 1973 SI No 2068, 2092; Electricity (Lighting) (Control) Order 1973 SI No 2080; Motor Fuel (Restriction of Acquisition) Order 1973 SI No 2087; Motor Fuel (Maximum Retail Prices) Order 1973 SI No 2119; Electricity (Industrial and Commercial Use) (Control) Orders 1973 SI Nos 2120, 2137, 2146, 2172, 1974 SI Nos 78, 117, 137. See further see Morris, GS, 'The Emergency Powers Act 1920' [1979] PL 317 at 340–1.

[60] Electricity (Industrial and Commercial Use) (Control) (Revocation) Order 1974 SI Nos 377, 511. [61] 1974 SI No 1893, 1975 SI No 1705.

[62] One exception might be the use of cordons under the Terrorism Act 2000, s 33.

[63] Dean, M, 'Entertaining the Troops' *The Guardian* 16 July 1984.

[64] Some examples have occurred such as driving ambulances in 1979, 1981, and 1982: Morris, GS, *Strikes in Essential Services* (Mansell, London, 1986) Ch 4.

2.20 The third trend, which arguably proved most decisive to the discontinuance of the Emergency Powers Act, was the successive reforms to industrial and union laws after 1979.[65] The effect has been to emasculate the opportunities for lawful strike action, especially secondary action. That trend coincided with more aggressive employment practices, such as no-strike agreements and no-union agreements. Though the Emergency Powers Act 1920 was not legally confined to industrial disruption, its usage was linked in practice.

2.21 The fourth trend was the break up of monolithic nationalized industries which were privatized after 1979. The result was often smaller units as well as new contractual terms.[66]

2.22 While the Emergency Powers Act was going out of fashion, the potential for emergencies and the need for emergency powers was felt to be growing. This contrast highlighted defects in the Emergency Powers Act. These included its emphasis upon dangers to the supply or distribution of food, water, fuel, light, or the means of locomotion, thereby ignoring new risks such as animal and human health diseases and terrorism or new vulnerabilities because of technology. Next, there was perceived to be the need to be responsive within more defined geographical bounds rather than waiting for the whole of Great Britain or Northern Ireland to be affected. There was also a need to take account of the reality of devolution.[67] It also did not embody the language of rights. In short, 'As currently constituted the Act does not serve a useful function in the early twenty-first century.'[68]

(c) Repeals

2.23 The Civil Contingencies Act, Schedules 2 and 3, repeals the Civil Defence Act 1948 (and regulations made under them), plus the Civil Defence Act (Northern Ireland) 1950, the Civil Defence (Armed Forces) Act 1954, the Civil Protection in Peacetime Act 1986, and the Civil Defence (Grant) Act 2002. It also terminates the Emergency Powers Acts of 1920 and 1964 (s 1 only) plus their Northern Ireland counterparts.

(d) The non-repeal of common law

2.24 Section 18(3) of the Civil Contingencies Act refers to the preservation of other statutory powers but does not mention residual case law. This stance leaves alone, lurking under the folds of pomp and tradition relating to the Crown, the effective and often draconian non-statutory powers to tackle episodes of crisis.

---

[65] See Ewing, KD, *The Right to Strike* (Clarendon Press, Oxford, 1991); Hepple, B, 'The Future of Labour Law' (1995) 24 ILJ 303; Lord Wedderburn, *Labour Law and Freedom: Further Essays in Labour Law* (Lawrence and Wishart, London, 1996).

[66] See Ernst, J, *Whose Utility?: The Social Impact of Public Utility Privatization and Regulation in Britain* (Open University Press, Buckingham, 1994).

[67] Cabinet Office, The Draft Civil Contingencies Bill, para 1.7.   [68] Ibid, para 5.15.

The precise scope of the Crown's powers to intervene in emergencies has been a perennial matter for fierce debate.[69] The contemporary cases amply demonstrate that prerogative powers continue to be vibrantly exercised in emergencies, such as the destruction[70] and requisitioning of property,[71] the banning of trade union membership,[72] the interception of communications,[73] the supply of weapons,[74] and the direction and disposition of the armed forces and acting in defence of the realm.[75]

As well as prerogative powers, vague common law powers can also be called in   **2.25** aid, such as 'the common law duty of every citizen to provide reasonable support to the police should they request it'.[76] There is precedent for such a duty being imposed on the citizen in olden days,[77] provided the request from the authorities is reasonably necessary. In addition, peace officers (including police constables but not soldiers)[78] must act without request under the threat of the penalty for the common law offence of misbehaviour in public office.[79] More recently, in *Albert v Lavin*, Lord Diplock opined that:

> . . . every citizen in whose presence a breach of the peace is being, or reasonably appears to be about to be, committed has the right to take reasonable steps to make the persons who is breaking or threatening to break the peace refrain from doing so; and those reasonable steps in appropriate cases will include detaining him against his will. At common law this is not only the right of every citizen, it is also his duty, although, except in the case of a citizen who is a constable, it is a duty of imperfect obligation.[80]

However, this opinion is without authority, and the final stanza points perhaps to a moral rather than legal duty in the mind of the judge.

---

[69] Evelegh, R, *Peacekeeping in a Democratic Society* (Hurst, London, 1978); Jeffery, K and Hennessy, P, *States of Emergency: British Governments and Strikebreaking since 1919* (Routledge & Kegan Paul, London, 1983); Winterton, G, 'The Prerogative in Novel Situations' (1983) 99 LQR 407; Greer, SC, 'Military Intervention in Civil Disturbances' [1983] PL 573; Peak, S, *Troops in Strikes* (Cobden Trust, London, 1984); Bonner, D, *Emergency Powers in Peacetime* (Sweet & Maxwell, London, 1985); Rowe, PJ and Whelan, CJ (eds), *Military Intervention in Democratic Societies* (Croom Helm, London, 1985); Campbell, C, *Emergency Laws in Ireland 1918–1925*, (Clarendon Press, Oxford, 1994); Walker, C and Reid, K, 'Military Aid in Civil Emergencies: Lessons from New Zealand' (1998) 27 *Anglo American Law Review* 133.

[70] *Burmah Oil Co Ltd v Lord Advocate* [1965] AC 75.

[71] Requisitioning of Ships Order 1982 SI No 1693.

[72] *Council of Civil Service Unions v Minister for the Civil Service* [1985] AC 374. See Walker, C, 'Review of the Prerogative' [1987] PL 62.

[73] See *Report of the Committee of Privy Councillors Appointed to Inquire into the Interception of Communications* (Cmnd 283, London, 1957); *Malone v Metropolitan Police Commissioner (No 2)* [1979] Ch 344.

[74] *R v Secretary of State for the Home Department, ex p Northumbria Police Authority* [1989] QB 26. See Bradley, AW, 'Comment' [1988] PL 298.

[75] *The Case of the King's Prerogative in Saltpetre* (1606) 77 ER 1294; *R v Hampden* (1637) 3 Cobb St Tr 826.   [76] HC Debs vol 400 col 437w 25 February 2003, Adam Ingram.

[77] See *R v Brown* (1841) Car & M 314; *R v Atkinson* (1869) 11 Cox CC 330.

[78] This duty is now set out in the Police Act 1996, Sch 4.

[79] *R v Dytham* [1979] QB 722. See also *R v Bowden* [1995] 4 All ER 505; Nicolson, D, 'The Citizen's Duty to Assist the Police' [1992] Crim LR 611.   [80] [1982] AC 546 at 565.

2.26 Whilst the use of the prerogative and common law to deal with matters of crisis and disturbance is well established, there is also wide agreement amongst observers as to the need for both clarification and reform.[81] To take one illustration, Evelegh asserts that:

> It is startling to reflect that in strict constitutional theory, a corporal with ten privates in a lorry who happened to drive through Grosvenor Square in London when a crowd of demonstrators had burst through a police cordon and were attacking an embassy, would have not merely a right to intervene and suppress the disorder with lethal weapons if necessary, but an absolute duty to do so, in spite of anyone from the Prime Minister to the senior policeman on the spot telling him not to.[82]

By contrast, Greer firmly refutes the suggestion that soldiers have a legal duty to intervene[83] but is forced to the observation that 'no clear conclusion on these matters can be derived'.[84] It is proper that there is constitutional concern arising through the 'democratic deficit'—that troops can be called out from their barracks by the Executive without the certainty of democratic approval or even notice. The deficit becomes most acute when the soldiers are ordered to engage in any activity under the heading of military aid to the civil powers (MACP) (maintaining or restoring public order)[85] which involves contact with civilians in a confrontational or coercive relationship.

2.27 From this debate,[86] it may be concluded that it would be helpful to clarify by statute the roles of the military and to set them out on a statutory basis.[87] Any reform process must involve the injection of constitutional precepts

---

[81] See also Royal Commission on Intelligence and Security, *4th Report* (Australian Government Publishing Service, Canberra, 1977) Vol 1, C, paras 122, 123; Commission of Inquiry concerning Certain Activities of the RCMP, *Second Report: Freedom and Security under the Law* (Ottowa, 1981) Pt V, Ch 4 para 2.

[82] Evelegh, R, *Peacekeeping in a Democratic Society* (Hurst, London, 1978) 8.

[83] Greer, SC, 'Military Intervention in Civil Disturbances' [1983] PL 573 at 595.

[84] Ibid, 599.

[85] See Bonner, D, *Emergency Powers in Peacetime* (Sweet & Maxwell, London, 1985) Ch 5; Rowe, P, *Defence: The Legal Implications* (Brassey's Defence Publishers, London, 1987) Ch 4. See further Ch 8.

[86] There is also a substantial debate about the applicability and meaning of martial law within the United Kingdom. See Townshend, C, *The British Campaign In Ireland 1919–1921* (Oxford University Press, Oxford, 1975); Lawson, FH and Bentley, DJ, *Keir and Lawson, Cases in Constitutional Law* (6th ed, Clarendon Press, Oxford, 1979) 217–50; Townshend, C, 'Martial Law: Legal and Administrative Problems of Civil Emergency in Britain and the Empire 1800–1940' (1982) XXV Hist Jo 167; Lee, HP, *Emergency Powers* (Law Book Co, Sydney, 1984) Ch 6; Campbell, C, *Emergency Laws in Ireland 1918–1925* (Clarendon Press, Oxford, 1994). For relevant cases, see *Egan v Macready* [1921] 1 IR 265; *R v Murphy* [1921] 2 IR 190; *R v Allen* [1921] 2 IR 241; *Re Clifford and O'Sullivan* [1921] 2 AC 570; *R (Garde) v Strickland* [1921] 2 IR 317; *R (Ronayne and Mulcahy) v Strickland* [1921] 2 IR 333.

[87] See Walker, C and Reid, K, 'Military Aid in Civil Emergencies: Lessons from New Zealand' (1998) 27 *Anglo-American Law Review* 133. The idea is rejected without argument by the Cabinet Office (Cabinet Office, *The Draft Civil Contingencies Bill*, para 37).

of both a substantive and a structural kind. It should: (i) specify the legitimate/illegitimate uses of military intervention (the principled parameters); (ii) provide a clear basis in detailed law for intervention and for the termination of intervention; (iii) clarify the chain of command; (iv) specify the powers available to soldiers which then arise; and (v) ensure accountability to democratic and judicial oversight. There are precedents for change along these lines in other common law jurisdictions which seek to follow very similar constitutional precepts. In the United States, for example, Congress has attempted to control Presidential external war-making through the War Powers Resolution 1973,[88] by the threat of Congress terminating a Presidential deployment of troops after 60 days, an exercise which has revealed substantial problems of drafting, enforcement,[89] and resistance to accountability. Likewise, there are problems in regulating the domestic deployment of military forces,[90] in part arising from uncertainty over the scope of the Posse Comitatus legislation.[91] It does not appear that there is much interest in emulation of these precedents from our common law cousin. The House of Commons Select Committee on Public Administration announced the objective of 'Taming the Prerogative', with particular focus on the power to engage in armed conflict which it felt should be approved by Parliament.[92] The Government gave a dismissive response—a pragmatic, case-by-case approach must be adopted and the Government felt it was already sufficiently accountable to Parliament.[93]

The Civil Contingencies Act should at least have ruled out reliance upon any **2.28** residual common law where the Act's remit was applicable (which is just about everywhere). A more thorough-going reform of prerogative powers is evidently and regrettably a task for tomorrow.

B. Legislative History of the Act

(a) Legislative formation

As well as remedying the inadequacies of prior legislation, the origins of the **2.29** Civil Contingencies Act 2004 reside within the experiences of dealing with

---

[88] 50 USC s 1541. For recent commentaries, see LeMar, AD, 'War Powers: What are they Good for?' (2003) 78 *Indiana Law Journal* 1045; Jinks, D and Goodman, R, 'International Law, War Powers and the Global War on Terrorism' (2005) 118 *Harvard Law Review* 2047.
[89] See *Crockett v Reagan* 467 US 1251 (1984); *Sanchez-Espinoza v Reagan* 770 F 2d 202 (1985); *Campbell v Clinton* 203 F 3d 19 (2000).
[90] The detailed regulation of military intervention is provided for in 32 CFR ss 215, 501. See further Engdahl, DE, 'Foundations for Military Intervention in the U.S.' and Sherman, EF, 'Contemporary Challenges to Traditional Limits on the Role of the Military in American Society' in Rowe, PJ and Whelan, CJ (eds), *Military Intervention in Democratic Societies* (Croom Helm, London, 1985). [91] 18 USC s 1385. See Ch 9 for details.
[92] 2003–04 HC 422. [93] Cm 6187, London, 2004.

a number of emergencies over the previous few years. Prime examples include the following.[94]

2.30 Weather emergencies were the most common. For instance, there were various incidents of severe flooding, especially in January and March 1999 and November 2000. These emergencies reflect the fact that five million people live in two million properties in flood risk areas in England and Wales and the prediction that the threat may be increasing with the onset of global climate change. Fortunately, there has been no disaster on the scale of the North Sea storm surge of February 1953, when there were 307 deaths on land, 234 at sea, and 30,000 people were rendered homeless.[95] Other severe weather is also a threat. The 'hurricanes' of January 1990 caused widespread damage, while Europe's summer heatwave in 2003 caused massive agricultural losses and killed around 20,000 people.[96]

2.31 Next, the Government was spurred into action by terrorist incidents, such as those experienced in Manchester city centre in 1996[97] and, most vividly, in New York on September 11, 2001, following which the Home Secretary declared that the Government's 'objective is to do everything that can be done to enhance our resilience'.[98] These events pointed to the need to widen the compass of contingency planning as well as putting some focus upon central government machinery.[99] There remains debate as to whether the emergence of the al Qaida terrorism network is a wholly new phenomenon and requires a new strategic disposition.[100] Certainly, both its goals and tactics are manifestly more geared towards catastrophe than most previous terrorist groups.[101] Terrorist events have dominated world politics, and so it is not surprising that they have given a substantial impetus to contingency planning, even if the need may be somewhat exaggerated by the emotion generated by the events on the basis that 'crashed aircraft, burning and subsequently collapsed high rise buildings...were not outside the emergency response preparedness parameters.'[102]

---

[94] Cabinet Office, *The Draft Civil Contingencies Bill: Explanatory Notes etc*, Annex A.

[95] One response was the Coastal Flooding (Emergency Powers) Act 1953.

[96] Civil Contingencies Secretariat, *Full Regulatory Impact Assessment*, para 16.

[97] See Walker, C, 'Political Violence and Commercial Risk' (2004) 56 CLP 531.

[98] HC Debs vol 375 col 1021 28 Nov 2001, Tim Collins.

[99] See further Cabinet Office, *The Draft Civil Contingencies Bill*, para 1.5.

[100] See Stern, J, *The Ultimate Terrorists* (Harvard University Press, Cambridge, Mass, 1999); Laqueur, W, *The New Terrorism: Fanaticism and the Arms of Mass Destruction* (Oxford University Press, New York, 1999); Raufer, X, 'New World Disorder and New Terrorism' (1999) 11 *Terrorism and Political Violence* 30; Tucker, D, 'What is New about the New Terrorism and How Dangerous is It?' (2001) 13 *Terrorism and Political Violence* 1; House of Commons Defence Committee, *The Threat from Terrorism* (2001–02 HC 348–1); Rohan, G, *Inside Al Qae'da* (C Hurst, London, 2002); Sedgwick, M, 'Al-Qaeda and the Nature of Religious Terrorism' (2004) 16 *Terrorism and Political Violence* 795.

[101] But it does resonate with the 'propaganda of the deed' and other nihilistic attacks of the nineteenth century; see Quail, J, *The Slow Burning Fuse: The Long History of the British Anarchists* (Paladin, London, 1978).

[102] Joint Committee on the Draft Civil Contingencies Bill, Appendix 7.

The next catastrophe of recent experience is pandemic disease relating to live- **2.32**
stock. The most serious was the Foot and Mouth Disease outbreak of 2001.[103]
The outbreak was the worst ever experienced by Britain since proper records
began. Around six million animals were culled at a cost of over £3bn. Better
contingency planning, especially at regional level, emerged as a key issue alongside
more specific powers taken in the Animal Health Act 2002.[104] As well as home-
generated problems, there is concern that international trade and travel will
introduce exotic problems, and such pandemic disease can also relate to humans.
The spread of Asiatic bird flu became a major preoccupation during 2005.[105]

Equally affecting Ministerial minds were the fuel price protests in September **2.33**
2000 which resulted in 'picket' lines at six out of nine oil refineries. The effect was
the closure of many petrol stations and a significant reduction in road traffic,[106] as
well as a threat to the viability of emergency services such as health.[107] The protest
was loosely organized and centred around the grievances of farmers and hauliers,
including a group called Farmers for Action. Though part of the problem could
be attributed to rising international oil prices, the Government's fuel duty 'escalator' also was denounced as producing some of the highest prices in Europe. The
escalator was discontinued in the Budget of spring 2000, but no decreases then
resulted. The protests were alleviated by Government pressure on oil companies
to ensure its drivers worked properly as well as promises of a review of fuel and
road taxes. This episode is rather less prominent in the official papers, perhaps
because it was too politically sensitive to mention. After all, it amounted to
legitimate protest as well as a threat to 'normality'.

In total, these crises demonstrated not only that risk was often unrelated to **2.34**
industrial conflict or civil defence but also that wide area or 'slow burn' emergencies were not being handled successfully, not least because of unclear relationships and responsibilities both within tiers of government and also between
public and private bodies.[108] Greater vulnerabilities were also felt to exist through
the emergence of '[g]reater dependency on supply and distribution networks,
on technology and upon increasingly complex interdependencies'.[109]

As a result, Ministers agreed at a meeting of the Central Local Partnership on 4 **2.35**
December 2000 that there should be a Review of Emergency Planning
arrangements in England and Wales.[110] On 13 February 2001, the Minister of

---

[103] See http://footandmouth.csl.gov.uk/. [104] See further Ch 4 for details.
[105] See ibid.
[106] See Hathaway, P, *The Effect of Fuel Protest on Road Traffic* (Transport Statistics Road
Traffic Division, DETR, London, 2000).
[107] The event is considered to be the closest to triggering what are now Part II powers in recent
years: HC Debs Standing Committee F col 147 3 February 2004, Fiona Mactaggart.
[108] Cabinet Office, *The Draft Civil Contingencies Bill: Explanatory Notes etc*, para 20.
[109] Civil Contingencies Secretariat, *The Government's Response*, 31.
[110] See http://www.ukresilience.info/epr/index.htm.

State at the Home Office announced the setting up of a steering group by the Home Office Emergency Planning Division.[111] The steering group was chaired by Charles Everett, the Director of the Home Office's Fire and Emergency Planning Directorate. A consultation document, Cabinet Office, *Emergency Planning Review: The Future of Emergency Planning in England and Wales*, appeared in August 2001. Even by that stage, the Government had concluded that 'the Civil Defence Act 1948, no longer provides an adequate framework for modern emergency planning in England and Wales particularly for local authorities.'[112] New risks and vulnerabilities were said to be to blame, along with the increasing demands of the public.[113]

2.36 As well as the broadening of types of emergency, the Cabinet Office was also firm about the need for broader conceptions of resilience which included a potential need to widen the remit to incorporate not just planning, preparation, maintenance, response, and recovery, but also prevention. The latter is underlined by the claim that 70 per cent of organizations experiencing a disaster cease trading within 18 months.[114]

2.37 Next, the Cabinet Office proposed that resilience planning would remain formally 'under the community leadership of the local authority' with the focus of the review being 'delivery of an emergency planning function responsive at a local level to the needs of all citizens and organisations'.[115] In this way, the pattern of diverting attention away from central responsibilities and facilities was reflected from the outset. As a result, it was local authorities which were to lead the multi-agency emergency planning process.[116] But there was recognition that a new factor is the extent of private organizational involvement in critical infrastructure and the need for local authorities to engage with others.[117]

2.38 An important issue which was raised in the paper was that of funding. From 1953 onwards, local authorities had received a Civil Defence grant from the Home Office. But, just as the 1948 Act was no longer viable, so 'The Government has now concluded that justification for specific grant to support wartime planning no longer exists.'[118] Instead, it proposed that emergency planning should be supported through the global grant alongside other locally delivered services, as had already occurred in Scotland. On other issues, policy remained fluid.

---

[111] HC Debs vol 363 col 100wa, Mike O'Brien. See further Cabinet Office, *Emergency Planning Review: The Future of Emergency Planning in England and Wales* (London, 2001) para 2.1.
[112] Cabinet Office, *Emergency Planning Review: The Future of Emergency Planning in England and Wales* (London, 2001) para 4.3.   [113] Ibid, paras 4.5, 4.6.
[114] Cabinet Office, *The Draft Civil Contingencies Bill: Explanatory Notes etc*, para 13.
[115] Ibid, para 3.2.   [116] Ibid, para 5.16.   [117] Ibid, para 4.7.
[118] Ibid, para 4.14. See further Ch 8.

The results of the consultation on these proposals were revealed by the Cabinet **2.39**
Office later in 2001.[119] There was amongst the responses (mainly from local
authorities) firm support for the broad sweep of the consultation paper. But
many had misgivings about the funding model and were unenthusiastic about
the involvement of the public.

In stark contrast to this consultation effort, there was no open debate on what **2.40**
became Part II of the Act. It was claimed in 2003 that 'Policy has been for-
mulated . . . in what has been an open and inclusive policy-making process.'[120]
In reality, that process was confined within the corridors of power; nothing was
revealed to the public nor had they been invited to participate.

The legislative parturition process was also impelled by the independent **2.41**
inquiries into disasters which had been undertaken in recent years.[121] Prominent amongst these were the reports on the sinking of the Herald of Free
Enterprise,[122] the Piper Alpha oil platform explosion in 1988,[123] the football
ground disasters at Bradford in 1985 and Hillsborough in 1989,[124] the sinking
of the Marchioness riverboat in 1989,[125] the Dunblane shootings in 1996,[126]
the underground rail fire at King's Cross in 1987[127] and the rail crashes at
Southall in 1997, Ladbroke Grove (Paddington) in 1999, and Hatfield in
2000,[128] the Foot and Mouth outbreak of 2001,[129] and arising from the BSE
Inquiry chaired by Lord Phillips.[130] Many of the reports were immediately sated

---

[119] Cabinet Office, *Emergency Planning Review: The Future of Emergency Planning in England and Wales: Results of the Consultation* (London, 2001) paras 2, 9, 15, 16.

[120] Cabinet Office, *The Draft Civil Contingencies Bill: Explanatory Notes etc*, para 8.

[121] The notion that catastrophes impel social change is associated with Samuel Henry Price: Scanlon, TJ and Handmer, J, 'The Halifax Explosion and the Port Arthur Massacre: Testing Samuel Henry Prince's Ideas' (2001) 19.2 *International Journal of Mass Emergencies and Disasters* 181.

[122] Sheen, B, *mv Herald of Free Enterprise: report of Court No. 8074 Formal Investigation* (Department of Transport, London, 1987).

[123] Lord Cullen, *Public Inquiry into the Piper Alpha Disaster* (Cm 1310, London, 1990).

[124] Lord Justice Popplewell, *Inquiry into Crowd Safety at Sports Grounds* (Cmnd 9585, IIMSO, London, 1985 and Cmnd 9710, HMSO, London, 1986); Lord Justice Taylor, *The Hillsborough Stadium Disaster, 15 April 1989* (Cm 962, HMSO, London, 1989).

[125] Lord Justice Clarke, *Thames Safety Inquiry* (Cm 4558, London, 2000).

[126] Lord Cullen, *Public Inquiry into the Shootings at Dunblane Primary School on 13 March 1996* (Cm 3386, HMSO, London, 1996) and Government Response (Cm 3392, HMSO, London, 1996).

[127] Fennell, D, *Investigation into the Kings Cross Underground Fire* (Cm 499, HMSO, London, 1988).

[128] Uff, J, *The Southall Rail Accident Inquiry Report* (HSE Books, Sudbury, 2000); Lord Cullen, *The Ladbroke Grove Rail Inquiry* (HSE Books, Norwich, 2001); Health and Safety Executive, *Hatfield Derailment Investigation* (http://www.hse.gov.uk/railways/hatfield.htm, London, 2002).

[129] Anderson, I, *Foot and Mouth Disease 2001: The Lessons to be Learned Inquiry* (2001–02 HC 888).

[130] BSE Inquiry, *Report* (1999–00 HC 887) and Government Response to the report on the BSE inquiry (Cm 5263, London, 2001).

by specific legislation,[131] but more general issues of planning, resilience, and coordination remained on the table.

(b) The draft Bill

2.42 The Civil Contingencies Act began to take firm shape with the publication by the Cabinet Office of the Draft Civil Contingencies Bill, issued in June 2003.[132] The Bill's key objectives were listed as:

> To create a modern framework for co-ordinating contingency planning and response at the local level, codifying existing arrangements.
>
> To enhance co-operation and understanding in support of a regional level capability.
>
> To modernise the legislation under which the Government can respond to extreme emergency situations to turn it into a usable tool fit for the twenty-first century.[133]

But the Bill was short on detail for it represented no more than 'an enabling Bill which seeks to establish in legislation a framework of powers which can be delivered through regulations.'[134] Therefore, much would depend on the contents of the regulations and guidance; draft regulations relating to Part I only (and only for England and Wales, as the Scottish versions did not appear for another year)[135] were produced in January 2004.

2.43 After the publication of the Draft Bill, the Government engaged in wide consultation[136] and also encouraged the formation of a Parliamentary Joint Committee on the draft Bill, chaired by Lewis Moonie MP.[137] That Committee reported in November 2003 and provided by far the most extensive and searching scrutiny of the Bill. Though broadly supportive, it concluded that the draft Bill lacked sufficient detail and provided inadequate safeguards against potential misuse. The Committee was critical of the fact that the draft regulations which were to accompany the Act had not been published and so required 'a leap of faith'.[138] It was concerned that the role and statutory responsibilities of central, regional, and devolved governments were not explicated or adequately defined.[139] There was criticism of the possible override of the Human Rights Act 1998 by cl 25 (described further in Chapter 7) and of the power to disapply or modify any Act of Parliament, which threatened

---

[131] See, for example, Public Order Act 1986, Pt IV; Football Spectators Act 1989; Football (Offences) Act 1990; Offshore Safety Act 1992; Firearms (Amendment) Act 1997; Animal Health Act 2002; Railways and Transport Safety Act 2003.   [132] Cm 5843, London.
[133] Ibid, para 4.   [134] Civil Contingencies Secretariat, *The Government's Response*, 15.
[135] See http://www.scotland.gov.uk/consultations/justice/crgcca-00.asp.
[136] For details, see Cabinet Office, *The Government's Response*, Annex A para 2.
[137] Joint Committee on the Draft Civil Contingencies Bill, *Draft Civil Contingencies Bill* 2002–03 HC 1074, HL 184. The Committee also had the benefit of memoranda from the Chairs of the House of Lords Select Committee on the Constitution (Appendix 1), the House of Commons Select Committee on Transport (Appendix 2), and the House of Lords Select Committee on Delegated Powers and Regulatory Reform (Appendix 3).
[138] Joint Committee on the Draft Civil Contingencies Bill, para 5   [139] Ibid, paras 10, 20.

constitutional law.[140] Concern was expressed about whether sufficient finance was being made available.[141] The Joint Committee also called for a dedicated civil contingencies inspectorate which should form part of a wider Civil Contingencies Agency.[142]

Additional to the Joint Committee, the draft Bill was scrutinized by three other **2.44** select committees. First, the Defence Committee had already considered planning for civil emergencies in its 2002 report, *Defence and Security in the UK*, which encouraged the publication of proposals on civil contingencies legislation.[143] After the draft Bill appeared, it quickly produced another report in July 2003,[144] which, *inter alia*, demanded the publication of draft regulations, questioned the legislative silence about central government, doubted the levels of funding proposed, and criticized the treatment of human rights.[145]

The Joint Committee on Human Rights also reported on the draft Bill.[146] It **2.45** concluded that Part I did not give rise to a significant threat of a violation of human rights, but cl 25 (discussed in Chapter 7) was strongly to be deprecated and Part II would give rise to a significant peril for Convention rights.[147]

Finally, the House of Lords Delegated Powers and Regulatory Reform Com- **2.46** mittee[148] offered a number of relatively technical improvements.

(c) The Bill

Significant differences emerged between the draft and the Bill introduced into **2.47** Parliament in January 2004.[149] The details were presaged by a paper issued by the Civil Contingencies Secretariat which distilled the latest thinking.[150] The key changes, amounting, according to taste, to a 'sober' response or to the Government having 'backed down',[151] between the draft of June 2003 and that of January were as follows: the definition of emergency in cl 1 was narrowed especially through the removal of reference to political, administrative, or economic stability; Part I was extended to the whole of the United Kingdom; the mechanism of promulgation in Part II was removed; the 'Triple Lock' (described in Chapter 5) was strengthened, as well as the need to submit to Parliamentary and other scrutiny, including the Council on Tribunals; cl 25 was dropped; and a power for devolved authorities to declare emergencies under Part II was omitted.

---

[140] Ibid, paras 11, 12, 13.   [141] Ibid, para 16.
[142] Ibid, paras 17, 18. These proposals are considered further in Ch 9.
[143] (2001–02 HC 518) para 158.
[144] Defence Committee, *Draft Civil Contingencies Bill* (2002–03 HC 557).
[145] Ibid, paras 13, 24, 58, 68.
[146] Joint Committee on Human Rights, *Scrutiny of Bills and Draft Bills; Further Progress Report* (2002–03 HC 1005, HL 149).   [147] Ibid, paras 3.4, 3.26, 3.35.
[148] *Twenty Fifth Report* (2003–04 HL 144).   [149] 2003–04 HC No 14.
[150] Civil Contingencies Secretariat, *The Government's Response.*
[151] *The Times* 8 January 2004 at 23.

2.48 The Bill was subsequently altered in a number of respects during passage: the definition of 'emergency' was again altered in various respects; the scope for urgent directions under s 7 was reduced; there was a greater recognition of the importance of voluntary groups in Part I; s 15, allowing for instructions about collaboration between responders across the borders of England/Wales/ Northern Ireland and Scotland, was added; s 22(5) was added so as to emphasize the importance of ensuring that Parliament and the courts are able to conduct relevant proceedings; by s 23(5), emergency regulations may not amend Part II of the Act or any part of the Human Rights Act 1998; a privy Counsellor will review any use of Part II; several road traffic laws (discussed later in this chapter) were altered in order to augment the prevention of terrorism.[152]

C. Outline, Miscellaneous Provisions, and Related Jurisdictions

(a) Outline

2.49 The 2004 Act comprises three parts. Part I (ss 1 to 18) covers local arrangements for civil protection against emergencies (the meanings are discussed in Chapter 3). Schedule 1 lists those persons and bodies—called 'responders'—which are subject to duties imposed under Part I. The Act imposes direct duties on local bodies ('Category 1 responders') to assess risks, to maintain response plans, and to provide advice and warnings. Other bodies ('Category 2 responders') must cooperate with, and provide information to, Category 1 responders. Many of the details are contained in the Civil Contingencies Act 2004 (Contingency Planning) Regulations 2005.[153] Part I is detailed in Chapter 4.

2.50 Part II (ss 19 to 31) deals with the powers to issue regulations in response to a wide range of emergencies. This Part of the Act contains long lists of the types of powers and provisions which may be required by emergency regulations, as well as a shorter list of what may not be included. Safeguards are included which reflect the 'Triple Lock' principles—that restraints will be imposed on the triggering definitions by reference to seriousness, necessity, and geographical proportionality. Part II is detailed in Chapter 5.

2.51 In addition to the legislation, Parts I and II can only be understood in the context of the two substantial and well-conceived guides issued by the Cabinet Office, one on *Emergency Preparedness*, the other on *Emergency Response and Recovery*.[154]

---

[152] HL Debs vol 665 col 1044 21 October 2004, Lord Bassam.

[153] See Civil Contingencies Act 2004 (Contingency Planning) Regulations 2005 SI No 2042; Civil Contingencies Act 2004 (Contingency Planning) (Scotland) Regulations 2005 SSI No 494.

[154] Cabinet Office, *Emergency Preparedness: Guidance on Part 1 of the Civil Contingencies Act 2004, its Associated Regulations and Non-statutory Arrangements* (http://www.ukresilience.info/ ccact/emergprepfinal.pdf, 2005), Cabinet Office, *Emergency Response and Recovery* (http:// www.ukresilience.info/ccact/emergresponse.pdf, 2005).

Part III (ss 32 to 36) deals with minor ancillary matters. Section 32 introduces **2.52** Schedules 2 and 3, detailing minor and consequential amendments and repeals and revocations. Section 33 allows for the reimbursement of any expenditure incurred by a Minister of the Crown.

The commencement (apart from ss 34 to 36) is to be set by statutory instrument **2.53** (s 34), with Scottish Ministers dealing with the measures exclusive to Scotland. The Act has come into force in five parts.

- Part II came into effect on 10 December 2004.[155]
- The minor amendments in Part 3 of Schedule 2 commenced on 19 January 2005.[156]
- The civil defence grant was terminated on 1 April 2005.[157]
- Most of Part I came into force on 14 November 2005.[158] Exceptions relate to the functions conferred on the Scottish Ministers ss 2(3) and (5), 4(2), (4), and (5), 6(1), 12, 15(3), and 17(6) which were commenced forthwith).
- Section 4 of the Act is commenced only for the purpose of permitting local authorities to provide advice and assistance to the public in relation to business continuity. The duty imposed by s 4(1) of the Act commences on 15 May 2006.

Section 35 makes it clear that the Act applies to all 'Parts' of the country— **2.54** England and Wales, Scotland, and Northern Ireland. No longer is there the separate legislative stream for Northern Ireland.

By s 36, the short title is the Civil Contingencies Act 2004. One could argue **2.55** that this title points more towards Part I than Part II. Some were therefore attracted to a title along the lines of the 'Civil Disasters and Emergencies Act'.[159]

(b) Miscellaneous materials

Within Schedule 2, a number of road traffic laws are changed in order to **2.56** introduce powers which might augment the prevention of terrorism. First, traffic-calming works regulations, which allow works such as pinch points and chicanes, under s 90H(2) of the Highways Act 1980 may include, by reference to an amended s 329, promoting safety in relation not only to road safety but also to avoiding or reducing, or reducing the likelihood of, danger connected with terrorism. In Scotland, the corresponding measure is s 63 of the Roads (Scotland) Act 1984.

---

[155] Civil Contingencies Act 2004 (Commencement No 1) Order 2004 SI No 3281.
[156] Ibid.
[157] Civil Contingencies Act 2004 (Commencement No 2) Order 2005 SI No 772.
[158] Civil Contingencies Act 2004 (Commencement No 3) Order 2005 SI No 2040. See also Civil Contingencies Act 2004 (Commencement) (Scotland) Order 2005 SI No 493 (C 26).
[159] See HL Debs vol 664 col 1232 15 September 2004, Lord Lucas.

2.57 Likewise, the Road Traffic Regulation Act 1984 is altered by the insertion of a new s 22C which allows for an order to be made under ss 1(1)(b), 14(1)(b), and 14(2)(b) for the purpose of preventing or reducing danger or damage connected with terrorism, such as by a ban on all, or all but specially authorized, vehicles. New s 22D is supplemental to s 22C and ensures that an anti-terrorist traffic regulation order may be made only on the recommendation of the local chief police officer. Section 22D further allows the placing in a road of obstacles and obstructions or authorization of works, such as blockers. Restrictions might also be varied according to the level of threat. In addition, the definition of 'extraordinary circumstances' in s 67 of the Road Traffic Regulation Act 1984 is expanded to include terrorism, so that the police can place traffic signs to control ordinary traffic in emergencies for 28 days. The use of these orders is subjected to para 1 of Schedule 9 to the Road Traffic Regulation Act 1984 which enables the Secretary of State (or with the consent of the Secretary of State, the National Assembly for Wales) to direct a local traffic authority to make or not make a permanent traffic regulation under ss 1 or 6 of that Act. In Scotland, the Scotland Act 1998 (Transfer of Functions to the Scottish Ministers etc) Order 2005[160] has since amended Schedule 2 and allows the powers to be exercisable by the Scottish Ministers, but with the agreement of the Secretary of State.

2.58 The House of Lords Delegated Powers and Regulatory Reform Committee[161] expressed concern that proposed s 22D(5)(d) would allow an order to provide that a constable may authorize an employee of a traffic authority to do anything that the constable could do by virtue of this subsection. This provision was 'seen as affording a degree of flexibility in the partnership between the police and local traffic authorities in dealing with terrorist threats.'[162] However, owing to the expressed doubts, it was later dropped.[163] The corresponding measure in Scotland is s 39BA of the Roads (Scotland) Act 1984.[164] In this way, the legislation gives birth to anti-terrorist traffic regulation orders—ATTROs—'to prevent or reduce the impact of vehicle-borne terrorist attacks, in particular no-warning vehicle suicide bombings'.[165] Temporary orders last for 18 months, permanent ones indefinitely.

(c) British Islands

2.59 Civil protection arrangements for the Channel Islands and Isle of Man are not covered in the 2004 Act. The Government consulted on the possible extension

---

[160] 2005 SI No 849 (S 2) Art 2. It follows that s 22C(7) has been repealed.
[161] *Thirtieth Report* (2003–04 HL 175) para 17.
[162] HL Debs vol 665 col 1049 21 October 2004, Lord Bassam.
[163] HL Debs vol 666 col 876 9 November 2004, Lord Bassam.
[164] This is also subject to the Scotland Act 1998 (Transfer of Functions to the Scottish Ministers etc) Order 2005 SI No 849 (S 2) Art 2 and so s 39BA((1) and (2) have been repealed, with consequent amendment to (3)).
[165] HL Debs vol 665 col 1047 21 October 2004, Lord Bassam.

of the draft Bill but no agreement was reached. The Privy Council has, though, requested these territories to 'plug into' the new arrangements as soon as possible, but the islands retain discretion as to their responses.

In the Isle of Man, emergency planning is a matter for the Department of Home **2.60** Affairs and its Emergency Planning Unit.[166] It is intended, by the vehicle of the Criminal Justice (Miscellaneous Provisions) Bill 2005, to alter the Emergency Powers Act 1936, s 3, by expanding the definition of 'emergency' to reflect those in ss 1 and 19 of the Civil Contingencies Act 2004.

In Guernsey, the Emergency Powers (Bailiwick of Guernsey) (Amendment) **2.61** Law 2005 has appeared as a Billet d'État in March 2005 to replace the Emergency Powers (Bailiwick of Guernsey) Law 1965. The Emergency Powers (Bailiwick of Guernsey) Law 1965 is thereby to be amended by widening the possible exercise of the emergency powers. There is also institutional change in the form of the establishment of the Emergency Powers Authority as an authority of the Policy Council and chaired by the Chief Minister. Current administration is dealt with by the Home Department and its Emergency Planning Officer.

In Jersey, the main law is the Emergency Powers and Planning (Jersey) Law **2.62** 1990. It sets up the Emergencies Council and the Emergencies Planning Officer. States of emergency can be declared by the Lieutenant-Governor.

D. Standards

The dilemma posed by emergencies is that they require swift and unencum- **2.63** bered executive action. Yet, by granting sweeping measures to achieve that goal, there is the grave danger that the crisis will be deepened by unforgiving or misplaced authoritarianism or that the state of normality thereby restored will be altered and damaged for all time. One must therefore consider, first, whether it is justifiable or desirable to resort to 'emergency' laws, and, second, assuming their existence, one must consider the normative standards which should govern those laws and the processes which should be inserted into the emergency powers legislation in order to ensure that those standards are observed.

(a) The resort to 'emergency'

Perhaps the most important lesson which can be learnt from the considerable **2.64** experience of one persistent form of response to the extraordinary, namely anti-terrorism legislation, is that the rule of law demands as much clarity in the

---

[166] Isle of Man Government, *Department Plans 2005–08* (Douglas, 2004) Vol II, 58, 59.

law, and consideration of checks and balances, before the crisis or emergency arises.[167] The broad approach implicit in the Terrorism Act 2000 is that there is a continuing need for extensive legislation against political violence now and for ever after. The Anti-Terrorism, Crime and Security Act 2001 reinforces that stance, though there are in both Acts 'sunset' clauses which ensure that some parts of the legislation must terminate after a set period. Amongst the disadvantages of these special laws are that they may be unnecessary (either because of the level of threat of the existence of other powers), there will be abuse of the wide powers, and there will be damage to the country's international reputation.[168] Therefore, this claim to a need for permanently based terrorism laws (and, on the same basis, the Civil Contingencies Act) should be examined at the outset. The 2004 Act might be justified in principle at two levels.

2.65 The first level concerns the powers and duties of states. In principle, it is justifiable for Liberal democracies to defend their existence and their values, even if this defence involves some limitation of rights. In the words of one American judge, a constitutional democracy is not a 'suicide pact', and proportionate measures can be taken against clear and present dangers.[169] This standpoint is also reflected in Art 17 of the European Convention for the Protection of Human Rights and Fundamental Freedoms:

> Nothing in this Convention may be interpreted as implying for any State, group or person any right to engage in any activity or perform any act aimed at the destruction of any of the rights and freedoms set forth herein or at their limitation to a greater extent than is provided for in the Convention.

It is also very much the point of the power of derogation from 'normal' Convention standards in time of emergency under Art 15(1):

> In time of war or other public emergency threatening the life of the nation any High Contracting Party may take measures derogating from its obligations under this Convention to the extent strictly required by the exigencies of the situation, provided that such measures are not inconsistent with its other obligations under international law.

Aside from the power to take action, there is a state responsibility to act against political or paramilitary violence. Each state has a duty, at least in international law, to safeguard the right to life and to family life of its citizens (as under Arts 2 and 8 of the European Convention on Human Rights). In addition, states should more generally ensure the enjoyment of rights and democracy (under

---

[167] See Walker, C, *A Guide to the Anti-terrorism Legislation* (Oxford University Press, Oxford, 2002) Ch 1. See further Simpson, AWB, *In the Highest Degree Odious* (Clarendon Press, Oxford, 1992) 408, 413.
[168] Lord Lloyd, *Inquiry into Legislation against Terrorism* (Cm 3420, London, 1996) paras 5.6–5.9.   [169] *Terminiello v Chicago* (1949) 337 US 1 at 37 per Douglas J.

Art 1).[170] International law duties to respond to terror were also reinforced by United Nations' statements in reaction to September 11, 2001.[171]

The second level of justification is more morally grounded. This argument **2.66** points to the illegitimacy of at least some kinds of threat to 'normal' life, such as terrorism. The fact is that many of its emanations are almost certainly common crimes, crimes of war, or crimes against humanity, even if the political cause of the terrorist is deemed legitimate.

The force of these arguments may be illustrated by jurisprudence under the **2.67** European Convention on Human Rights. It has been asserted on several occasions that the suppression of terrorism may protect individual rights as well as collective interests.[172] Moreover, the duty of the state to safeguard the right to life of its citizens was one of the main concerns of the European Court of Human Rights in the *McCann* case which arose out of the fatal shooting by soldiers of three IRA suspects in Gibraltar in 1988.[173] In the Court's view, the duty of the state to prevent the danger from IRA bombs to the population of Gibraltar should have taken precedence over the interests of criminal prosecution which influenced the security forces not to arrest at the border the three IRA suspects. A duty of protection has also been applied against non-terrorist assailants.[174]

The argument in relation to Art 8 has been asserted by the European Court of **2.68** Human Rights in *López Ostra v Spain*, concerning dire environmental pollution from unlicensed leather industries.[175] The same principle has been asserted in relation to English law arising from environmental emergencies. In *Marcic v Thames Water Utilities Ltd*,[176] the House of Lords concluded that the refusal of Thames Water to take measures to prevent the flooding of Marcic's property,

---

[170] This point is also reflected in the view of the Joint Committee on Human Rights, Scrutiny of Bills and Draft Bills (2002–03 HC 1005) para 3.4.

[171] See UN Security Council Resolutions 1269 (1999) of 19 October 1999 and 1368 (2001) of 12 September 2001, 1373 of 28 September 2001. These instruments added to the United Nations General Assembly Resolutions 40/61 of 9 December 1985 and 49/60 of 9 December 1994.

[172] *X v Ireland*, App No 6040/73, CD 44, 121; *W v UK*, App No 9348/81, DR 32, 190; *W v Ireland*, App No 9360/81, DR 32, 211; *X v UK and Ireland*, App No 9825/82, 8 EHRR 49; *G v United Kingdom and Ireland*, App No 9837/82, 7 March 1985; *X v UK and Ireland*, App Nos 10019, 10020, 10023, 10024/82, 8 EHRR 71; *X and Y v Ireland*, App Nos 9837, 9839/83.

[173] *McCann, Farrell and Savage v UK*, App No 18984/91, Ser A vol 324.

[174] *Osman v United Kingdom*, App No 23452/94, Reports 1998-VIII.

[175] App No 16798/90, Ser A 303-C, 1995. See further *Moreno Gomez v Spain*, App No 4143/02, 16 November 2004; *Fadayeva v Russia*, App No 55723/00, 09 June 2005; Sands, P, 'Human Rights, Environment and the *López-Ostra* Case' [1996] EHRLR 597; De Merieux, M, 'Deriving Environmental Rights from the European Convention for the Protection of Human Rights and Fundamental Freedoms' (2001) 21 *Oxford Journal of Legal Science* 521.

[176] [2003] UKHL 66. See Howarth, W, 'Flood Defence Law after *Marcic*' (2003) 5 *Environmental Law Review* 23.

arising repeatedly from overloaded sewers, did not breach his right to respect for his private and family life (as well as his property rights under Art 1 of Protocol 1) since the scheme of priority for the carrying out of works under the Water Industry Act 1991 was viewed in this case as consistent with Convention values. But the possibility of conflict is confirmed in the decision of the European Court of Human Rights in *Hatton v United Kingdom*,[177] albeit that the claim (concerning noise pollution from Heathrow airport) failed in that case. The Court stated, 'Article 8 may apply in environmental cases whether the pollution is directly caused by the State or whether State responsibility arises from the failure to regulate private industry properly.'[178]

2.69 It follows that the weight of arguments are against those who oppose on a platform of concerns for human rights and constitutionalism all conceivable forms of special laws. Rather our collective concern for constitutional values should lead us to the conclusion that the state should take proportionate action to protect citizens or indeed the state itself, including its essential attributes of democracy and the protection of rights, against disaster.[179] Accordingly, recent United Kingdom governments have been correct to reject the call, inspired by events in Northern Ireland, 'No emergency, no emergency law.'[180] A model of legislation, based on a stance of 'Break glass in case of emergency legislation' is to be preferred.[181] This contingency model of a permanent legislative code reflects the philosophy of constitutionalism and democratic accountability—that the legislature can secure an important input for itself and the courts if it can speak in advance of a crisis. One cannot coherently complain about 'panic' legislation but at the same time deny to the state the principled and refined means to defend itself when a genuine crisis arises and to allay its genuine fears (and often those of the general public and in Parliament). It is foolish not to plan for contingencies, especially as the planning process can allow the legislature to safeguard its future role as well as allowing expert consideration to be brought to bear on this specialist area of law. As argued by the New Zealand Law Commission:

> The choice will be between legislation carefully prepared in advance, conforming to the principles and safeguards... and hastily drafted legislation conferring wider powers than are necessary and omitting appropriate protections against abuse. Moreover, New Zealand and overseas experience suggests that emergency legislation passed in haste is likely to remain on the statute book long after its immediate purpose has been served.[182]

---

[177] App No 36022/97, 2003-VIII.   [178] Ibid, para 98.
[179] This conclusion was also reached by the *Inquiry into Legislation against Terrorism* (Cm 3420, London, 1996) para 5.15.
[180] Committee on the Administration of Justice, *No Emergency, No Emergency Law* (Belfast, 1995).
[181] See further Walker, C, 'Terrorism and Criminal Justice' [2004] Crim LR 311; Walker, C, 'Anti-terrorism Laws for the Future' (1996) 146 NLJ 586.   [182] Ibid, para 4.12.

There are several advantages flowing from this approach. The most important is **2.70** that it can reduce the dangers of the passage of badly designed and dangerous emergency laws, so long as it contains within it mechanisms for continued scrutiny. Secondly, one can seek to build upon the experience of permanent legislation and to impose effective scrutiny:

> Prolonged and sustained exposure to the asserted security claims may be the only way in which a country may gain both the discipline necessary to examine asserted security risks critically and the expertise necessary to distinguish the bona fide from the bogus.[183]

The conclusion here reached appears consistent with the observations in **2.71** Chapter 1 concerning the embedded nature of risk as a preoccupation of late modern society. As a result, those who posit a clear 'normalcy–emergency' dichotomy[184] are perhaps living in the past. It is not that the emergency laws should be in force and pervasive at all times. The US 'war' model in relation to terrorism is to be deprecated, for that approach is conducive to a lack of accountability and proportionality and threatens an everlasting departure from civil society.[185] But to ignore or deny the need to address by law risk in a risk society involves either forlorn imprudence or misplaced piety to rights.

(b) The standards in 'emergency'

The principal danger associated with the permanent availability of special laws is **2.72** the inclination towards overuse—that there will be too much smashing of the glass to take out the special laws and they will be utilized inappropriately or disproportionately. There must be an adherence to limiting principles which reflect the values of individual rights,[186] democratic and legal accountability and review, and constitutionalism (respect for the rule of law and proportionality between emergency and measures used).[187]

One may ask next how we can be sure that the emergency powers will adhere to **2.73** these standards? More pertinently, how can we be sure that Civil Contingencies Act powers are only triggered by serious threats, are proportionate to them, and will not be inimical to individual rights or to other constitutional features such as

---

[183] Justice Brennan, 'The American Experience', in Shetreet, S (ed), *Free Speech and National Security* (Nijhoff, Dordrecht, 1991) 16–17.

[184] For discussion, see Gross, O, 'Once More into the Breach' (1998) 23 *Yale Journal of International Law* 440; Lowe, V, ' "Clear and present danger": Responses to Terrorism' (2005) 54 ICLQ 185.

[185] See Allen, FA, *The Habits of Legality* (Oxford University Press, New York, 1996) 37–40.

[186] See especially Council of Europe, *Guidelines on Human Rights and the Fight Against Terrorism* (Strasbourg, 2002).

[187] See Walker, CP, 'Constitutional Governance and Special Powers against Terrorism' (1997) 35 *Columbia Journal of Transnational Law* 1; Bingham, T, 'Personal Freedom and the Dilemma of Democracies' (2003) 52 ICLQ 841.

democracy, accountability, and devolution? These issues will be discussed further in the following chapters, with special focus in Chapters 6 and 7. However, it is clear from what has already been said that attention must be paid to processes such as the possibility of informed and effective Parliamentary debate, the extent to which adjudication remains available, whether on human rights grounds or otherwise, and finally the scope for independent scrutiny of executive action.

2.74 In addition, to the constitutional principles here adduced, laws must in substance achieve effectiveness (secure their aims), economy (do not take up unnecessary resources), and efficiency (provide cost-effective solutions to problems). Many of these points are captured by the Cabinet Office's Better Regulation Task Force, which publishes the *Principles of Good Regulation*;[188] good regulation should be: 'Proportionate', 'Accountable', 'Consistent', 'Transparent', and 'Targeted'.

2.75 In the light of this discussion, the advent of the Civil Contingencies Act can be welcomed in principle. The building of resilience in Part I reduces the ultimate need to resort to emergency powers and seeks to address the concerns of risk averse publics. Even the emergency provisions in Part II have been delineated with greater attention to detail than their predecessors. At the same time, many measures are broadly worded. A legislative scheme so dependent on guidance and so silent about important aspects of the civil contingency universe (which includes central government) may eventually be found wanting.

E. Conclusions

2.76 Civil resilience had not been ignored until 2004. Local planning and inter-agency working had taken place for many years in regard to issues such as flooding, industrial hazard, and otherwise, sharpened in places (such as the City of London)[189] by the experience of terrorism. Nevertheless, deficiencies had become apparent by the new Millennium, and it was a fair comment that 'The current framework for civil emergency planning was not designed with the needs of modern society in mind.'[190] The breadth and depth of perceived risks in a late modern society were certainly not captured by the previous legislation. The 2004 Act is therefore to be commended as appropriate for its era, and few will bemoan the passing of the previous regime. The Act is not, however, a comprehensive statement, and many other laws, some inconsistent, remain beyond its scope. Nor has it entirely captured the governance debate. While, as shall be seen in Chapter 4, some aspects of the private sector are invited into the world of public planners, many others are, at best, offered advice or, at worst,

---

[188] http://www.brtf.gov.uk/reports/principlesentry.asp, 2003.
[189] See Walker, C, 'Political Violence and Commercial Risk' (2004) 56 CLP 531.
[190] HC Debs vol 416 col 1096 19 January 2004, Douglas Alexander.

wholly ignored, with the general public largely in the latter category of spectators rather than players. Nor, by maintaining silence over the roles of central and regional players, can we be confident that the Act will ensure that adequate standards of constitutionalism must be observed by law, though the Cabinet Office guidelines assist in practice.

As befits the notion of reflexivity in the risk society, legislation about risk carries **2.77** risk—the danger of excessive or inappropriate invocation. These dangers must be taken seriously as they have bedevilled most other 'emergency' legislation. The 2004 Act has the advantage over many failed examples in that, while dealing with risk and emergency, it was not passed in circumstances of dire crisis. Its performance on the basis of standards set here will be estimated in the remaining chapters.

3

THE MEANINGS OF 'EMERGENCY'

Much of the controversy surrounding the new Act is triggered by the impre- **3.01** cision of its definitions. Prominent amongst these debates is the scope of an 'emergency'. The definitions adopted are consciously more encompassing than formulae in prior Acts and thereby provide 'programmatic declarations'[1] of an interventionist stance towards potential crisis which also reflect globalizing and politicizing tendencies.[2]

A. Part I Definitions

Central to the operation of Part I are the three meanings of 'emergency', set out **3.02** in s 1(1). Each is accorded an expansive definition to cater for a variety of catastrophic events.

(a) Human welfare and the environment

The threats relating to human welfare and the environment include both: an **3.03** event or situation which 'threatens serious damage to human welfare in a place in the United Kingdom' (s 1(1)(a)); and an event or situation which 'threatens serious damage to the environment of a place in the United Kingdom'

---

[1] See Dombrowski, WR, 'Again and Again' in Quarantelli, EL (ed), *What is a Disaster?* (Routledge, London, 1998) 20.
[2] See Rosenthal, U, 'Future Disaster, Future Definitions' in ibid.

(s 1(1)(b)). By these clauses, a wide variety of disasters can be addressed. Though consequences rather than causes are here mentioned in the Act,[3] the following sources can be addressed—epidemics affecting humans or their food chain, fuel shortages, chemical/biological/radiological/nuclear or major industrial accident, and fires, floods, storms, droughts, and land movements.

3.04 Further elucidation of s 1(1)(a) is offered in s 1(2). Fearful of the breadth of the term, 'welfare', which might be taken to include such esoteric and immeasurable aspects of the human condition as spiritual welfare, it is specified that the term is confined to: (a) loss of human life, (b) human illness or injury, (c) homelessness, (d) damage to property, (e) disruption of a supply of money, food, water, energy, or fuel, (f) disruption of a system of communication,[4] (g) disruption of facilities for transport, or (h) disruption of services relating to health.[5]

3.05 Two concerns have been voiced about the meaning of the 'supplies' in (e). First, it has been argued that the definition might be overbroad as catching 'ordinary', manageable failures of 'utility' networks and associated services. Yet, one can hardly see the harm in requiring some planning under Part I of the Act for such eventuality, and perhaps the real complaint here is that other agencies are being allowed to poke their noses into the business of utilities. The argument about overreach is more forceful as regards the need for Part II powers in those circumstances, but the operation of the Triple Lock (described later in this chapter and in Chapter 5) is meant to safeguard against overuse. Furthermore, the original list in the draft Bill of June 2003 was longer still. To heading (e) was tagged 'or another essential commodity'. The vagueness of this phrase was criticized during consultation,[6] so it was dropped in January 2004.[7] In the same sub-heading (e), the inclusion of 'money' means that aspects of the banking system become 'essentials' which are protected by the Act.[8]

3.06 Next, heading (h) encompassed in the original draft 'medical, educational or other essential services'. This aspect was again the subject of much criticism and was dropped.[9] Education is not listed as a basic essential of life in the European

---

[3] The Government resisted the alternative approach (HC Debs Standing Committee F col 30 27 January 2004, Douglas Alexander) as it could become dated.

[4] The system may be physical, such as a roadway (though transport is covered under (g)), or ethereal, such as broadcasting: HL Debs vol 663 col 1224 15 September 2004, Lord Bassam. The version of the Bill as tabled referred to 'an electronic or other system' of communication, but the Government was convinced to remove any qualifying categories: HL Debs vol 666 col 1326 16 November 2004.

[5] 'Health' is intentionally wider than 'medical services': HC Debs Standing Committee F col 32 27 January 2004, Douglas Alexander.

[6] Joint Committee on the Draft Civil Contingencies Bill, para 72.

[7] Cabinet Office, *The Government's Response to the Report of the Joint Committee*, para 2.

[8] Civil Contingencies Secretariat, *The Government's Response*, 33.

[9] See Cabinet Office, *The Government's Response to the Report of the Joint Committee*, paras 9, 14.

Convention on Human Rights, nor are educational services considered 'essential' in international labour law.[10] Furthermore, the phrase, 'other essential services' was rejected as unduly broad and vague.[11]

The corresponding elucidation of s 1(1)(b) is set out in s 1(3). Damage to the **3.07** environment is confined to: (a) contamination of land, water, or air with biological, chemical,[12] or radio-active matter,[13] or (b) disruption or destruction of plant life or animal life. The original list of environmental hazards in the draft of June 2003 also mentioned 'fuel oils and flooding'.[14] The Joint Committee on the Draft Civil Contingencies Bill expressed puzzlement as to why only fuel oils and not, say, lubricating oils and edible oils, were mentioned.[15] The solution was at first to change 'fuel oil' to 'oil',[16] but the bolder step was later taken to remove any specific reference to oils being per se threats so that they could be covered by the more general formulae in (a) or (b).[17] The same applied to flooding.[18] The word, 'water' in (a) can be taken to include 'river water, estuarial water and sea water', 'lake water, the water table or rain water', or indeed any other type of water which can be said to occur within the environment.[19] Finally, it was not felt necessary to add 'noxious' to the description of matter within s 1(3)(a), as the existing terminology is broad enough.[20]

Those emergencies within s 1(1)(a) and (b) must affect a 'place' in the United **3.08** Kingdom. The intention is to permit planning for localized threats and situations. This degree of refinement differs from the position under Part II (discussed later in this chapter). There is no definition of 'place', but it can readily apply to a town, say, where flooding is a threat, or even to a district or a street within a district.[21]

---

[10] Joint Committee on the Draft Civil Contingencies Bill, para 53

[11] Cabinet Office, *The Government's Response to the Report of the Joint Committee*, paras 2, 14. See further Bonner, D, *Emergency Powers in Peacetime* (Sweet & Maxwell, London, 1985) 212.

[12] A chemical may be in solid, liquid, or gaseous form: HC Debs Standing Committee F col 38 27 January 2004, Douglas Alexander.

[13] The word 'harmful' was dropped as unnecessary: HL Debs vol 666 col 1326 16 November 2004, Lord Bassam.   [14] Cabinet Office, *The Draft Civil Contingencies Bill*, cl 1(3).

[15] Draft Civil Contingencies Bill (2002–03 HC 1074, HL 184, para 78). Compare Merchant Shipping Act 1995 s 151(1).

[16] Cabinet Office, *The Government's Response to the Report of the Joint Committee*, para 17.

[17] HL Debs vol 666 col 1326 16 November 2004, Lord Bassam.

[18] HL Debs vol 666 col 755 9 November 2004, Lord Bassam.

[19] Cabinet Office, *The Government's Response to the Report of the Joint Committee*, para 17.

[20] HC Debs Standing Committee F col 38 27 January 2004, Douglas Alexander; HL Debs vol 664 col 1230 15 September 2004, Lord Bassam. Whether a substance is 'noxious' requires the jury to consider 'quality and quantity and to decide as a question of fact and degree in all the circumstances whether that thing was noxious' (see Archbold, *Criminal Pleading Evidence and Practice* (49th ed, Sweet & Maxwell, London, 2001) para 19–229).

[21] HL Debs vol 664 col 1211 15 September 2004, Lord Bassam.

(b) Security

3.09 By s 1(1)(c), a security 'emergency' means 'war, or terrorism, which threatens serious damage to the security of the United Kingdom'. The mention of terrorism makes clear that sub-state threats may be the subject of contingency planning, an understandable response to the attacks of September 11, 2001. 'Terrorism' is defined by s 18 by adopting the meaning given by s 1 of the Terrorism Act 2000,[22] while 'war' is there stated to include armed conflict. This term is broader than war,[23] but it cannot encompass domestic criminal gang conflicts, even if guns are used.[24] Security threats do not include serious or organized crime, even though they are treated as security threats for other purposes.[25] Such crimes could of course be a threat to human welfare under s 1(1)(a) if of sufficient potential impact.[26]

3.10 The definition of security threats in terms of war or terrorism represents a narrower conception than originally set out in cl 1(1) of the draft Bill produced in June 2003, which listed instead '(c) the political, administrative or economic stability of a place in England or Wales, or (d) the security of a place in England or Wales.'[27] Clause 1(4) went on to define threats under (c) as those which cause or may cause 'disruption of (a) the activities of Her Majesty's Government, (b) the performance of public functions, or (c) the activities of banks or other financial institutions.' Public functions were in turn defined in cl 1(6) as comprising (a) functions conferred or imposed by or by virtue of an enactment, (b) functions of Ministers of the Crown (or their departments), and (c) functions of persons holding office under the Crown. These wider aspects were criticized in the consultations which led up to the Bill in January 2004. The inclusion of the phrase, 'political, administrative or economic stability', was especially picked out as unnecessarily vague, duplicative, or wide.[28] It was feared that the phrase could cover a perplexing range of disruptions. More seriously still, it might be invoked more for the protection of the essentials of life for governments and to ensure their survival rather than the essentials of life for the community.[29] The formulation thereby prompted the criticism, used against prior legislation that, 'In Britain, the idea of "civil" defence has been turned on its head. Home defence is about the protection of government—if need be,

---

[22] See Walker, C, *A Guide to the Anti-terrorism Legislation* (Oxford University Press, Oxford, 2002); EC Framework Decision on Combating Terrorism OJ 2002 L164/3. 'Terrorism' therefore includes 'threat of terrorism': HC Debs Standing Committee F col 44 27 January 2004, Douglas Alexander.

[23] See *Spinney's (1948) Ltd and others v Royal Insurance Co Ltd* [1980] 1 Lloyd's Rep 406; *IF P & C Insurance Ltd (Publ) v Silversea Cruises Ltd and others* [2004] EWCA Civ 769.

[24] HL Debs vol 663 col 1224 15 September 2004, Lord Bassam.

[25] See Security Service Act 1996.

[26] HC Debs Standing Committee F col 45 27 January 2004, Douglas Alexander.

[27] Cabinet Office, *The Draft Civil Contingencies Bill*, cl 1(1).　　[28] Ibid, para 63.

[29] Joint Committee on the Draft Civil Contingencies Bill, para 49.

against the civil population.'[30] The Government accepted these criticisms and dropped those elements.[31] It accepted that 'while political, administrative and economic stability is a means to an end, it should not be an end in itself in the Bill.'[32] At the same time, the definition of 'threat to human welfare' was extended to a disruption of the supply of money (see above).

A further debate arose in January 2004, when the Bill introduced into Parliament simply referred to the threat of serious damage to 'the security of the United Kingdom or a place in the United Kingdom', with cl 1(4) thereafter categorizing war, armed conflict, or terrorism as being threats to security. But they are not stated to be the only threats. The protean nature of 'security' could, if left unqualified, have reinserted much of the ground covered by the deleted cl 1(1)(c). For example, s 1(2) of the Security Service Act 1989 defines the functions of the Security Service as being 'the protection of national security and, in particular, its protection against threats from espionage, terrorism and sabotage, from the activities of agents of foreign powers and from actions intended to overthrow or undermine parliamentary democracy by political, industrial or violent means.' It was therefore to be commended that the words 'war or terrorism' remained as the only sources relevant to a threat to 'security'. **3.11**

(c) Common elements

All three types of emergency must 'threaten' serious damage of one kind or another. The word 'threaten' reflects a prime purpose of Part I which concerns the encouragement of preparedness and resilience. When an emergency has materialized rather than been threatened, then Part II of the Act comes into play, though it may equally deal with a disruptive challenge which threatens serious consequences as well as catastrophic events which have already inflicted serious impacts. **3.12**

A further common element is that all threats must be capable of producing 'serious' adverse outcomes: 'This makes more transparent the fact that it is not just the threat that must be serious but the potential or actual consequences to human welfare, the environment or security.'[33] There is no definition of 'serious', though the phrase is commonly used in law (for example, within the definition of 'terrorism' in s 1 of the Terrorism Act 2000). The insertion of 'serious damage' represents a change from the draft Bill released in June 2003, which merely required a serious threat. The condition of seriousness caused some consternation in the responder community. Many wondered how they **3.13**

---

[30] Campbell, D, *War Plan UK* (Burnett Books Ltd, London, 1982) 15.
[31] Civil Contingencies Secretariat, *The Government's Response*, 5, 6, 32; Cabinet Office, *The Government's Response to the Report of the Joint Committee*, para 2.
[32] HC Debs vol 416 col 1105 19 January 2004, Douglas Alexander.
[33] Cabinet Office, *The Government's Response to the Report of the Joint Committee*, para 10.

were to assure themselves that the emergency was serious enough.[34] Many preferred the definition of 'major emergency' in *Dealing with Disaster* because it contained an outcome test:

> . . . any event or circumstance (happening with or without warning) that causes or threatens death or injury, disruption to the community, or damage to property or to the environment on such a scale that the effects cannot be dealt with by the emergency services, local authorities and other organisations as part of their normal day-to-day activities.[35]

In response, the Government claimed to be persuaded of the need for change but deprecated the cited definition since it was conditional upon the capacity of emergency services in the area and such capacity is variable.[36] Thus, it proposed in 2003 that the existing definition would be included as a trigger in the Bill but only in the form of a requirement in regulations that responders maintain a procedure for determining whether an emergency has occurred—not, in other words, a change to the definition in s 1 itself.[37] This position was modified in January 2004, when a condition was inserted into s 2(2):[38] duties under Part I only arise if the emergency would be likely to seriously obstruct the responder or the responder would consider it necessary or desirable to intervene in the emergency and would be unable to take that action without changing the deployment of resources or acquiring additional resources. The drawback remains that the result will be a trigger dependent upon the nature of the capabilities of the particular responder—this is why the provision is set out in s 2, thus preserving s 1 as more 'objective'. It would have been possible to change the level of seriousness mentioned in s 1(1) to, say, 'very serious' or 'severe',[39] but the Government believed that such wording would not make much difference.[40]

3.14 For Part II purposes, it is felt that the Triple Lock devices already provide sufficient safeguards to ensure seriousness, especially the requirement as to necessity under s 21. Comparison was again made to the formulae in *Dealing with Disasters*, in this case with the definition of 'major incident' given in Annex A:

> A major incident is any emergency that requires the implementation of special arrangements by one or more of the emergency services, the NHS or the local authority for:
>
> - the initial treatment, rescue and transport of a large number of casualties;
> - the involvement either directly or indirectly of large numbers of people;

---

[34] Civil Contingencies Secretariat, *The Government's Response*, 4.

[35] Revised 3rd ed, Cabinet Office, London, 2003, para 1.5.

[36] Joint Committee on the Draft Civil Contingencies Bill, para 60.

[37] Civil Contingencies Secretariat, *The Government's Response*, 6.

[38] Cabinet Office, *The Government's Response to the Report of the Joint Committee*, para 11.

[39] The latter was suggested a number of times during the Parliamentary passage, for example HL Debs vol 663 col 1214 15 September 2004, Baroness Buscombe.

[40] Ibid, col 1221, Lord Bassam.

- the handling of a large number of enquiries likely to be generated both from the public and the news media, usually to the police;
- the need for the large scale combined resources of two or more of the emergency services;
- the mobilisation and organisation of the emergency services and supporting organisations, e.g. local authority, to cater for the threat of death, serious injury or homelessness to a large number of people.

This definition is widely used by the emergency services, the NHS, local authorities, and others for planning for localized emergencies. The Government considered that most, if not all, 'major incidents' would satisfy the definition of emergency in Part I of the Act but would not be sufficiently serious to trigger special legislative measures under Part II.[41]

Finally, a threat to human welfare is not a common element to all definitional **3.15** subsections. Some representations were made that human welfare, as the essential core of any 'emergency' (as opposed to 'hazard')[42] and the only conceivable justification for special measures, should invariably be in its immediate jeopardy.[43] The requirement would further corroborate the seriousness of the situation and would take the focus away from narrower political or governmental interests. It was an essential condition in the Emergency Powers Act 1920 and ensures that emergency action is the interests of 'the people' and not 'the state'. The Government accepted that the disjunction between (a), (b), and (c) in s 1(1) means that the threat to human welfare may not be apparent or immediate in all cases. Nonetheless:

> ...a situation such as an environmental catastrophe may have a serious effect with very serious consequences, yet the impact upon human welfare could be minimal or might not be direct or immediate in the initial stages. Similarly, with a terrorist attack on the stock exchange. As the impact on human welfare may only become clearer as the situation unfolds, it is essential to retain depth and flexibility.[44]

In all cases, the 'emergency' must have impact in the United Kingdom. **3.16** However, by s 1(5), the event or situation relevant to subsection (1) may occur or be inside or outside the United Kingdom. In this way, the threat could be internal—for example, the fall-out on 25 April 1986 from the Chernobyl nuclear power plant, located 80 miles north of Kiev, affected farming in Cumbria for several years and so potential catastrophes of that kind can fall within Part I. Even broader is the situation under s 1(1)(c), relating to security.

---

[41] Joint Committee on the Draft Civil Contingencies Bill, Appendix 9 q 12.

[42] Quarantelli, EL (ed), *What is a Disaster?* (Routledge, London, 1998) 259.

[43] Joint Committee on the Draft Civil Contingencies Bill, para 47; Oliver-Smith, A, 'Global Changes and the Definition of Disaster' in Quarantelli, EL (ed), *What is a Disaster?* (Routledge, London, 1998) 195.

[44] Cabinet Office, *The Government's Response to the Report of the Joint Committee*, para 7.

It has been held in the context of immigration laws that the security of the United Kingdom can be affected by threats to the security of its foreign allies.[45] An important antidote to this coverage is that the duties imposed under Part I of the Act are conferred primarily on localities who are not encouraged to develop foreign policies.[46]

3.17 Despite the breadth of the definitions of 'emergency', a power of augmentation is afforded by s 1(4) in two ways. The first way, s 1(4)(a), is to allow a definitive interpretation. A Minister of the Crown, or, in relation to Scotland, Scottish Minister, may by order provide that a specified event or situation, or class of event or situation, is to be treated as falling, or as not falling, within any of paragraphs (a) to (c) of subsection (1). This power could offer a summary form of resolution of any doubt as to whether a given situation falls within an existing definition as well as presenting an opportunity for debate in Parliament in advance of the exercise of powers. Thus:

> ...it may be possible to identify impending events and situations that might become emergencies. An example of a situation that turned out not to be an emergency, but that we all thought might become one, was the millennium bug. That was an example of an apparently impending potential emergency that turned out not to be an emergency.[47]

Section 1(4)(b), an addition in January 2004, is more radical, for it allows the Minister to amend s 1(2) so as to provide that in so far as an event or situation involves or causes disruption of a specified supply, system, facility, or service, (i) it is to be treated as threatening damage to human welfare, or (ii) it is no longer to be treated as threatening damage to human welfare. Thus, the order can recognize the changing profile of the essentials of society, such as the emergence of new sources of energy or the exhaustion or marginalization of others, whereupon the Act can be updated:

> The Government considers that the list of essential services and commodities in the Bill may become out of date at some point in the future. Experience with the Emergency Powers Act 1920 shows that it is not possible to predict what may become an essential service or commodity in ten, thirty or eighty years. (For example, the 1920 Act did not include computers or means of communication.)[48]

The breadth of this power is very disturbing:

> This sub-clause does not require that reasons be given or that the Minister is satisfied that such action is necessary. It simply states that he may do it. In other

---

[45] *Secretary of State for the Home Department v Rehman* [2001] 3 WLR 877.
[46] *Wheeler v Leicester City Council* [1985] 1 AC 1054. Compare the misleadingly wide list at http://www.ukresilience.info/contingencies/localrisk.htm.
[47] HC Debs Standing Committee F col 167 3 February 2004, Fiona Mactaggart. This example was given in the context of Part II.
[48] Cabinet Office, *The Government's Response to the Report of the Joint Committee*, para 14.

words, the definition of emergency still contains a catch-all element. If what can be done by order is included, the new definition may even increase the scope of the definition relative to that in the draft Bill.[49]

Following criticism, changes were later made so that any orders under subsection (4) are now subject to the affirmative procedure (s 17(2) and (3)).[50]

B. Part II Definitions

'Emergency' is defined by s 19(1) to include (a) events and situations which **3.18** threaten serious damage to human welfare[51] or (b) the environment, or (c) war or terrorism which threaten serious damage to security.[52] Similar to s 1, by s 19(6), the event or situation mentioned in subsection (1) may occur or be inside or outside the United Kingdom.

By comparison, the Emergency Powers Act 1920 referred to 'depriving all or **3.19** part of the community of the essentials of life', which may be viewed as a narrower definition which is largely equivalent to just s 19(1)(a). As a result, s 19 is, on the one hand, broader and encompasses, for instance, mass technological failure, electronic attack, environmental disaster, or a 9/11 style attack.[53] In this way, the definition is apt 'to cover all forms of disruptive challenges—natural, accidental or deliberate—where the measures flowing from their use would aid response or recovery'.[54] Some have argued that a 'threat to existence for significant numbers of people should be the only scenario that is sufficient' for emergency powers and the wording of s 19 should have reflected this requirement,[55] but the widening of emergency jurisdiction is the pronounced policy of the Civil Contingencies Act. On the other hand, s 19 is no

---

[49] House of Commons Library, *Civil Contingencies Bill* (Research Paper 04/07, House of Commons, London) 17.

[50] Joint Committee on the Draft Civil Contingencies Bill, Appendix 3.

[51] At drafting stage, Part II referred to a threat to 'welfare', rather than 'human welfare' as in Part I. Furthermore, a threat to welfare under Part II was not inclusively defined, since the words 'in particular' were included in the first sentence of cl 17(2) (now s 19(2)). These differences were removed in January 2004, following adverse comment by the Joint Committee on the Draft Civil Contingencies Bill. See Draft Civil Contingencies Bill, paras 43, 69; Cabinet Office, *The Government's Response to the Report of the Joint Committee*, paras 6, 13.

[52] Section 19 does not replicate the various wider formulations in cl 17 of the original draft Bill of June 2003, corresponding to those mentioned in connection with Part I, cl 1. In particular, the definition of security threats no longer contains cl 17(1)(c), which also listed '(c) the political, administrative or economic stability of a place in England or Wales, or (d) the security of a place in England or Wales.' These terms were heavily criticized during consultation on Part II just as they ran into trouble in the context of Part I. See House of Commons Library, *Civil Contingencies Bill* (Research Paper 04/07, House of Commons, London) 16, 17.

[53] Cabinet Office, *The Draft Civil Contingencies Bill: Explanatory Notes etc*, para 3.

[54] Cabinet Office, *The Draft Civil Contingencies Bill*, para 5.17.

[55] Joint Committee on the Draft Civil Contingencies Bill, paras 46, 48.

broader than the 1920 Act in that it does not directly protect against damage to the economy.[56]

3.20 There follows in s 19(2) a list of types of emergencies relevant to s 19(1)(a). That list is identical to that in s 1(2). Correspondingly, s 19(3) details those threats relevant to s 19(1)(b). That list is identical to that in s 1(3). It means that the likely fare of Part II will be sufficiently serious industrial strikes, disruptive political protests, terrorist outrages, disasters arising from storms or epidemics, and incursions on national infrastructure such as computer networks.

3.21 The differences from the Part I definition in s 1 are very few, which, on the face of it, is remarkable since they serve quite distinct purposes. A potential emergency under Part I requires planning and resources which, if adequate, should avert a state of emergency under Part II. One would also expect the thresholds in Part II to be much more demanding since its powers are more intrusive.[57]

3.22 In response to these concerns, the scheme of the Act is to maintain parity in the two sets of definitions but to insert into Part II more pre-conditions before use (s 21) and more limitations over use (s 23)—all reflecting the Triple Lock strategy of demanding seriousness, necessity, and geographical proportionality—rather than refining the definition in s 19.[58] An example of the alternative strategy might be the substitution of the word 'severe' for the word 'serious' in s 19(1) and by the addition of the word 'immediate'.[59] Calls were additionally made to heighten the level of threat by changing 'threatens serious damage' to 'seriously threatens', thereby requiring a higher level of credibility in the threat;[60] but that proposal would also, out of keeping with the rest of the defintion, have shifted the emphasis away from the significance of its consequences and onto the seriousness of the threat itself which is not helpful. Rather than such changes, the Home Department Minister emphasized the brakes applied by the Triple Lock.[61] It was also pointed out that it is conceivable for a threat not to be immediate but for the need for a response to be immediate: 'For example, in the case of a biological element that could cause birth defects, the consequences would not happen for nine months, but urgent intervention of a distinctive peculiar kind might be required.'[62] More generally, the Government preferred a 'highly inclusive' approach so as to allow for flexibility

---

[56] The point was debated in relation to the invocations of the 1920 Act in 1948, 1949, and 1966: Morris, GS, 'The Emergency Powers Act 1920' [1979] PL 317 at 322. Compare Industrial Relations Act 1971, s 138(1) (see Ch 6).

[57] Joint Committee on the Draft Civil Contingencies Bill, para 27.

[58] Compare Joint Committee on the Draft Civil Contingencies Bill, paras 28, 42.

[59] See HC Debs Standing Committee F col 127 3 February 2004, Richard Allan.

[60] HL Debs vol 666 col 1319 16 November 2004, Lord Lucas.

[61] HC Debs Standing Committee F col 145 3 February 2004, Fiona Mactaggart.

[62] Ibid.

and adaptability,[63] despite the worrying drawback of the potential for overuse of Part II. Thus, the Triple Lock demand of 'seriousness' is not adequately reflected in s 19. It could have been more fully secured, for example, by demanding proof of a threat to a portion of the community or the community itself[64] alongside the 'serious damage' which can arise in everyday life. A related objection is that the various legs of the Triple Lock are not drawn together so that they are readily visible and underscored.[65]

One explicit difference which does exist between Part I and II—the most **3.23** important of all—concerns the scale of events which they address. In Part II, the words, 'Part or region' are used instead of the Part I word, 'place'. This phrase makes it clear that responses can be tailored to emergencies which do not affect the whole jurisdiction (in other words, England/Wales or Scotland or Northern Ireland or the United Kingdom as a whole). At the same time, the Part II powers cannot be triggered by localized problems, such as might only affect a specific 'place' within a Part or region. Thus, the principal difference from the definition in Part I is geography—that, for the purposes of Part II, the situation must threaten serious damage in the United Kingdom or in a Part or region (rather than a local place in the United Kingdom). The goal is to ensure a level of seriousness before these grave powers are invoked.

The definitions of 'Parts' and 'regions' are set out in s 31(2). A 'Part' means **3.24** England, Northern Ireland, Scotland, and Wales. A 'region' means a region for the purposes of the Regional Development Agencies Act 1998, which in turn, in Schedule 1, lists nine regions; East Midlands, Eastern, London, North East, North West, South East, South West, West Midlands, and Yorkshire and the Humber.[66] It is further specified in s 31(3) that a Part or a region is to be taken to include: (i) any part of the territorial sea that is adjacent to that Part or region,[67] (ii) any part of the area within British fishery limits that is adjacent to the Part or region, and (iii) any part of the continental shelf that is adjacent to the Part or region. The choice of regions as the basic administrative unit for these purposes ties in with the boundaries of bodies such as Army Land Command and not, in the main, with political boundaries.

The geographical distinction is not entirely clear-cut for two reasons. First, the **3.25** rule of geographic extent can be traduced by the severity of a local emergency

---

[63] Cabinet Office, *The Government's Response to the Report of the Joint Committee*, para 1.
[64] Joint Committee on the Draft Civil Contingencies Bill, para 41.
[65] Ibid, para 37.
[66] It follows that doubts concerning the legality of the invocation of the Emergency Powers Act 1920 in 1924 in relation to the London transport strike (see Morris, GS, 'The Emergency Powers Act 1920' [1979] PL 317 at 322) will no longer arise.
[67] By the Territorial Sea Act 1987, s 1 the breadth of the territorial sea adjacent to the United Kingdom shall for all purposes be 12 nautical miles.

which then requires a more extensive response. Examples might be the Lockerbie Pan Am aircraft bombing in 1988.[68] The aircraft fell upon a small Scottish town with no significant regional or national infrastructure being affected; but the event was beyond the capacities of the small Dumfries and Galloway Constabulary and required national police resources as well as the involvement of the military in the search for wreckage and evidence. No proclamation was made under the Emergency Powers Act, but one might expect that if the event had been scaled up to the cataclysmic 9/11 attack in New York, which affected even fewer acres than Lockerbie but caused reverberations across the nation, that Part II would be engaged.[69] In this way, a very localized impact can still have consequences which engage Part II.

3.26 Secondly, one can understand that a certain achievement of scale is required before one can claim that an 'emergency' exists and that barriers against the use of potentially draconian powers are to be encouraged rather than deprecated. Yet, the choice of large administrative boundaries, while potentially more discerning than the all-or-nothing application of the Emergency Powers Acts, does cut across the Triple Lock strategy of geographical proportionality. Emergencies, especially environmental disasters, do not always befall neat administrative areas.[70] The result is therefore a compromise between the promotion of proportionality during use and the avoidance of the excessive invocation of Part II.

3.27 A further demarcation between the two Parts is that there is no equivalent to the related qualification in s 2(2), by which duties under Part I only arise if the emergency would be likely to seriously obstruct the responder or the responder would consider it necessary or desirable to intervene in the emergency and would be unable to take that action without changing the deployment of resources or acquiring additional resources. For Part II purposes, it was felt that the Triple Lock devices already provided sufficient safeguards to this effect, especially the requirement as to necessity under s 21.[71]

3.28 Another difference is that the powers in s 19(4) to amend the lists in relation to emergencies affecting human welfare under s 19(2) is in some respects more constrained than under s 1. Only a Secretary of State may issue an order and not any Minister of the Crown nor (since it is a reserved matter) a Scottish Minister. Next, there is no equivalent to the provision in s 1(4)(a) that an order may provide that a specified event or situation, or class of event or situation, is to be treated as falling, or as not falling, within any of paragraphs (a) to (c) of

---

[68] See Klip, A and Mackarel, M, 'The Lockerbie Trial' (1999) 70 *Revue Internationale de Droit Penal* 777; *Lockerbie Trial*, http://www.scotcourts.gov.uk/library/lockerbie/index.asp.

[69] Even Lockerbie might fall within both Parts of the Act: HC Debs Standing Committee F col 14 27 January 2004, Douglas Alexander.

[70] Joint Committee on the Draft Civil Contingencies Bill, para 270.

[71] Cabinet Office, *The Government's Response to the Report of the Joint Committee*, para 11.

subsection (1). The equivalent clause was taken out during parliamentary passage as being less vital than under Part I, given the lack of time for issuance and consideration in an unfolding emergency;[72] it also avoids the provision being used as 'a back-covering operation to enable Ministers to take difficult decisions'.[73] On the other hand, s 19(5)(a) goes further than s 1: an order under (4) may make a consequential amendment of Part II itself. For instance:

> The rapid development of, and our ever growing dependence upon, new technology, changes in lifestyle and patterns of employment, the growth of new means of production and the networks of supply and delivery that have developed since 1920 are startling.... With the pace of change seemingly ever increasing it would be irresponsible not to allow for the updating of the list of specified means of supply, systems facilities and services to ensure it reflects future developments whose disruption may threaten human welfare. Clause 19(5)... allows for that and only for that, not for wholesale changes to the definition of the emergency to be made.[74]

3.29 On its face, s 19(5) may appear to be 'the Henry VIII clause to end them all.'[75] But, as indicated, this power is not to be understood as allowing substantive change to the existing definitions in s 19(1),[76] but it may affect the list in s 19(2)(e) to (h) with reference, for example, to new computer viruses,[77] new forms of hostile weapons, or an invasion of aliens. The House of Lords Delegated Powers and Regulatory Reform Committee interpreted the role of the clause as being to 'particularise, so far as is consistent with what is in the bill' and accepted it as such.[78] Any order under s 19(4) is subject to the affirmative resolution of each House of Parliament (s 19(5)(b)).

3.30 It can readily be recognized that the definition in s 19 is extremely broad in comparison with the 1920s legislation and also in the light of the very expansive regulation-making power which is triggered. The Government claims that limits are imposed by reference to the Triple Lock concept—that restraints will be imposed on the triggering definitions by reference to seriousness (the situation must be serious enough in nature to warrant the use of Part II regulations), necessity ('there is a genuine need to take such special legislative measures'), and geographical proportionality (the minimum geographical extent required is affected).[79] The

---

[72] HL Debs vol 666 col 850 9 November 2004, Baroness Scotland.

[73] HC Debs Standing Committee F col 170 3 February 2004, Oliver Heald.

[74] HL Debs vol 665 col 745 14 October 2004, Baroness Scotland. One 'consequence' might be to the drafting of s 22: HL Debs vol 666 col 851 9 November 2004, Baroness Scotland.

[75] HL Debs vol 665 col 740 14 October 2004, Lord Stoddart.

[76] HC Debs Standing Committee F col 167 3 February 2004, Fiona Mactaggart.

[77] It might be said that these examples are events rather than the consequences of events which is the fare of the Act: ibid, col 749 Lord Lucas.

[78] House of Lords Delegated Powers and Regulatory Reform Committee, *Twenty Fifth Report* (2003–04 HL 144) paras 6, 7, 8, 19.

[79] Cabinet Office, *The Draft Civil Contingencies Bill*, para 5.19. See further s 23(1)(b).

effectiveness of these concepts should be dissected. There are two issues to consider. First, are the concepts sufficiently reflected in Part II; secondly, are there other criteria which are missing?

3.31 On the first point, the main problem relates to the concepts of seriousness and necessity. The concepts are reflected in s 19(1) in the case of seriousness and s 21(3) in the case of necessity and are fortified by the concept of proportionality in s 23(1). But the basis on which powers can be exercised in Part II is chronically subjective—the word 'satisfied' appears in ss 20 and 23(1) without qualification as to reasonableness. This does not mean that Ministers may act unreasonably or without factual grounding—to do so may fall foul of judicial review which has long sought to imply forms of objectivity into powers which are apparently subjective.[80] But the setting of an explicitly objective standard could produce a more demanding standard, requiring more proof of compliance rather than an absence of proof of non-compliance. The Government remarkably felt that the setting of objective standards would be a bad precedent:

> Both Her Majesty (in this capacity) and a senior Minister of the Crown are subject to a duty to act reasonably by virtue of public law. It is therefore unnecessary to provide specifically that they must do so in the Bill. To make express provision of this kind in this Bill could be positively harmful; it might imply that action might be taken under other enactments that do not specify that the decision-maker must act reasonably which is unreasonable.[81]

On the other hand, the Act does secure geographical proportionality effectively by requiring under s 23(2) that emergency regulations must specify the Parts of the United Kingdom or regions in relation to which the regulations have effect.

3.32 On the second point, many important considerations are integrated:

> Ministers must be satisfied that there is a serious threat before the use of special legislative measures is possible. They must also be satisfied that special legislative measures are genuinely necessary. In addition, they must be satisfied that existing legal provision is not sufficient for the purpose for which they wish to make emergency regulations, or believe that relying on existing legal provision would occasion serious delay. Finally, they must believe that the proposed provision, and not some less intrusive provision, is necessary for the purpose of preventing, controlling or mitigating a serious aspect or serious effect of the emergency.[82]

The misgiving in these respects is again more the low standard of proof than the omission of relevant considerations.

---

[80] See *Secretary of State for Education and Science v Tameside Metropolitan Borough Council* [1977] AC 1014. See further Ch 7.

[81] Cabinet Office, *The Government's Response to the Report of the Joint Committee*, para 4.

[82] Cabinet Office, *The Draft Civil Contingencies Bill*, para 5.29.

C. The Concept of 'Emergency'—Overall Assessment

(a) Single definition

The approach of adopting what is effectively the same definition (with a **3.33** tweaking in Part II affecting geographical proportionality) in both Parts of the Act offers insufficient distinction between the two Parts of the Act which serve different, albeit related, purposes. It is arguable that a much narrower definition should be used in the circumstances of acute emergency and draconian powers in Part II compared to the much wider range of imaginable contingencies but gentler planning processes in Part I. To put it another way, one expects responders to prepare for, and respond to, a far wider range of situations under Part I than one would expect to trigger the extraordinary powers under Part II.

(b) The triggers in the definitions

The meanings of 'emergency' in ss 1 and 19 have been found to be highly **3.34** inclusive. The advantage of such a conceptualization is, of course, that flexibility and adaptability are built into the legislative framework which can then respond effectively even to novel or unforeseen events or where the extent of threat or damage is unascertainable because of the confusion of the crisis.[83] However, the very generality of this overarching definition does give rise to the drawbacks of allowing ample scope for ambiguity and misinterpretation of events and for the abuse of powers for political convenience rather than public protection. For example, the section entitled, 'When Emergency Powers May be Used', in the Cabinet Office Consultation Document uses terminology such as 'Major accidents', 'Major health crisis, such as a flu pandemic', 'Serious Economic Crises', and 'major acts of terrorism'[84] without further explanation of the underlying quantitative or qualitative terms which will determine whether a given event or events should be labelled an emergency.

Events could have been categorized using clearly understood (rather than simply **3.35** implied) quantitative terms of reference relating to severity of impact (financial loss, fatalities, etc) and/or qualitative assumptions that differentiate between terrorist attack, viral pandemic, failure of critical infrastructure/services, and so on.[85] Importantly, these qualifications could have provided guidance as to where responsibility for managing/responding to emergencies should reside.

As regards the quantification of risk, the Consultation Document states: 'The aim **3.36** of building resilience is to reduce susceptibility to challenges by reducing the

---

[83] Joint Committee on the Draft Civil Contingencies Bill, Appendix 9 q 3.
[84] Cabinet Office, *The Draft Civil Contingencies Bill*, para 5.17.
[85] See further Joint Committee on the Draft Civil Contingencies Bill, Appendix 4.

probability of their occurrence and their likely effects.'[86] This is a fairly standard definition of risk which is widely used to underpin quantified approaches to risk assessment. That some form of quantitative differentiation is being utilized is also clearly implied by the statement that: 'Disruptive challenges exist along a spectrum of severity ranging from local flooding to massive terrorist attack'.[87] However, neither the Act nor the accompanying documentation address where key thresholds lie on this 'spectrum of severity'. At what point will a 'local' emergency become a 'regional' or 'national' emergency? How are Ministers to make such a judgement? Indeed, the purpose of Part II is to 'enable' Ministers to respond to an emergency, but the framework proposed appears to assume that Ministers, in the face of novel or unforeseen events, will be vested with perfect information and a comprehensive understanding of the substantive essence of the problem they face. Yet, during periods of crisis, decision-makers are subject to severe limitations, asymmetries, and distortions of information as well as being subject to highly stressful, threatening, and surprising events requiring rapid response. There would therefore seem to be a case for building in to the Act a clearer structure or clarification of what is meant by 'severity' and better guidance as to how different levels of severity should be tackled.

3.37 Moving from quantification of risk to qualitative issues, there is scant attempt in the Act to make qualitative distinctions between 'disruptive challenges' such as 'local flooding' and 'terrorist attack'. The conjoining of issues in this way may not be helpful. In considering the appropriateness of the Act, attention should not solely be paid to exploring merely whether an 'increase' in powers is commensurate with the 'severity' of future disruptive challenges. This restricts our thinking to simplistic (and partial) terms of reference that over-emphasizes quantification at the expense of important qualitative concerns about the substantive nature of 'emergency' and the context in which emergency is perceived and crisis decision-making occurs. An insight into the problem is revealed in reference to Margaret Thatcher's recollection of the Government's reaction to the Miners' Strike of 1984. In *The Downing Street Years*, Mrs Thatcher notes that in response to the declaration of a national dock strike:

> We mobilized the Civil Contingencies Unit to prepare to meet the crisis but avoided proclaiming a state of emergency, which might have meant the use of troops. Any sign of overreaction to the dock strike would have given the miners and other union militants new heart.[88]

3.38 Clearly, for Mrs Thatcher, the problems of 1984 were a complex mix of economic, social, and political problems and calculations. What is certainly apparent

---

[86] Cabinet Office, *The Draft Civil Contingencies Bill, Executive Summary*, para 1.
[87] Ibid.
[88] Thatcher, M, *The Downing Street Years* (HarperCollins, London, 1993) 356.

in 1984 is that, faced by the extension of industrial unrest beyond the mining sector, Ministers were not assisted by the unitary structure of emergency powers legislation. If anything, the possibility of military involvement, as well as the political signals which would be sent by using such powers, were disincentives to invoke such legislation. The invocation could easily be interpreted as a sign of weakness or a repressive overreaction. It is not certain that the Part II of the Civil Contingencies Act, equally lacking in nuanced responses, will improve matters.

Instead of refining the definitions, the Government opted for the Triple Lock. **3.39** There are, however, several problems with this tactic. One is that the Triple Lock relates only to Part II. Its prominence there is understandable as the threats of overuse, misuse, and abuse are more acute with Part II. But there remains a danger of public interference in everyday life and the wasting of public money if Part I is not interpreted properly. Secondly, the effectiveness of the Triple Lock can be doubted because it is translated into subjective rather than objective criteria. Thirdly, the Triple Lock applies to the exercise of powers rather than to their definitions per se. In this way, the definitions do not work to rule out the application of powers—after all, many of the listed problems such as loss of life and human illness or injury are commonplace.

(c) **Practical application**

Some indications of the attitudes of government to the likely use of Part II have **3.40** already been glimpsed. For example,[89] neither the Manchester nor Docklands bombings by the IRA in 1996 would have led to special legislative measures because of their very limited geographical impact. This implies that single locus events do not generally meet the criteria of Part II (taking ss 19 and 21 together) unless one of two factors comes into play, either of which disrupts effective reactions within 'normal' laws.

The first would be that the single locus incident is so serious of itself or is in such **3.41** a strategic location that it could satisfy the tests. If the 9/11 attacks in New York were to be replicated in the centre of London, though the geographic area of impact might be narrow, the impact would be enormous in terms of 'serious-ness' and 'extent'. Of course, it is possible that Part II powers would still be averted as not fulfilling the 'necessary' test on the basis that the emergency would be over within a very short period and the consequences could be dealt with by the Part I resilience plans of the local authorities. Nevertheless, one could imagine that powers might be used to restrict movement in a blighted area or to restrict access to the vicinity of other potential targets.

The second example of a single locus event which triggers Part II is where the **3.42** single locus has serious consequential effects elsewhere. For example, an event

---

[89] Joint Committee on the Draft Civil Contingencies Bill, Appendix 9 q 14.

such as September 11, 2001 might overwhelm all NHS resources or might disrupt telecommunications networks on a national basis.

3.43 It is easier to imagine that multi-seated terrorist attacks, outbreaks of animal disease, or flooding (for example, across the whole of the Thames Estuary) might trigger Part II. There might be a need to direct the allocation of resources and also to restrain movement. The Energy Act 1976 can deal with the immediate issues relating to the distribution and supply of fuel, but the Civil Contingencies Act might deal with the consequential impacts on essential public services, such as by relaxing standards so that a wider variety of personnel or vehicles can be utilized. Another example might be the fuel crisis of 2000, which was both serious and wide in extent, the threatened repetition of which in September 2005 did give rise to murmurings about the invocation of Part II.[90] On the other hand, there remains evidence of reluctance to resort to Part II, especially because of the adverse signal it sends of loss of normality. For instance, following the bombings in London in July 2005, there was no announcement of an intention to invoke the Civil Contingencies Act.

(d) Final conclusions

3.44 There may be two conflicting findings from this chapter. The first is that the broadening of the definition of 'emergency' is to be expected. The range of risks with which the legislation is expected to contend is far wider than for its pre-decessors. Society may now be more risk aware and averse and expects thought to be given to the broader range of dangers, even if the responsibility for action does not necessarily reside with the state. The second is that the way in which risk has actually been defined does not adequately address the concerns of constitutionalism or practical implementation since it fails to discern sufficiently within and between Part I and II as to what should be actionable.

---

[90] *The Guardian* 14 September 2005 at 1.

4

LOCAL ARRANGEMENTS FOR CIVIL
PROTECTION: PART I OF THE ACT

A. Introduction

An overview will be given here of the planning duties in Part I of the Act, their **4.01**
effect on different public and private bodies, their fit with other legal provisions
about planning and emergencies, and their meaning within a wider theoretical
and critical context. The essence of Part I is to plan for 'emergency' (as defined
in Chapter 3) by requiring preparations and protections to be undertaken and
advice to be given (ss 2 and 4). The 'Category 1' responders form the core
contingency workforce, and so they have specified responsibilities 'based on key
themes—risk management, emergency planning, business continuity of the
responder itself, and warning and informing the public.'[1] Bodies in the public
and private sectors whose involvement is crucial to facilitating the efforts of

---

[1] Cabinet Office, *The Draft Civil Contingencies Bill: Explanatory Notes etc*, para 3.

Category 1 responders will also be brought into the fold as 'Category 2' responders; their contribution demands 'communication, co-operation and information sharing'.[2]

4.02 As well as a structural focus on public bodies, Part I adopts a geographic preference for contingency planning by local authorities. Part I also delimits their duties:

> The Bill does not place an express duty on local responder bodies to respond to an emergency. But it does require them to maintain *plans* for response and to maintain *arrangements* to warn, inform and advise.[3]

Ministers of the Crown remain in the background—able to require a Category 1 responder to perform a Part I function but not bearing statutory responsibilities themselves. Their role will be teased out later, in Chapter 8. Regions are barely mentioned but will be covered here.

B. Duty Holders—'Responders'

4.03 The duty holders under Part I of the Act are designated as Category 1 responders and Category 2 responders under s 3. The emergence of distinct categories of responders might be traced to the Cabinet Office consultation in 2001, when there was comment about 'a two-status system—of 'core' and 'optional statutory partners'.[4]

(a) Category 1 responders

4.04 The 'general' range of Category 1 responders is set out in Part 1 of Schedule 1 (covering primarily UK-wide or England and Wales bodies, with Part 2 of Schedule 1 relating a corresponding list for Scotland).[5] As shall be described in the next segment of this chapter, they shoulder a wide range of duties of planning and preparation:[6] to assess the risk of emergencies occurring; to put in place planning arrangements; to put in place business continuity management; to warn and inform the public in the event of an emergency; (for local authorities alone) to give advice to businesses and voluntary groups about resilience; and to cooperate with each other through local resilience forums. Furthermore, regulations might require them to discharge established statutory

---

[2] Ibid.   [3] Civil Contingencies Secretariat, *The Government's Response*, 11.
[4] Cabinet Office, *Emergency Planning Review: The Future of Emergency Planning in England and Wales: Results of the Consultation* (London, 2002) para 5.
[5] In so far as bodies are national in coverage (such as the Maritime and Coastguard Agency), the phrase used is 'general', rather than 'United Kingdom': HC Debs Standing Committee F col 124 29 January 2004, Douglas Alexander.
[6] Civil Contingencies Secretariat, *Full Regulatory Impact*, para 29.

powers for the purpose of responding to an emergency.[7] The Category 1 responders fall into four categories.

The first are local authorities in England, Wales, and Scotland. These are **4.05** defined in England as (a) a county council, (b) a district council, (c) a London borough council, (d) the Common Council of the City of London, and (e) the Council of the Isles of Scilly. In total, there are 34 county councils, 46 unitary authorities, and 239 district councils. Category 1 local authority responders are defined in Wales as (a) a county council, and (b) a county borough council (there are 22 such unitary authorities). In Scotland, they comprise any council constituted under s 2 of the Local Government etc (Scotland) Act 1994 (there are currently 32 unitary council areas). The 26 district councils in Northern Ireland are not scheduled, and their limited involvement is described later in this chapter. Though English district councils are listed, it was envisaged from the outset that county (shire) councils will take 'full' responsibility for local authority civil protection planning in their area,[8] as has historically prevailed. The residual role for the district councils will relate mainly to housing and environmental health,[9] but this demarcation is not so stated on the face of the Act. A possible alternative would have been to specify only counties, with duties imposed on districts to cooperate with them. Service-level agreements by which the districts contract with their counties to deliver their functions will in practice point in that direction. A further problem is that there is no uniform structure within local government, so devolved regions and London merit distinct treatment, as discussed later.

The second category of Category 1 responders are the emergency services. **4.06** Within this group is, first of all, any 'police force maintained by a police authority' within the definition of s 101 of the Police Act 1996 or under the under the Police (Scotland) Act 1967 (43 in England and Wales, 8 in Scotland), the Police Service of Northern Ireland,[10] and the British Transport Police Force.[11] The formula does not include the Civil Nuclear Constabulary.[12] Though it is a police force maintained by a police authority, it is not linked to a police area listed in Schedule 1 of the Police Act 1996. This omission arises from their involvement in existing radiological preparedness networks, so that the Civil Nuclear Constabulary does in fact participate in local, regional, and national resilience structures.

---

[7] Cabinet Office, *The Draft Civil Contingencies Bill*, para 3.16.   [8] Ibid, para 3.10.
[9] Joint Committee on the Draft Civil Contingencies Bill, para 104.
[10] Police (Northern Ireland) Act 2000. It can be included since policing is a reserved matter: Northern Ireland Act 1998 Sch 3 para 11.
[11] See Railways and Transport Safety Act 2003 Pt III.
[12] Energy Act 2004, s 51. See *Managing the Nuclear Legacy—A Strategy for Action* (Cm 5552, London, 2002) Ch 8.

4.07 Fire and rescue authorities (there are 50 fire authorities in England and Wales, 8 in Scotland, and one in Northern Ireland) are the next Category 1 responders if, in England and Wales, within the meaning of s 1 of the Fire and Rescue Services Act 2004 or, in Scotland, under s 1 of the Fire (Scotland) Act 2005.[13] The Northern Ireland Fire Brigade is left outside the scope of the Act.[14] For England and Wales, Fire and Civil Defence Authorities (FCDAs), set up under s 26 of the Local Government Act 1985, will not exercise any aspects of civil protection duties on behalf of local authorities. In most metropolitan areas, the metropolitan districts councils carry out the whole range of emergency planning tasks, and so no change is required, but those FCDAs which did so will continue only by agreement with the constituent local authority. FCDAs will be subject to the duties of Category 1, but only in their capacity as fire authorities. In addition, they are retitled 'fire and rescue authorities' (Schedule 2, para 10). These arrangements were questioned by the Joint Committee which felt there should be a continuing role in contingency planning for FCDAs.[15] Some will still be involved in planning under the Control of Major Accident Hazards (COMAH) Regulations 1999.[16] After reflection, the Government concluded that 'local authorities will be given powers of delegation in the Bill, so those that wish to delegate their functions to the Fire and Rescue Service or some other body will be permitted to do so.'[17] This outcome can be achieved under s 2(3).[18]

4.08 Next are health authorities within Britain. For England and Wales, these are specified as including National Health Service trusts[19] (which include primary care[20] and hospital trusts,[21] ambulance trusts,[22] and mental health trusts[23]),

---

[13] Their primary role is rescue and health and safety issues: Cabinet Office, *Dealing with Disaster* (rev 3rd ed, Cabinet Office, London, 2003) paras 2.17, 2.18. Under s 9 of the Fire and Rescue Services Act 2004, the Secretary of State may by order confer on a fire and rescue authority functions relating to a wide range of emergencies (as defined by s 58).

[14] It is a transferred matter: Northern Ireland Act 1998, s 4. See Fire Services Order (NI) 1984 SI No 1821 (NI 11). A draft Fire and Rescue Services (Northern Ireland) order is pending.

[15] Joint Committee on the Draft Civil Contingencies Bill, para 107.

[16] 1999 SI No 743 as amended by 2005 SI No 1088. See also Control of Major Accident Hazards Regulations (Northern Ireland) 200 SR No 93, as amended by 2005 SR No 305.

[17] Civil Contingencies Secretariat, *The Government's Response*, 11; see also Cabinet Office, *The Government's Response to the Report of the Joint Committee*, para 21.

[18] Cabinet Office, *The Government's Response to the Report of the Joint Committee*, para 21.

[19] See National Health Service and Community Care Act 1990, s 5.

[20] National Health Service Act 1977, s 16A.

[21] Strategic health authorities are in Category 2: Civil Contingencies Act 2004 (Amendment of List of Responders) Order 2005 SI No 2043.

[22] Ambulance services (there are 32 trusts in England plus one each for Scotland, Wales and Northern Ireland) are expected to offer the on-site National Health Service response: Cabinet Office, *Dealing with Disaster* (rev 3rd ed, Cabinet Office, London, 2003) para 2.20. Ambulance trusts were solely listed in the first draft of the Bill: Joint Committee on the Draft Civil Contingencies Bill, paras 108–113; Cabinet Office, *The Government's Response to the Report of the Joint Committee*, para 22, 23.

[23] The importance of mental needs of victims of disasters is stressed: Hodgkinson, P.E. and Stewart, M, *Coping with Catastrophe: A Handbook of Post-disaster Psychological Aftercare*

NHS foundation trusts[24] covering the same functions, Local Health Boards in Wales,[25] the Health Protection Agency (in so far as its functions relate to Great Britain), and port health authorities.[26] In Scotland, the Scottish Ambulance Board (a unitary body though with six divisions) and any Health Board constituted under s 2 of the National Health Service (Scotland) Act 1978 are the relevant bodies. The position of the Health Protection Agency (HPA), established as a special health authority (SpHA) in 2003 and then, by the Health Protection Agency Act 2004, transformed into a non-departmental public body, is especially significant since it handles chemical, biological, and radiological substances.[27] It was listed as a responder after initial hesitation.[28] There is no health authority listed for Northern Ireland. Albeit that health is a transferred rather than excepted or reserved matter under the Northern Ireland Act 1998, it is not seen as appropriate to include a Northern Ireland Minister within the (Northern Ireland) Department of Health, Social Services and Public Safety as a Category 1 responder. These devolution arrangements are further reflected in a statutory order, issued in 2005, to limit the Act's duties to HPA functions only in relation to Great Britain.[29] At national level, the National Blood Service (plus the Northern Ireland Blood Transfusion Service and the Welsh Blood Service),[30] has been omitted from both categories, even though Category 2 status was recommended by the Joint Committee on the Draft Civil Contingencies Bill.[31] The most plausible reason is that it operates within the NHS fiefdom, since its services are delivered via NHS or ambulance trusts.[32]

(Routledge, London, 1998); Cabinet Office, *Dealing with Disaster* (rev 3rd ed, Cabinet Office, London, 2003) para 4.11; FEMA, *Responding to Incidents of National Consequence: Recommendations for America's Fire and Emergency Services Based on the Events of September 11, 2001, and Other Similar Incidents* (Washington, 2002) 65; Executive Session on Domestic Preparedness, *Beyond the Beltway: Focusing on Hometown Security* (John F Kennedy School of Government, Harvard University, 2002); Wood-Heath, M and Annis, M, *Working Together to Support Individuals in an Emergency or Disaster* (British Red Cross, London, 2004).

[24] See Health and Social Care (Community Health and Standards) Act 2003, s 1.
[25] National Health Service Act 1977, s 16BA. There are 22 LHBs.
[26] See Public Health (Control of Disease) Act 1984, s 2(4). These are separately constituted local authorities (see Local Government Act 1972, s 263) whose primary duties relate to the control of infectious disease, environmental protection, imported food control, and hygiene on vessels.
[27] See Ministry of Defence, *Operations in the UK*, para 875; Walker, C, 'Biological Attack, Terrorism and the Law' (2004) 17 *Journal of Terrorism and Political Violence* 175.
[28] Cabinet Office, *The Draft Civil Contingencies Bill*, para 3.11. See also Ch 1.
[29] Civil Contingencies Act 2004 (Amendment of List of Responders) Order 2005 SI No. 2043. See further HL Debs vol 673 col 1171 13 July 2005, Lord Bassam; House of Commons Seventeenth Standing Committee on Delegated Legislation, 14 July 2005.
[30] The National Blood Service is a Special Health Authority (National Blood Authority (Establishment and Constitution) Order 1993 SI No 585). Welsh Blood is constituted under the Velindre National Health Service Trust (Establishment) Amendment Order 1999 SI No 826.
[31] Draft Civil Contingencies Bill (2002–03 HC 1074, HL 184) para 114.
[32] Ibid, para 23.

ScotBlood—the Scottish National Blood Transfusion Service—is within the Common Services Agency included within Category 2.[33]

4.09 Finally, there are two other responders, the Environment Agency (and its equivalent, the Scottish Environment Protection Agency)[34] and 'the Secretary of State, in so far as his functions include responding to maritime and coastal emergencies (excluding the investigation of accidents)'.[35] The latter laborious phrase means the Maritime and Coastguard Agency, but, because it has the status of an Executive Agency within the Department for Transport, it lacks a legal personality to be named in its own right. The relevant operations of the Maritime and Coastguard Agency, which are UK-wide, are divided between the HM Coastguard (Search and Rescue Response) and the Counter Pollution Unit.[36] The Environment Agency can be proudly named on the face of the Act since it was established under the Environment Act 1995 as a body corporate.[37] Its function is the environmental protection of water, land, and air in England and Wales. The Scottish Environment Protection Agency has equivalent status but some variant powers (for example, Scottish local authorities are responsible for flood defence). In Northern Ireland, the Environment and Heritage Service is a Next Steps (non-statutory) agency within the Department of the Environment. It is not a responder.

4.10 These responders can be added to or removed from the list or transferred from one category to the other under s 13 by Ministerial order, which can relate to the responder as a whole or for specified functions. The order may also may make incidental, transitional, or consequential provision including the amendment of the 2004 Act or another 'enactment', which term includes (by s 18) Acts of the Westminster Parliament, Acts of the Scottish Parliament, Northern Ireland legislation, or instruments made under Scottish or Northern Irish legislation). The amending power could be necessitated, for example, by the addition as a Category 1 responder of a body which has functions in Wales which then correspondingly needs to be added to the list of Category 1 responders and which must therefore be mentioned in s 16(4) (see below).

4.11 By way of assessment, there are some initial points to note. First, the list of bodies goes much further than the previous legislation, reflecting the wider definition of 'emergency', though its extent still omits many socially and financial important institutions as well as the general public. The second point is

---

[33] See NHS (Functions of the Common Services Agency) (Scotland) Order 1974 SI No 467.
[34] See Environment Act 1995, ss 1, 20.
[35] Schedule 1 para 12. See further Ministry of Defence, *Operations in the UK*, para 870.
[36] See further Ministry of Defence, *Operations in the UK*, para 862.
[37] Its inclusion as a category 1 provider can replace the DEFRA Ministerial Direction on Emergency and Security Measures issued in 1996 (power to issue flood warnings) under the Water Resources Act 1991, s 5.

that the list is confined to local bodies. Conversely, most national bodies—government departments and their regional offices, plus the Army[38]—are not listed, with the exception of the Health Protection Agency, the Environmental Agency, and, by a roundabout way, the Maritime and Coastguard Agency. Had they been incorporated within the scheme, then central bodies would have been categorized as Category 1, since 'all [Government] departments have a responsibility to plan, prepare, train and exercise for handling incidents and emergencies that might occur within their field of responsibility'.[39] But, rather than applying the scheme, the Government promised that it is working to 'establish standards against which departmental contingency planning activities can be monitored and audited'.[40] The result is that the Act is 'very "bottom heavy", with all statutory duties being accorded to local providers and a cloak of invisibility being drawn over the regional and central tiers.'[41] The position of central government will be further explored in Chapter 8.

(b) Category 2 responders

Forcibly allied to the Category 1 responders are the 'cooperating bodies', **4.12** Category 2 responders, who are 'less likely to be involved at the heart of planning work but will be heavily involved in incidents that affect their sector'.[42] They are conferred with duties to cooperate with, and to provide information to, Category 1 responders in connection with their civil protection duties. As for the composition of the list which follows:

> There are two principal criteria for inclusion in category 2. They are, first, being an organisation that operates a risk source or, secondly, being an organisation that has a role to play in responding to emergencies.

> There are also two key criteria for excluding bodies from category 2 responders. The first is that some bodies are already covered by other arrangements. The Bill is focused on local arrangements for responder bodies that have an operational role in emergencies. It does not directly concern itself with national arrangements in relation to the critical national infrastructure, which fall outside its local focus. They are a matter for the lead government department to manage. For example, the Department of Trade and Industry, as the lead government department, co-ordinates a national fuel contingency plan. . . .

> Secondly, inclusion in the framework is governed by practical considerations. While we might wish to bring in more bodies than we have currently, we need to be careful not to overburden the local arrangement infrastructure. Priority has been given to those bodies that local responders themselves judged to be

---

[38] See Ch 9. [39] Joint Committee on the Draft Civil Contingencies Bill, para 92.
[40] Ibid. [41] Ibid, para 101.
[42] Cabinet Office, *The Draft Civil Contingencies Bill*, para 38. See further HC Debs vol 416 col 1106 19 January 2004, Douglas Alexander.

important, and some bodies that we might like to have included—for example, buses and road freight—have been excluded for this reason.[43]

4.13 Category 2 responders are listed for the whole of the United Kingdom or for England and Wales in Schedule 1, Part 3, and for Scotland in Part 4 (Scotland), with none specific to Northern Ireland. They fall into three categories.

4.14 First, there are utilities—including the holders of electricity or gas transmission, distribution and interconnector licences;[44] water sewerage undertakers;[45] and public electronic communications networks which make telephone services available (whether for spoken communication or for the transmission of data).[46] The utilities, often nationally based, fear practical and financial difficulties arising from the obligation to liaise with several local resilience forums at the same time. As a result, they would have preferred a greater emphasis upon regional and national planning forums. The Government has sought to ensure that they should not be required to be involved below the level of police force area.[47]

4.15 Secondly, there are transport undertakers, including railway operators (both passenger and freight),[48] Transport for London and London Underground Limited,[49] any British airport operator within the meaning of s 82(1) of the Airports Act 1986,[50] British harbour authorities within the meaning of s 46(1) of the Aviation and Maritime Security Act 1990,[51] and the Secretary of State, in so far as his functions relate to matters for which he is responsible by virtue of s 1 of the Highways Act 1980 (in other words, the Highways Agency). The latter was included because the Agency has assumed the role on trunk roads, including motorways, of responding to emergencies at a local level. Passenger transport authorities are not listed.[52]

---

[43] HL Debs vol 666 col 794 9 November 2004, Lord Bassam.

[44] Electricity Act 1989, s 6; Gas Act 1986, ss 7, 7ZA.

[45] Water Industry Act 1991, s 6; Water Industry (Scotland) Act 2002, s 20, Sch 3.

[46] See Communications Act 2003, ss 32, 151.

[47] Joint Committee on the Draft Civil Contingencies Bill, para 118.

[48] Railways Act 1993, s 8. Network Rail will generally represent rail freight companies at Local Resilience Forums: HL Debs vol 666 col 796 9 November 2004, Lord Bassam.

[49] It was included after consultation in January 2004: Civil Contingencies Secretariat, *The Government's Response*, 48. Transport for London encompasses London Buses, Docklands Light Railway, Tramlink, River Services, Streets, Public Carriage Office, Woolwich Ferry, and Victoria Station and manages London Underground.

[50] The Act is applicable only to those airports through which, in the most recent year for which data is available, at least 50,000 passengers or 10,000 tonnes of freight and mail were transported: Civil Contingencies Act 2004 (Contingency Planning) Regulations 2005 SI No 2042, reg 3.

[51] The Act is applicable only to those harbours through which the average annual maritime traffic, calculated by reference to the most recent three years for which data is available, is at least 1.5 million tonnes of cargo or 200,000 passengers: Civil Contingencies Act 2004 (Contingency Planning) Regulations 2005 SI No 2042, reg 3.

[52] Cabinet Office, *The Government's Response to the Report of the Joint Committee*, para 28. See Traffic Management Act 2004 Pt I which sets up a system of 'traffic officers' within the Highways Agency.

Thirdly, there are health bodies—the Health and Safety Executive (HSE) is **4.16** specified 'because of the importance of its sharing information and cooperating fully with the other partners.'[53] Despite its importance in dealing with the assessment of risk and failures in the observance of regulatory standards, the HSE does not in the main have any direct emergency response capability.[54] In Scotland, there is listed the Common Services Agency (commonly called the NHS National Services Scotland) established by s 10 of the National Health Service (Scotland) Act 1978.[55] Strategic health authorities in England (there are 28) have also been added to the list.[56]

The logic behind Category 2 is to capture those entities, nowadays as often **4.17** private as public, which perform 'functions vital to the life of the community' or are 'key parts of the local infrastructure which maintain the life of the community'.[57] Their proposed duties are depicted as benefiting them as much as the community.[58] Adding them all together, the network of Category 2 responders comprised the following in June 2003:

Table 4A Category 2 responders, June 2003[59]

Type	Number
Water	24
Gas	1
Electricity	10
Telecoms	18
Rail	25
Harbour authorities	259
Airports	21
Total	358

[53] Cabinet Office, The *Draft Civil Contingencies Bill*, para 3.12.

[54] But it could be argued that the HSE's Nuclear Safety Directorate, which provides a 'Government Technical Adviser', fits better with Category 1 as having prime responsibility.

[55] The NSS includes the Scottish National Blood Transfusion Service and Health Protection Scotland: Scottish Executive Health Department, *National Health Service in Scotland Manual of Guidance: Responding to Emergencies* (http://www.show.scot.nhs.uk/emergencyplanning/guidance.htm, Edinburgh, 2004) para 2.6.

[56] They were created in England by the National Health Service Reform and Health Care Professions Act 2002, s 1, and they must fall within the National Health Service Act 1977, s 8: Civil Contingencies Act 2004 (Amendment of List of Responders) Order 2005 SI No 2043. In relation to Wales, the same Act provides for Local Health Boards, which take responsibility alongside each Local Authority in Wales but these are already listed under Category 1.

[57] Cabinet Office, *The Draft Civil Contingencies Bill: Explanatory Notes etc*, para 17.

[58] Ibid, para 25.

[59] Source: Civil Contingencies Secretariat, *Full Regulatory Impact*, Annex C. There are doubt about the figure given for harbour authorities: Cabinet Office, *The Draft Civil Contingencies Bill: Explanatory Notes etc*, para 36; Civil Contingencies Secretariat, Full Regulatory Impact, para 42.

(c) General provisions and assessment

4.18 A prolonged debate took place as to whether there might be further responders, whether in Categories 1 or 2. A huge range of organizations were nominated, some of which have already been mentioned.[60] For example, in Category 1, probably the most important addition would be the central and regional government ministries. Their lack of involvement is discussed in Chapter 8. Basically, the Act is silent in this regard, save for the regulation-making powers granted to Ministers and the creation of Regional Nominated Coordinators (or Emergency Coordinators).[61] Another candidate was the armed forces, especially as some have been attending at county resilience forums, but it was said that 'Their participation could not be guaranteed in the face of operational requirements.'[62] There is also no mention of coroners, though they are advised to devise emergency plans and to consult with other responders.[63]

4.19 Category 2 does not include such movers and shakers of UK society as financial institutions or major employers in the distribution, retail, and catering industries.[64] The Defence Committee also suggested the inclusion of major industrial sites and even private security providers.[65] Another suggestion was to embrace all operators of establishments subject to the Control of Major Accident Hazards (COMAH) Regulations (such as chemical factories).[66] The prospect of one planning umbrella and one regulatory framework is alluring, but the Government was discouraged from consolidation[67] due to the administrative complications engendered by involving such a wide range of bodies, including universities, road transport hauliers, and industrial companies. More positively, a degree of coordination is assured because COMAH is managed by the HSE and the Environment Agency and both bodies are responders under the Act. Though the direct relationship between the Local Resilience Forums (LRFs) under the 2004 Act and the COMAH arrangements was not addressed in the 2004 Act, the Civil Contingencies Act 2004 (Contingency Planning) Regulations 2005, reg 12 (reg 9 in Scotland), states expressly that general Category 1 responders need not perform a duty under s 2(1) in relation to a variety of emergencies covered by other specified legislation (which are described later in this chapter).[68]

---

[60] See Joint Committee on the Draft Civil Contingencies Bill, Appendix 10.

[61] See Ch 5.

[62] Cabinet Office, *The Government's Response to the Report of the Joint Committee*, Annex A para 11.

[63] Cabinet Office, *Dealing with Disaster* (rev 3rd ed, Cabinet Office, London, 2003) para 2.29.

[64] See Nasima Begum, 'Employment by occupation and industry' (2004) 112 No 6 Labour Market Trends.

[65] Defence Committee, *Draft Civil Contingencies Bill* (2002–03 HC 557) paras 46, 47, 50.

[66] Joint Committee on the Draft Civil Contingencies Bill, para 131.

[67] See Cabinet Office, *The Government's Response to the Report of the Joint Committee*, para 27.

[68] But they are encouraged to receive reports from those of its members involved in the activities of these separate emergency planning groups: Cabinet Office, *Emergency Preparedness: Guidance on Part 1*, para 2.62.

A further suggested supplement to Category 2 concerned road transport **4.20**
undertakers, such as bus and coach companies, who might be vital to evacuation
and the transportation of materials over a much wider area of the country than
the railways, airports, and harbour authorities who are listed in Category 2.[69]
But the Government once again felt queasy about imposing statutory obligations on such a 'numerous and...disparate' range of the private sector.[70] A
special case can be made for the 'life-line' ferry operators in western and
northern Scotland, and further consideration is pending.[71]

Similar arguments and responses were made on behalf of the fuel distributors **4.21**
and producers[72] and also the food and drink industry and the relevant regulatory body, the Food Standards Agency.[73] In these cases, the argument was
that there were alternative regulatory regimes already in existence.

A specific proposal was that the British Broadcasting Corporation (BBC) should **4.22**
be listed. In an emergency, it might act as a vital, and singularly authoritative,
source of information. The BBC has devised a project, known as 'Connecting in
a Crisis',[74] which envisages public service announcements via local radio. But
the Government was not keen on formal incorporation into Category 2 for fear
that it would compromise the broadcaster's independence.[75] The official fears
may hark back to the criticism of the involvement of the BBC during the
General Strike of 1926, quickly followed by a Charter providing for independence.[76] The official publication during the same crisis of the *British Gazette*
was equally controversial.[77] Further systems for warning the public and the
resilience of communications networks are discussed later in this chapter.

The most radical suggestion was that Category 2 responders should not be **4.23**
specified at all. Instead, Category 1 responders should be able to designate
Category 2 bodies from their locality in the light of local circumstances.[78] Just as
community safety partnerships remain flexible under the Crime and Disorder

---

[69] See Joint Committee on the Draft Civil Contingencies Bill, para 132 and Appendix 2; Civil
Contingencies Secretariat, *The Government's Response*, 13.

[70] HL Debs vol 665 col 492 14 October 2004, Lord Bassam. There are 8,300 licensed public
service vehicle operators: ibid col 503. See also Cabinet Office, *The Government's Response to the
Report of the Joint Committee*, para 28. Rail freight provides 11.5% of total traffic: Department of
Transport, *The Future of Transport* (Cm.6234, London, 2004) Ch 8.

[71] Letter from Scottish Executive Justice Department, 30 June 2005, http://www.scotland.
gov.uk/Resource/Doc/55971/0015631.pdf.

[72] HL Debs vol 665 col 499 14 October 2004, Baroness Buscombe.

[73] Ibid, col 503, Lord Bassam.

[74] See Connecting in a Crisis, http://www.bbc.co.uk/connectinginacrisis/index.shtml.

[75] Joint Committee on the Draft Civil Contingencies Bill, Appendix 9 q 58.

[76] O'Brien, TH, *Civil Defence* (HMSO, London, 1955) 30.

[77] Edited by Winston Churchill, it was published between 5–13 May 1926 under the imprint
of the HMSO but on printers donated by the press owners: Jenkins, R, *Churchill* (MacMillan,
London, 2001) 408–9.

[78] Joint Committee on the Draft Civil Contingencies Bill, para 90.

Act 1998, s 1, so local circumstances might dictate that, for example, the operators of major shopping, leisure, and sporting complexes in some areas and marsh wardens in others, could be included as appropriate.[79] This proposal was rejected because of the burden of new regulatory costs, as well as inconsistencies developing between areas and too many bodies formally 'involved'.[80] But, since local authorities are obliged to give advice and assistance under s 4 (described later) on Business Continuity and Risk Management—why not go a step further and cement relationships with a formal 'statutory Partnership' which could further foster 'risk awareness' in the private sector?

4.24 Some have suggested an extra 'Category 3', the prime candidates for which were voluntary organizations. It has been recognized that, 'Major emergencies can overstretch the resources of the emergency services and local authorities. The value of additional support from the voluntary sector has been demonstrated on many occasions.'[81] Prominent candidates include the Association of Lowland Search and Rescue, the British Association for Immediate Care Schemes, the British Cave Rescue Council, the Maritime Volunteer Service, the Mountain Rescue Council of England and Wales, and the Mountain Rescue Committee of Scotland,[82] RAYNET (The Radio Amateurs' Emergency Network), the Red Cross, St John Ambulance and St Andrew Ambulance, the Salvation Army, Victim Support, and the Women's Royal Voluntary Service (WRVS).[83] The Red Cross at one stage argued for inclusion in Category 1, 'given the legal functions established under our Royal Charter and the Geneva Conventions.'[84] In very specialized circumstances, where the Geneva Conventions are brought into force because of a state of 'armed conflict', the Red Cross could have a tenable case. Red Cross organizations must then be allowed to carry out humanitarian protection, including visits to prisoners of war and civilian internees. It would be a breach of international law for the public authorities to obstruct these functions. Of course, the notion of 'armed conflict' is not a common contingency. It probably did not apply, for example, even during the height of 'The Troubles' in Northern Ireland in 1971 and 1972 (though it is true that the Red Cross did then perform humanitarian functions, including visits to internees).[85] While planning for civil war or invasion by a foreign power

---

[79] Ibid, para 127.

[80] Civil Contingencies Secretariat, *The Government's Response*, 14. See also Cabinet Office, *The Government's Response to the Report of the Joint Committee*, para 18.

[81] Cabinet Office, *Dealing with Disaster* (rev 3rd ed, Cabinet Office, London, 2003) para 6.1. See further Wood-Heath, M, and Annis, M, *The Role of Non-Governmental Organisations' Volunteers in Civil Protection in European Member States & European Economic Area Countries* (British Red Cross, London, 2002).

[82] See Ministry of Defence, *Operations in the UK*, para 886.

[83] The latter claims 12,000 available volunteers: http://www.wrvs.org.uk/.

[84] Civil Contingencies Secretariat, *The Government's Response*, p.12.

[85] See Walker, C, 'Irish Republican Prisoners' (1984) 19 *Irish Jurist* 189.

will probably not be on top of the 'to do' list of emergency planners, the Red Cross should be included in planning processes when and if 'armed conflict' is considered. Further, one might argue that the contingency of armed conflict is not as remote as here suggested. President George W Bush has notoriously conjured 'the first war of the twenty-first century',[86] and martial powers include the internment without trial of terrorism suspects,[87] with the American Red Cross being allowed access for humanitarian purposes. Thus, major terrorism assaults might be construed as 'armed conflict' or at least something akin to it which triggers humanitarian intervention de facto.

Another voluntary organization with a unique role, approaching that of **4.25** Category 1 responder, is the Royal National Lifeboat Institute, which operates over 200 stations around the British Isles.[88] Furthermore, it acts as a primary emergency service in a way which most other volunteer groups do not.[89]

The view of the Government was that, while volunteers could always be invited **4.26** to join in the planning process without compulsion:

> Voluntary organisations are not covered by the new duty. This is because voluntary organisations rely on the goodwill of their members and supporters to provide the services that they do, and because those services are not in themselves based on statutory obligations. As a consequence, the skills and expertise available to the voluntary sector may vary from place to place.[90]

Thus, if there is no obligation to act and no certainty of ability to deliver services **4.27** (not least because key volunteers such as medical staff will be required in priority by Category 1 responders), then it is better not to impose statutory requirements, the failure of which might entail dire consequences in an emergency.[91] However, a formal recognition of the role of voluntary bodies was signalled by the addition at the House of Lords' Report Stage[92] of s 2(5)(k): Ministerial regulations may permit or require a person or body responsible for maintaining contingency plans to have regard to the activities of bodies (other than public or local authorities) whose activities are not carried on for profit to the lists of statutory requirements placed upon Category 1 responders (see further below). This met or exceeded the wishes of most of the voluntary sector who wanted to have a consultative role in emergency planning.[93] Though

---

[86] *The Guardian* 14 September 2001 at 5.
[87] See *Hamdi v Rumsfeld* [2004] US LEXIS 4761; *Rasul v Bush* [2004] US LEXIS 4760.
[88] http://www.rnli.org.uk/. See further Ministry of Defence, *Operations in the UK*, para 883.
[89] HL Debs vol 664 col 1277 15 September 2004, Lord Bassam.
[90] Cabinet Office, The Draft Civil Contingencies Bill, para 3.13, See also Cabinet Office, *The Government's Response to the Report of the Joint Committee*, para 25. See further Ch 8.
[91] HC Debs vol 421 col 1358 24 May 2004, Hazel Blears.
[92] HL Debs col 666 col 798 9 November 2004, Lord Bassam.
[93] See Joint Committee on the Draft Civil Contingencies Bill, para 126.

the Government stood firm against a net-widening tendency during the passage of the Bill, it has since relented to a degree. It now advises that:

> ... organisations which are not required to participate under the Act should be encouraged to take part in forums and co-operate in planning arrangements wherever this is appropriate. The statutory nature of the framework is not a consideration which in itself should imply their exclusion.[94]

4.28 Finally, in the outer reaches of the constellation of responders, there was even a suggestion of a Category 4, namely the public:

> One of the great strengths of the 30 year campaign against Irish Terrorism was the engagement of the public, both the commercial sector and individual citizens, thereby enabling them to provide information that prevented or disrupted attacks, to alert the authorities of the possible presence of explosive device or other suspicious item or activity and to minimise the probability of becoming victims of terrorist action. While the prevailing official view is that to initiate greater public engagement will increase anxiety there is a counter argument based on the premise that an informed and involved society is a resilient society and their inclusion would facilitate mobilisation of the full complement of resources.[95]

In a late modern society, where levels of welfarism are increasingly problematized rather than merely increasing, self-reliance is vital. Thus, the passive statutory status accorded to the public, both pre-emergency and post-emergency, disregards the deficiencies and limits in core responders which could be alleviated by communal involvement. But it would probably dilute the concept of 'responder' if all became responders without discernment. Thus, the legislation depicts the public primarily as passive consumers of contingency planning, in counterpoint to a 'securitocracy' who have to be trusted and obeyed. Their position would be more palatable were there to be greater duties imposed upon that securitocracy to inform the public in order to make the Act more 'public-facing'.[96] Consistent with the rejection of public rights to involvement, the legislation imposes no special duties on the public:

> The proposal that a civil protection duty should fall directly on the general public itself to co-operate in civil protection was examined in some depth at an early stage of policy development, but considered to be unfeasible and not pursued, partly on civil liberties grounds.[97]

---

[94] Cabinet Office, *Emergency Preparedness: Guidance on Part 1*, paras 15.3–15.4. The relevant bodies comprise the armed forces; retail companies, including supermarkets; insurance companies; bus and road haulage companies; taxi firms; airlines; shipping companies and ferries; media companies; private communications networks; offshore oil and gas industry; security firms; internal drainage boards; and general practitioners and chemists.

[95] Joint Committee on the Draft Civil Contingencies Bill, Appendix 7.

[96] See HL Debs vol 663 col 628 5 July 2004, Lord Lucas.

[97] Civil Contingencies Secretariat, *The Government's Response*, 50.

C. Power Holders

The Act mentions the 'Secretary of State' at some points and a 'Minister of **4.29** the Crown' elsewhere. Reference to 'Secretary of State' means any of Her Majesty's Principal Secretaries of State.[98] A 'Minister of the Crown' is defined as the 'holder of an office in Her Majesty's Government in the United Kingdom, and includes the Treasury, the Board of Trade and the Defence Council'.[99] In this way, 'Minister of the Crown' is a wider category. An important constitutional and political distinction is that Secretaries of State are all members of the Cabinet but only a few Ministers are so appointed. This distinction between Secretaries of State and Ministers of the Crown plays out as follows in the Act:

> Functions under Part 1 of the Bill are conferred on Ministers of the Crown. This reflects the range of Ministers who may need to take action under that Part of the Bill. In particular, it should be noted that the Minister currently responsible for the Civil Contingencies Secretariat is not a Secretary of State. Functions under Part 2 of the Bill are conferred on the Secretary of State. This reflects the Government's view that action under Part 2 is a major undertaking and so should only be taken by a senior Government Minister.[100]

D. Preparedness of Responders

(a) General requirements

The Act imposes sweeping duties upon the specified responders. The duties are **4.30** imposed individually—though there are duties to act cooperatively, the responders do not amount to a single body corporate and the notion that those in each locality should be treated as one collective duty holder was rejected.[101] There are broadly seven civil protection duties, all of which affect their functions: risk assessment, emergency planning, business continuity management, cooperation, information sharing, advice to business, and communicating with the public.[102] By s 18, the 'function' of a responder means any power or duty whether conferred by virtue of an enactment or otherwise. 'Enactment' has already been defined. The power or duty 'otherwise' would presumably include prerogative or other common law based powers. For example, the police have broad powers to uphold the Queen's Peace.

---

[98] Interpretation Act 1978, Sch 1.   [99] Ministers of the Crown Act 1975, s 8.
[100] Joint Committee on the Draft Civil Contingencies Bill, Appendix 9 q 36.
[101] Civil Contingencies Secretariat, *The Government's Response*, 7.
[102] Cabinet Office, *Emergency Preparedness: Guidance on Part 1*, para 1.30. See further Ch 8.

4.31 The duties to assess, plan, and advise in s 2 are extensive and amount to 'a strong emphasis'.[103] In order to achieve excellence in Comprehensive Performance Assessments, introduced in 2002,[104] local authorities already need to demon-strate effective risk management processes,[105] so these demands of local authority Category 1 responders are not entirely new. But s 2 encourages new levels of consistency and compliance, and it extends the duty to non-local government responders.

4.32 The first stage in the process is for Category 1 responders to assess from time to time the risk of an emergency occurring (s 2(1)(a)) which involves identifying those circumstances which would amount to an emergency as well as their chances of occurrence. These risks should, however, be confined to an emer-gency which affects or may affect the geographical area for which the Category 1 responder is responsible (by reg 13 (reg 9 in Scotland) of the Civil Contin-gencies Act 2004 (Contingency Planning) Regulations 2005). To further allay the danger of planning for fanciful or irrelevant risks and to concentrate minds, reg 14 (reg 10 in Scotland) allows a Minister of the Crown to issue in writing to a 'general' (meaning, non-Scottish)[106] Category 1 responder guidance (Local Risk Assessment Guidance) or even a precise assessment as to (a) the likelihood of a particular emergency or an emergency of a particular kind occurring; or (b) the extent to which a particular emergency or an emergency of a particular kind would or might (if it occurred) cause damage to human welfare in a place in the United Kingdom, the environment of a place in the United Kingdom, or the security of the United Kingdom. For Wales, there must first be consultation with the Assembly, and the Assembly can, reciprocally with the consent of a Minister of the Crown, issue corresponding guidance or assessments to a general Category 1 responder. The same powers are also granted to the Office of the First Minister and Deputy First Minister in Northern Ireland and to the Scottish Ministers. Any existing plans must then be revised in the light of the guidance or assessment (reg 26/20 in Scotland). A Local Risk Assessment Guidance, covering risk priorities for a five-year period was issued by the Cabinet Office in July 2005; the document, which is understood to cover, *inter alia*, terrorist threats, is categorized as confidential.

4.33 Category 1 responders must next assess the ways in which it might be necessary or expedient to perform their functions (s 2(1)(b)). No set periods are detailed to explain the phrase 'from time to time', which may not impose a tight enough duty: 'If one asks a child to clean his bedroom from time to time, it simply will

---

[103] Cabinet Office, *The Draft Civil Contingencies Bill*, para 3.17.
[104] Http://www.audit-commission.gov.uk/cpa/index.asp?page ='index.asp&area = hpcpa.
[105] Http://www.audit-commission.gov.uk/cpa/index.asp.
[106] See Civil Contingencies Act 2004 (Contingency Planning) Regulations 2005 SI No 2042, reg 3.

not happen.'[107] But the Government rejected a fixed review period, such as three months, because of the variation in circumstances, because of the check of audit and guidance and also because s 2(1)(e) specifically requires review if (and whenever) circumstances change.[108]

The outcome of the assessment under s 2(1)(a) should be a 'Community Risk **4.34** Register', the product of a collaborative effort required by reg 15 (reg 12 in Scotland) of the Civil Contingencies Act 2004 (Contingency Planning) Regulations 2005.[109] Responders must reflexively have regard to any relevant Community Risk Register in performing their duties under s 2(1)(a) and (b). The purpose of the register is thus to assist with planning rather than as an objective in itself. The Community Risk Register is additional to the duties of responders to develop documented plans within their own sphere of operations—for example, the Environment Agency assesses the risk of flooding, the Health and Safety Executive deals with industrial risk, and so on. That the register is for the benefit of responders rather than the public is emphasized by the absence in reg 15 of a direct duty of publication,[110] but the register will be published by reference to s 2(1)(f) below and also there could be an application for disclosure under the Freedom of Information Act 2000 (or the Freedom of Information (Scotland) Act 2002),[111] subject to the need in both cases to protect sensitive information. A very limited form of publication is specified under reg 16—any general Category 1 responder which has functions which are exercisable in a particular local resilience area must from time to time provide the Category 1 responders which have functions which are exercisable in any neighbouring local resilience area with a copy of its local Community Risk Register. In addition, the Secretary of State shall be informed by responders in England and the National Assembly by Welsh responders. Those UK-wide Category 1 responders operating in Scotland must also cooperate with Scottish responders to produce Community Risk Registers (reg 17). A more proactive form of publication must be considered where appropriate for UK-wide responders. By reg 18, which requires such 'general' Category 1 responders to provide information about risk to other general Category 1 responders or Scottish Category 1 responders where the information relates to an emergency which would affect that other responder because it is (a) likely to be seriously obstructed in the performance of its functions if that emergency or an emergency of that particular kind occurred; or (b) likely to consider it necessary or desirable to take action to prevent, reduce, control, or

---

[107] HC Debs Standing Committee F col 47 27 January 2004, Nigel Evans.
[108] See also Cabinet Office, *Emergency Preparedness Guidance on Part 1*, para 4.8; HL Debs vol 664 col 1239 15 September 2004, Lord Bassam.
[109] It is suggested that it is acceptable to include COMAH risks: Cabinet Office, *Emergency Preparedness: Guidance on Part 1*, para 1.37. [110] See further ibid, para 3.31.
[111] Note also the Data Protection Act 1998, the Environmental Information Regulations 2004 SI No 3391, and the Environmental Impact (Scotland) Regulations 2004 SSI No 520.

mitigate that type of emergency or otherwise take action, and would be unable to take that action without changing the deployment of resources or acquiring additional resources.

4.35 Next, there must be the devising and maintenance of 'business continuity plans'[112] and emergency plans. These will enable the Category 1 responder itself to remain viable so far as is reasonably practicable[113] in an emergency in its normal operations (s 2(1)(c)) as well as seeking to prevent that form of emergency from affecting its own abilities to prevent or ameliorate an emergency or to take action in connection with it (s 2(1)(d)). Some argued that Business Continuity Management plans or emergency plans should equally be required of Category 2 responders, since many are obligated by existing sector-specific rules, but this duty has not been imposed.[114] Section 2(1)(e) obliges a Category 1 responder to keep under consideration whether to change its plans under (c) or (d) as a result of a new risk assessment. Section 2(1)(f) requires the publication of aspects of business continuity plans so far as necessary or desirable for the purposes of preventing, controlling, or mitigating the effects of an emergency or responding to the emergency.

4.36 The Civil Contingencies Act 2004 (Contingency Planning) Regulations 2005 echo these requirements. Regulations 19 and 20 (regs 13 and 14 in Scotland) tie in this duty to the duties under s 2(1)(a), (b), or (g), which must be considered at the same time. There are likely to be multiple plans, both generic and specific (reg 21/15 in Scotland), and it will often be appropriate to participate in multi-agency plans where duties in relation to a type of emergency are shared with other Category 1 responders (reg 22/16 in Scotland). Consideration must equally be given to multi-level plans, covering not only local responses but equally regional and/or central responses.[115] In addition, and reflecting the concession in s 2(5)(k) for the benefit of voluntary agencies, reg 23 (reg 17 in Scotland) reminds responders of the need in performing their duty under s 2(1)(c) or (d), to have regard to the relevant activities of voluntary organizations.

4.37 Having determined the format of plans and the players in the planning process, reg 24 (reg 18 in Scotland) requires Category 1 responders to include in the plans a procedure for determining whether an emergency has occurred which seriously obstructs the ability of the Category 1 responder to continue to perform its functions. To ensure that decision-making is certain and transparent,

---

[112] Cabinet Office, *The Draft Civil Contingencies Bill: Explanatory Notes etc*, para 14.
[113] Cabinet Office, *Emergency Preparedness: Guidance on Part 1*, para 6.10: 'Ideally, Category 1 responders would be able to continue all of their functions at ordinary service levels in the event of an emergency. In practice, this may not prove possible, and therefore the duty is qualified.'
[114] Civil Contingencies Secretariat, *The Government's Response*, 9.
[115] For further details, see Cabinet Office, *Emergency Preparedness: Guidance on Part 1*, para 5.69.

a person (who can be determined by office rather than specific name) shall be identified as responsible for determining whether an emergency has occurred, and the plans must also lay down the procedures for decision-making and how the decision shall be made known.

Completing the details, reg 25 (reg 19 in Scotland) requires that the plans under **4.38** s 2(1)(c) or (d) must include arrangements for exercises and training.[116]

In terms of the public dissemination of information, each Category 1 responder **4.39** must arrange for the publication to the public of the assessment and plans in so far as consistent with its objectives (s 2(1)(f)). It must also maintain arrangements to warn the public and to provide information and advice to the public about a forthcoming or extant emergency (s 2(1)(g)). An important message in the Civil Contingencies Act 2004 (Contingency Planning) Regulations 2005 about the performance of duties under s 2(1)(f) and (g) is to highlight the importance of not alarming the public unnecessarily (regs 27, 30/21, and 24 in Scotland).[117] As regards s 2(1)(g) specifically, advice is given to bear in mind the concluded plans under s 2(1)(d) (reg 28/22 in Scotland), and a reminder is also given of the need for training and exercises (reg 31/25 in Scotland). The arrangements for warning may relate either to generic or specific emergencies (reg 29/23 in Scotland). Thus, there could be a generic warning to householders living in a flood-plain or there could be a specific warning about a looming flood. A Category 1 responder, which has functions exercisable in a particular local resilience area and which is subject to a duty under s 2(1)(g) in relation to a particular emergency or an emergency of a particular kind, should be identified as the 'lead' responder, and there should be clear procedures for revealing which agency that might be in an emergency (reg 32/26 in Scotland). That lead agency must then work with the others of relevance in delivering the duties under s 2(1)(g) (reg 33/27 in Scotland), and the others must cooperate (reg 34/28 in Scotland). For example, the Environment Agency has long maintained a flood warning system; one would expect that Agency to become the leader for that purpose, though others might also have duties to issue warnings when a flood occurs (such as by touring an area). In order to avoid duplication and confusion, consideration should also be given not only to the duties to warn and inform the public of dangers of other responders but also to specified bodies beyond the Act (reg 35/29 in Scotland). Examples include COMAH arrangements, by which

---

[116] See futher ibid, para 5.153 (which uses the typology of discussion-based, table top and live).
[117] The draft version of the Regulations issued in January 2004 also included rules that the responder must ensure, so far as reasonably practicable, effective communications with vulnerable persons, and persons for whom English is not their first language (regs 18 and 22). These matters have now been left to guidance: Cabinet Office, *Emergency Preparedness: Guidance on Part 1*, para 5.98; *Scottish Guidance on Preparing for Emergencies* (http://www.scotland.gov.uk/consultations/justice/crgcca-03.asp, 2005) paras 5.4, 7.57.

local authorities must warn about hazardous industries. Other illustrations include the Meteorological Office (which operates a National Severe Weather Warning Service Office), announcements by the Secretary of State (likely to include terrorism and other public security issues), the Highways Agency (which is developing a National Traffic Control Centre for the collection and dissemination of traffic and travel information, with seven regional centres in England), and the Food Standards Agency.[118]

4.40 Digressing further into official policy on warnings, there exists under the aegis of the Cabinet Office the National Steering Committee for Warning and Informing the Public (NSCWIP).[119] It was created in 1995 and deals with the formulation of strategy concerning the effective use of new technology, the media, and a programme of public education and information as part of its objective of improving systems of public warnings. Before the 2004 Act, there had been no national peacetime warning system in place, following the end of the wartime national siren network in 1992–93. A further working group set up in 2003 by the Home Office Emergency Planning Division agreed that responsibilities for alerting the public in England and Wales in the event of emergencies should fall into three generic groups: incidents at hazard sites covered by regulations (such as COMAH), where responsibility is placed on the site operator; incidents where advance warning is likely such as flooding, severe weather, overseas nuclear accidents, and satellite incidents, where responsibility is accorded to the Lead Government Department or nominated agency such as the Environment Agency; incidents where advanced warning is unlikely, where responsibility is accorded to emergency responders, probably the police. The Government was reluctant to add to this panoply of bodies.[120] It wished to retain the flexibility of 'the long-established practice of saying that members of the public should follow the advice of the emergency services on the ground', often summed up as the all-encompassing 'go in, stay in and tune in'.[121] Rather less convincing in the era of resilience is the undercurrent view that trained officials know best and the duty of the public is to follow instructions rather than be trained to think for themselves.[122] This approach is contradicted by US research by Lasker in 2004.[123] Using scenarios in which smallpox and dirty

---

[118] See Food and Environmental Protection Act 1985, s 1.

[119] Http://www.nscwip.info/.

[120] HC Debs Standing Committee F col 102 29 January 2004; HL Debs vol 665 col 447 14 October 2004, Baroness Buscombe.

[121] HL Debs vol 665 col 448 14 October 2004, Lord Bassam.

[122] HC Debs Standing Committee F col 107 29 January 2004, Douglas Alexander.

[123] Lasker, RD, *Redefining Readiness: Terrorism Planning Through the Eyes of the Public* (Center for the Advancement of Collaborative Strategies in Health, New York Academy of Medicine, New York, 2004). See also Burkhart, FD, *Media, Emergency Warnings and Citizen Response* (Westview Press, Boulder, 1991).

bomb attacks are announced, the study found a high level of reluctance to follow instructions which could be reduced by the disclosure of further information, such as overall preparations which affect relatives or 'significant others' and specific advice about the nature of counter-measures. Taking the perspective of the public is essential but can only be achieved through dialogue. Aside from ss 2(1)(f) and 4, the Act is silent about public engagement. Unlike, say, planning processes with regard to land use,[124] there is no formal stage for public consultation or objection. This deficiency was mentioned during consultation,[125] and it was proposed that a 'Emergency Public Education and Training Board' should ensure the public had information and could respond with effective resilience measures.[126] Those unsympathetic to the idea conjured an image of 'hordes of ill-trained and semi-informed actors milling about trying to be helpful.'[127]

Returning to the meat of s 2, as part of the safeguards against overuse, the need for **4.41** the situation to become serious before these duties are triggered is underlined in s 2(2) which was added in to the Bill in January 2004 as an attempt to reflect to some extent the definition of 'major emergency' in *Dealing with Disaster*.[128] Duties under s 2[129] only arise if either the emergency would be likely seriously to obstruct the responder or the responder would consider it necessary or desirable to intervene in the emergency and would be unable to take that action without changing the deployment of resources or acquiring additional resources. It follows that routine tasking of the emergency services, such as dealing with a single residential fire as opposed to a large industrial fire in a densely populated area[130] or a huge moorland blaze,[131] should not trigger the inter-agency planning under Part I which should be confined to disruptive challenges of such a scale that they cannot be dealt with as part of normal day-to-day activities of a single responder. It also follows that not all types or levels of emergency demand intervention in the same way by all responders. The responder must consider their responses in relation to

---

[124] Compare the local inquiries into development plans: Town and Country Planning Act 1990, s 16. [125] Civil Contingencies Secretariat, *The Government's Response*, 49.

[126] HC Debs vol 421 col 1366 24 May 2004, Patrick Mercer.

[127] Ibid, col 1373, Lewis Moonie.

[128] (3rd ed, Cabinet Office, London, 2003) para 1.5. See further, Cabinet Office, *The Government's Response to the Report of the Joint Committee*, para 11; HC Debs vol 416 col 1105 19 January 2004, Douglas Alexander; HL Debs vol 665 col 394 14 October 2004, Lord Bassam.

[129] It follows that duties under s 4 are not affected by this qualification: Cabinet Office, *The Government's Response to the Report of the Joint Committee*, para 11.

[130] HC Debs Standing Committee F col 18 27 January 2004, Douglas Alexander.

[131] This event was said to be outwith the whole Act by the Home Office Minister (HL Debs vol 665 col 728 14 October 2004, Baroness Scotland), but it is surely a legitimate subject of planning and, if the blaze suddenly flares and affects urban areas, Part II could arise. States of emergency were declared in Portugal in summer 2003 and 2005 because of widepread forest fires, and 'Bellwin' payments were made in connection with the Thorne Moor fire in August–September 1995 (see Ch 8).

the functions they provide, their modes of delivery, how they have allocated or concentrated their resources (in that some authorities will be more capable of dealing with some challenges than others in the normal course of events), and their interconnection with other responders.

4.42 Further clarification as to the extent of duties can be offered under s 2(3) and (4). A Minister of the Crown (or the Scottish Ministers in the case of Part 2 Category 1 responders) may make regulations about the extent of a duty under subsection (1) and the manner in which it is to be performed. The purpose is 'to avoid duplication of effort by Category 1 responders and to ensure that responders do not waste resources on assessing risks that fall outside their area of responsibility and expertise'.[132] As mentioned earlier, the s 2(3) power could be used to allow a local authorities to delegate s 2 duties to FCDAs.

4.43 These regulation-making powers are very extensive and are set out under 19 different categories in s 2(5). Despite the length of the list, it is not expressed to be exhaustive but merely itemizing specific instances 'in particular'. The powers focus on the planning process and so do not in the main produce burdens or obligations for individual citizens or even for emergency workers (such as an obligation to undergo vaccination against smallpox).[133] Of course, in a live emergency, the powers under Part II of the Act become relevant. Despite the extent of s 2(5), the regulation-making power under s 2(3) and (4) remains subject to the negative procedure in Parliament, not so much because of its extent but because it is depicted as 'detailed' and 'technical'.[134] Since the powers allow regulations to take precedence over provisions in an Act of Parliament, the Chair of the House of Lords Select Committee on Delegated Powers and Regulatory Reform suggested that the affirmative procedure would be preferable, though his Committee did not echo this idea.[135]

4.44 Regulations under s 2(5) may specify the performance of duties, including (a) the kinds of emergency relevant or not relevant to a specified person or body, (b) the circumstances in which functions may not be exercised by a specified person or body, and (c) the timing of performance of a duty. Both (a) and (b) may involve specifying who is to be the 'lead' responder in a given type of emergency,[136] and sometimes it will be a non-responder who is specified, such as

---

[132] Cabinet Office, *The Draft Civil Contingencies Bill: Explanatory Notes etc*, para 13.

[133] HL Debs vol 665 col 398 14 October 2004, Lord Jopling. It is predicted that one per million of those not previously vaccinated might die as a result of the vaccine, with a further 20,000 suffering severe complications. Those vaccinated must be excluded from giving blood for a defined period, so additional deaths could also be expected due to a shortage of blood: col 409, Lord Bassam.

[134] Cabinet Office, *The Government's Response to the Report of the Joint Committee*, Annex A para 22. [135] Joint Committee on the Draft Civil Contingencies Bill, Appendix 3.

[136] HL Debs vol 665 col 394 14 October 2004, Lord Bassam. Section 2(3) is also relevant to this point.

a Minister in relation to issuing risk assessments for national threats such as terrorism.[137] Under (a) 'Category 1 responders might be required to prepare a mass decontamination plan. Correspondingly, they might be specifically required not to prepare a plan that the Minister thought was not needed, for example an earthquake plan.'[138]

Regulations may specify the relations between responders, including (d) duties **4.45** to consult a specified person or body or class of person or body (such as a requirement to consult with the Meteorological Office in relation to weather emergencies),[139] (e) the subsuming of the duties of a district council by the relevant county council (as was the case under the Civil Defence Act 1948, s 2(2)(c)), (f) requiring or prohibiting collaboration between responders or (g) delegation, (h) permitting or requiring cooperation or (i) the provision of information by both Category 1 and 2 responders (such as electricity companies' planning arrangements for restoring loss of supply),[140] or (j) the reliance upon work undertaken by another specified person or body. Items (g) and (h) were not in the draft 2003 Bill. Joint or delegated working is envisaged by regs 4 to 6 (regs 3 and 4 in Scotland) of the Civil Contingencies Act 2004 (Contingency Planning) Regulations 2005.

The possibility of delegation under (g) is echoed in reg 8 (reg 5 in Scotland) of **4.46** the Civil Contingencies Act 2004 (Contingency Planning) Regulations 2005. Each general Category 1 responder may perform a duty under s 2(1) jointly with another responder or may make arrangements with another responder for that responder to perform such a duty on behalf of the general Category 1 responder. As a result of these options, many district councils are being tempted to enter deals with county councils so as to allow the counties to transact the new functions for them. It remains to be seen how widespread this practice becomes and whether the Act will eventually be redesigned to reflect emergent patterns. Furthermore, guidance suggests that, for most LRFs, the British Transport Police will be represented by the local area police, port health authorities will be represented by the local authority or the HPA, and the NHS should have a single representative (but not including the ambulance service or the HPA), as should each category of utility.[141]

Regulation 6 simply reiterates the duty under s 2(5)(h) of Category 1 responders **4.47** to cooperate with each other in connection with the performance of their duties under s 2(1) and the duty of Category 2 responders to cooperate. Regulation 7 provides for more formal 'protocols' to be developed in order to facilitate

---

[137] Ibid, col 396.
[138] Cabinet Office, *The Draft Civil Contingencies Bill: Explanatory Notes etc*, para 21.
[139] Ibid, para 22.   [140] Ibid, para 24.
[141] Cabinet Office, *Emergency Preparedness: Guidance on Part 1*, paras 2.70–2.96.

cooperation. Guidance suggests that these can be legally binding or not.[142]
Cooperation is also to take the form of 'Local Resilience Forums' (LRFs) which
are described below.

4.48 Coordination between the overlapping responsibilities of Category 1 responders
is important. Regulation 9 (reg 6 in Scotland) deals with the situation where
more than one Category 1 responder in a particular local resilience area is
subject to the same relevant civil protection duties under (a), (b), (c), (d), (e), or
(f) of s 2(1) in relation to a particular emergency or an emergency of a particular
kind. Such responders must identify between them which is to be conferred
with the 'lead responsibility'. The lead Category 1 responder must then, under
reg 10 (reg 7 in Scotland), take the lead under s 2 for that function, though
continuing consultation, both inside and outside the LRF, is required. Non-
lead Category 1 responders must, under reg 11 (reg 8 in Scotland), cooperate
and provide information but should avoid unnecessary duplication.

4.49 Next, regulations may specify the relations between responders and other bodies,
including, under s 2(5), by (k) permitting or requiring regard to the activities of
bodies (other than public or local authorities) whose activities are not carried on
for profit, and (o) permitting a person or body to make arrangements with
another person or body, as part of planning undertaken under subsection (1)(c) or
(d). The provision in (k) was inserted during the parliamentary process in order to
assure voluntary bodies of their important role.

4.50 Regulations under s 2(5) may next specify the details of plans, including (l) the
degree of detail to be contained in a plan, requiring (m) a plan to include
provision for the carrying out of exercises, and (n) a plan to deal with the
training of staff or other persons.

4.51 Powers for central government or other intervention, especially if responders are
in default, may follow from regulations which specify the conferment of duties.
Under (p), a function may be conferred on a Minister of the Crown, on the
Scottish Ministers, on the National Assembly for Wales, on a Northern Ireland
department or on any other specified person or body. Regulations under (q)
may even specify the overriding of other enactments.

4.52 These varieties of possible regulations can then be further refined by (r) applying
them to a specified person or body or only in specified circumstances or by (s)
making different provision for different persons or bodies or for different circumstances. In this way, Category 1 responders do not have to assess the
likelihood of every conceivable type of emergency occurring. An example might
be that 'regulations may be applied to harbour authorities only to the extent that

---

[142] Ibid, para 2.29.

the annual volume of passengers and freight handled by them exceeds a specified threshold.'[143] These provisions are adapted to the situation of the Scottish Ministers under s 2(6).

Section 3(1) ((2) for Scotland) augments the regulation-making powers with **4.53** Ministerial powers to issue less formal guidance on the same issues as in s 2 to Category 1 or 2 responders. The relevant responder must 'have regard to guidance' (s 3(3)) which will make it a relevant consideration for administrative law purposes. In *Laker Airways Ltd v Dept of Trade*,[144] Lord Justice Roskill stated that 'guidance is assistance in reaching a decision proffered to him who has to make the decision, but guidance does not compel any particular decision'. Lord Denning in *De Falco v Crawley BC*,[145] said that 'the [public body] of course had to have regard to the code...but, having done so, they could depart from it if they thought fit'. This formula may be contrasted with the stronger impact under the Police and Criminal Evidence Act 1984, s 67, and the Trade Union and Labour Relations (Consolidation) Act 1992, s 207, which make a failure to observe the code not actionable in itself but admissible in evidence, and any provision of the Code which appears to be relevant to any question arising in proceedings must be taken into account.

(b) Local resilience forums

'Cooperation' under s 2(5)(h) enables the setting up of multi-agency LRFs. The **4.54** Minister of State for the Cabinet Office emphasized that 'The local response capability is the foundation of our response to all emergencies.'[146] These LRFs will replace and rationalize any existing structures, such as Local Civil Protection Committees, Local Emergency Planning Groups, or Strategic/Senior Coordinating Groups (SCGs) at county level:

> Category 1 responders will take the lead in the business of these forums, which may meet every six months, with Category 2 organisations required to attend or to be represented as appropriate.[147]

The LRFs are be established on the basis of police force areas,[148] though the **4.55** chairing of the body might be located within local authority bodies rather than the police.[149] The LRFs can also manage sub-county level multi-agency groups[150] and subsume existing arrangements under the European-derived COMAH planning for major accident hazards at industrial sites or under the

---

[143] Cabinet Office, *The Draft Civil Contingencies Bill: Explanatory Notes etc*, para 28.
[144] [1977] QB 643 at 714.   [145] [1980] QB 460 at 478.
[146] HC Debs vol 416 col 1105 19 January 2004, Douglas Alexander.
[147] Cabinet Office, *The Draft Civil Contingencies Bill: Explanatory Notes etc*, para 24. See further ibid, para 31; Defence Committee, Draft Civil Contingencies Bill (2002–03 HC 557) para 35.   [148] Cabinet Office, *The Draft Civil Contingencies Bill*, para 3.24.
[149] Civil Contingencies Secretariat, *The Government's Response*, 18.   [150] Ibid.

Health and Safety at Work Act or procedures for radiation emergencies and oil and gas pipelines (Pipeline Regulations).[151] But some multi-agency working will remain beyond the 2004 Act such as under the Crime and Disorder Act 1998 and the Local Government Act 2000.[152] The fact that the relations are statutory is thought to be helpful to encourage the sharing of sensitive information,[153] though there is no plan to change the laws or practices relating to official secrecy or security clearance.[154] LRFs will not have a role in the response phase—the lead agency identified within the plans will at that point take charge.[155] In this way, LRFs are planning teams not action teams (the exception being Scotland, where the functions coalesce). When an emergency arises, the response and recovery phases will be dealt with by Strategic Coordinating Groups—these will strongly overlap in terms of membership, but are constituted differently and form a part of a different chain of command (as described in Chapter 8).

4.56 LRFs are elaborated in reg 4 (reg 5 applies to those UK-wide 'general' responders with responsibilities in Scotland and reg 6 likewise in relation to Northern Ireland) of the Civil Contingencies Act 2004 (Contingency Planning) Regulations 2005,[156] which establishes that there must be a forum and confers its title. The title is 'local resilience forum' in England and Wales but is to remain as the 'Strategic Coordinating Group' in Scotland[157] (for ease of reference in this chapter, 'LRF' will normally be used to cover both). A few other substantive working rules are there specified. One is that Category 1 responders in a particular local resilience area must hold a meeting at least once every six months. Next, every Category 1 responder must, so far as reasonably practicable, attend such a meeting or arrange for representation at that meeting. Category 2 responders have the same level of duty of attendance whenever specifically invited to a specific meeting, but in the case of other meetings simply have to consider whether it is 'appropriate' to attend the meeting or to be represented. This weaker level of commitment may address the concerns of those national bodies, such as utilities, who might otherwise have to attend

---

[151] See Radiation (Emergency Preparedness and Public Information) Regulations 2001(REPPIR) SI No 2975. See Health and Safety Executive, *Guide to the Pipelines Safety Regulations 1996* (HSE Books, Sudbury, 1996); Cabinet Office, *The Draft Civil Contingencies Bill*, para 3.28.   [152] Cabinet Office, *The Draft Civil Contingencies Bill*, para 3.29.
[153] Ibid, para 3.26.
[154] Joint Committee on the Draft Civil Contingencies Bill, Appendix 9 q 59. See further Ch 8.   [155] Cabinet Office, *The Draft Civil Contingencies Bill*, para 3.29. See further Ch 8.
[156] SI No 2042.
[157] Civil Contingencies Act 2004 (Contingency Planning) (Scotland) Regulations 2005 SSI No 494, reg 3. Model terms of reference are set out in the Scottish Guidance on Preparing for Emergencies (http://www.scotland.gov.uk/consultations/justice/crgcca-03.asp) Ch 2 Annex A. See Scottish Executive, *Civil Contingencies Bill: Scottish Consultation Report* (Justice Department, Edinburgh, 2004) para 5.

many LRFs. Central government departments or agencies (unless specified in the Act) are not present at LRFs, but one notable exception is the recognition of the need to maintain a link with the armed forces through the Regional Brigade, though attendance is not formally required.[158] Furthermore, in Scotland, it is promised that the 'Scottish Executive Civil Contingencies Division will endeavour to attend all SCG meetings'.[159] Because of the very limited impact of the 2004 Act in Northern Ireland, more discretion as to liaison is allowed under reg 6 for those UK-wide responders with responsibilities in the Province.

(c) Regional structures in England

Part I of the Act is silent about regional structures in England. However, they **4.57** will be related here since they tie in closely with local resilience structures.

The need for this tier of response was brought home during the Fuel Crisis in **4.58** 2000 and the Foot and Mouth outbreak in 2001, as well as the fire-fighters' strike of 2002:

> These situations highlighted two key issues:
> —With increasingly complex social and economic networks, it was difficult to handle the volume of information coming directly to the centre from so many local areas and flowing back out again. This was leading to ad hoc regional arrangements.
> —Regional arrangements put in place in an ad hoc fashion did not always perform as well as established structures.[160]

The regional tier is meant to respond to local coordination problems. It equally **4.59** acts as a conduit to and from the centre, such as for the local mobilization of central resources and the central reception of local information. It also reflects the policy announced in the Office of the Deputy Prime Minster's White Paper, *Our Fire and Rescue Services* which stated that, 'The regional level is acknowledged to be the right operational level for many functions, in particular securing the safety of the community in the event of terrorist attack or other major emergencies.'[161] The regional tier has also been strengthened with Part II of the Act in mind.

Since it is primarily a facet of central government, this regional tier is not at all **4.60** visible within the Act[162] but has certainly existed in one form or another for much

---

[158] Cabinet Office, *Emergency Preparedness: Guidance on Part 1*, para 2.110.

[159] *Scottish Guidance on Preparing for Emergencies* (http://www.scotland.gov.uk/consultations/justice/crgcca-03.asp, 2005) paras 2.30–2.31.

[160] Cabinet Office, *The Draft Civil Contingencies Bill*, para 4.1 See further Joint Committee on the Draft Civil Contingencies Bill, Appendix 9 q 75. FEMA (Responding to Incidents of National Consequence: Recommendations for America's Fire and Emergency Services Based on the Events of September 11, 2001, and Other Similar Incidents (Washington, 2002) 23) likewise found the need to develop a regional capacity to cater for situations where the locality is over-whelmed.   [161] (Cm 5808, London, 2003) para 4.14.

[162] It is mentioned in the guidance: See further Cabinet Office, *Emergency Preparedness: Guidance on Part 1*, paras 17.2–17.4, Annex C.

of the previous century.[163] At the time of enactment, Regional Resilience Teams (RRTs) were already in place in the nine outpost Government Offices for the English Regions,[164] having been operational since 1 April 2003. Designated regions are North East, North West, Yorkshire and Humberside, East, East Midlands, West Midlands, South East, South West, and London. Each is lead by a senior official with around half a dozen staff (with more in London).

4.61 Associated with RRTs are Regional Resilience Forums (RRFs), likewise nowhere mentioned in the Act, which will exclusively discuss planning:

> Regional Resilience Forums have been formed to bring together the key players, including central government agencies and the Armed Forces, and representatives of local responders including the emergency services and local authorities. The Forums will work to improve the co-ordination of planning at a regional level and improve communications between the centre and the region and between the region and the local response capability.[165]

4.62 The purposes of the RRF comprise:[166] the compilation of an agreed regional risk map (which will include a generic response plan for activating Regional Civil Contingencies Committees (RCCCs) as well as a framework for regional awareness and planning regimes); ensuring a business continuity plan for the Government Office itself; consideration of policy initiatives; the facilitation of information-sharing and cooperation between local responders, between regional partners, and between the locality, the region, and central Departments; the coordination of multi-agency exercises and other training; and the sharing of lessons learned from emergencies and exercises; and support for the preparation of multi-agency plans and other documents, including a Regional Capability Co-ordination Plan to develop consistency of approach at the local level and to act as a mechanism whereby local plans can be 'scaled up' where necessary.[167] Improving and securing coordination is therefore the key role,[168] but the process is not overtly hierarchical in that LRFs are not immediately subordinate to regional structures which, in turn, are not commanded daily by central government.[169] They also act as a regional voice to the media and as cheerleaders for central government's response at the regional level.[170]

---

[163] See Ch 2.

[164] See http://www.gos.gov.uk/national/. See Cabinet Office, *Emergency Preparedness: Guidance on Part 1*, para 17.6; Cabinet Office, *Emergency Response and Recovery*, Ch 7. The Offices have a liaison role themselves in emergencies that could generate ministerial interest or national/regional press coverage.

[165] See http://www.odpm.gov.uk/stellent/groups/odpm_civilres/documents/page/odpm_civilres_30673.hcsp.

[166] Cabinet Office, *Emergency Preparedness: Guidance on Part 1*, paras 18.3–18.15; Cabinet Office, *Emergency Response and Recovery*, Ch 8. [167] Ibid, paras 17.8, 18.1.

[168] Cabinet Office, *The Draft Civil Contingencies Bill*, para 4.5.

[169] Cabinet Office, *Emergency Preparedness: Guidance on Part 1*, para 17.10.

[170] Cabinet Office, *The Government's Response to the Report of the Joint Committee*, para 50.

Furthermore, they have a part to play in leading a regional response when an emergency arises, assisting with recovery.

As for membership, existing forums consist of, by way of illustration, one or two **4.63** lead officials from the Government Office in the Region, a Government News Network representative, a police chief, ambulance and fire chiefs, representatives from the NHS and the Health Protection Agency, the Maritime and Coastguard Agency and the Environment Agency, an armed forces representative, a local authority leader, the chief emergency planning officer, the Army regional brigade commander, and a voluntary sector representative, with others as specifically required.[171] The guidance suggests some cross-membership between LRFs and RRFs,[172] that the chair will be the director of the relevant Government Office (save for London, as described below) and that there should be sub-groups such as a Regional Working, Risk, Sectoral, capabilities, specialist, and project groups, plus a Media Emergency Forum (with possible links in a live emergency to the News Co-ordination Centre established by the Cabinet Office Communications Group) and a Mass fatalities group. Small teams of civil servants headed by a senior official and supported by four or five staff have been provided from each Government Office to support the RRFs.[173]

When RRTs are exceptionally engaged in response mode, there may excep- **4.64** tionally arise the RCCC to coordinate the regional response.[174] Their invocation occurs in response to a request from a strategic coordinating group or a member of the RRF in 'very exceptional circumstances when the response to an emergency would benefit from co-ordination at a regional level—such as in gradual and widespread so-called "rising tide" or "slow burn" emergencies'.[175] 'Such circumstances could include where the local response, including locally agreed mutual aid arrangements, is overwhelmed, or where an emergency affects the majority of localities within a region.'[176] The membership of the RCCC will be drawn from representatives of the emergency services, local authorities, central government departments, and the Government Office (or the London Resilience Team), agencies with a regional presence, with others being invited as necessary. Meetings of the RCCC are to be convened by the Government Offices, with three modes of response, depending on the seriousness. Level one meetings would be convened by a senior person in the Government Offices at

---

[171] Ministry of Defence, *Operations in the UK*, para 724.
[172] See Cabinet Office, *Emergency Preparedness: Guidance on Part 1*, paras 17.28–17.32.
[173] Ibid, para 17.6.
[174] Cabinet Office, *The Draft Civil Contingencies Bill*, para 4.10; Cabinet Office, *Central Government Arrangements*, Annex C. Regional Emergency Committees were in existence during the 'Winter of Discontent' of 1978–79: Jeffery, K and Hennessy, P, *States of Emergency: British Governments and Strikebreaking since 1919* (Routledge & Kegan Paul, London, 1983) Ch 8.
[175] Cabinet Office, *Emergency Response and Recovery*, paras 8.1, 8.24.
[176] Ibid, para 8.1.

the time of an incipient emergency at the request of any RCCC member. Level Two meetings would be convened when a region or nationwide disruptive challenge actually occurs in, or affects, the region and would involve contact with the Cabinet Office. At Level Two, the RCCC might be chaired by a 'shadow' Regional Nominated Coordinator.[177] Level Three will be reached if Part II of the Act comes into operation, in which case the RNC is recognized in s 24 and can be empowered under Part II (see Chapter 5).[178]

4.65 There was apparently considerable support during consultation for this regional tier of planning.[179] It allows 'joint management of risk' as well as a 'systematic and holistic approach'.[180] It may be especially beneficial to responders, such as utilities, which operate across the whole of a region or are national. A regional tier, if properly managed, can lead to the establishment of a 'cellular' framework of emergency response that would provide the entire country with a robust, flexible, interoperable, and interdependent system of emergency planning and response. It could also ameliorate the bottom-heavy emphasis of the present framework, which might otherwise suffer from the duplication of effort by responders, as well as ensuring a degree of consistency and support for them.

4.66 Amongst the problems which the regional focus encounters are first that it still leaves room for divergence of practice which a more centralized form of control through central government might avoid.[181] There remains some danger that the regional tier will simply act as another layer of bureaucracy, which does not itself have any significant power or authority and which will not be able to provide support either in terms of resource allocation or strategic guidance.[182] Ultimately, is there 'value added' by inserting a regional level in contingency management?[183]

4.67 A second problem is that the key figure of the Regional Nominated Coordinator (RNC) is primarily and legally associated with Part II crisis response and not Part I crisis planning. Thus, in normal times, there is no clearly defined or legally constituted regional leader, even during a Level Two crisis.[184] The Government envisages appointees who can show 'specialist subject knowledge, strong leadership and effective emergency management'.[185] But, in the absence of an institutional structure, it is not clear how such persons will emerge. RNCs do not exist for the purposes of Part I, though training will be given to

---

[177] Cabinet Office, *The Draft Civil Contingencies Bill*, para 4.13.   [178] Ibid, para 4.14.
[179] Civil Contingencies Secretariat, *The Government's Response*, 29.
[180] Cabinet Office, *The Draft Civil Contingencies Bill: Explanatory Notes etc*, para 19.
[181] House of Commons Defence Committee, *Draft Civil Contingencies Bill* (2002–03 HC 557) para 30.   [182] Joint Committee on the Draft Civil Contingencies Bill, Appendix 4.
[183] It is claimed that substantial improvements have been made: Home Affairs Committee, *Homeland Security* (2003–04 HC 417-I) Q19, Sir David Omand.   [184] Ibid, para 263.
[185] Ibid, Appendix 9 q 60. See further ibid, para 254, Appendix 7.

pre-nominees (the nature of whom is described in Chapter 5) during 'quiet' or pre-emergency periods and such nominees could act as shadow RNCs by chairing relevant meetings even without formal authority; much as happened before the Act came into force. A statutorily designated cadre would be more likely to achieve their objectives. One suspects that hesitation about costs and the sensitivities of departmental boundaries conduced against up-front appointments.

A third problem is that the sponsoring Government Offices for the English **4.68** Regions remain rather shadowy operations. These bodies are essentially administrative constructs which reflect ten Whitehall departments (the Office of the Deputy Prime Minister, the Department for Education and Skills, the Department of Trade and Industry, the Department for Environment, Food and Rural Affairs, the Home Office, the Department for Culture, Media and Sport, the Department for Work and Pensions, the Department for Transport, the Department of Health, and the Cabinet Office) in nine regions. Their function is to represent central Government (especially by managing spending departmental programmes), to gather local knowledge and expertise to be fed back into departmental administration and to coordinate with local and regional stakeholders. As a result, RRTs amount to the 'regional presence' of national government and are not a part of 'regional government' in the sense either of reporting to regional assemblies or representing a means of devolving power to the regions from the Centre. They have limited public visibility and are not democratically accountable by any direct instrument. The Government sought to address this objection by developing elected regional assemblies,[186] but it suffered a mighty rebuff with the adverse vote against such an assembly for the North East in November 2004.[187] Thereafter, similar plans for the North West and Yorkshire and the Humber have been shelved. Consequently, there is left the regional footprint of national government but not a structure of regional government which is accountable to the locality or exercises defined powers.[188] Accordingly, the RRTs will report to the Civil Contingencies Secretariat in the Cabinet Office and to whichever lead Government Department is acting in a given emergency. They will also report to the Regional Coordination Unit (the Whitehall centre of operations for the Government Office Network) within the Office of the Deputy Prime Minister (ODPM). The ODPM also contains the Regional Resilience Division (now part of the Civil Resilience Directorate) which was set up in April 2003 to implement the regional 'resilience' tier by checking that current response

---

[186] See Office of the Deputy Prime Minister, *Your Region — Your Choice: Revitalising the English Regions* (Cm 5511, Stationery Office, London, 2002). There are still unelected assemblies (such as the Yorkshire and Humber Assembly) whose membership is drawn from local councils and elsewhere.　　　　　　　　　　　　　　[187] *The Times* 6 November 2004 at 6.
[188] Joint Committee on the Draft Civil Contingencies Bill, para 265.

capabilities are understood, that gaps in preparedness are identified and that plans are put in place to address contingencies by regional and local authorities.[189]

4.69   In summary, there arises an indistinct chain of command and control.[190] Outside the eventuality of a Part II emergency, the RRTs have almost no authority over the local tier in order to carry out the coordination expected of them. In essence, the regional arrangements reflect the Lead Government Department concept (discussed in Chapter 8)—leadership is to emerge from the relevant area of expertise related to the crisis which has arisen. It is not suggested that the regional tier should be abandoned. But a regional tier which is almost entirely disembodied from the Act is not a model for success.

E. External Actions Demanded of Responders

(a) Business continuity advice and assistance to sectors of the public

4.70   Of those bodies qualifying as Category 1 responders, local authorities alone[191] (not including port authorities) have duties[192] under s 4(1) to provide business continuity management (BCM) advice and assistance[193] to the public in connection with the making of arrangements for the continuance, in the event of an emergency, of commercial activities by the public, or the continuance of the activities of bodies other than public or local authorities whose activities are not carried on for profit.

4.71   As for the former aspect, there is no definition of 'commercial'. It is suggested that:

> It should not be taken narrowly to mean only private sector businesses operating for a profit. Charities, for example, may undertake commercial activities in their fundraising work (eg charity shops). Building societies and credit unions, too, carry out commercial activities—although not privately owned, they do operate as a business and generate financial benefits for their members.[194]

---

[189] Cabinet Office, *The Government's Response to the Report of the Joint Committee*. Annex A para 7.   [190] Joint Committee on the Draft Civil Contingencies Bill, para 266.
[191] But other responders may have other statutory duties. Police constables must prevent crime (Police Act 1996, Sch 4), and specialist crime prevention and counter-terrorism advice is available. The Fire and Rescue Service Act 2004, s 6 requires fire authorities to promote fire safety. Local authorities sponsor community safety teams under the Crime and Disorder Act 1998, s 6.
[192] The word 'shall' is used in the section and not 'may' as wrongly indicated by the Minister of State for the Cabinet Office who contended that 'Government do not propose to impose an obligation': HC Debs Standing Committee F col 87 29 January 2004, Douglas Alexander.
[193] See Barnes, J, *A Guide to Business Continuity Planning* (John Wiley, Chichester, 2001); Hiles, A, and Barnes, P, *The Definitive Handbook of Business Continuity Management* (John Wiley, Chichester, 2001).
[194] Cabinet Office, *Emergency Preparedness: Guidance on Part 1*, para 8.24. There is no definition in the Act.

The latter aspect of non-commercial activities represents a late addition to the Civil Contingencies Bill.[195] But the main focus and legislative intention seems to be on 'keeping the economy moving in the event of an emergency'.[196] Regulation 39 (reg 33 in Scotland) demands that the responder should confine its efforts to members of the business community who are present or resident in the area.[197] The duty to voluntary organizations is more hedged. A relevant responder need only provide advice and assistance to those voluntary organizations which it considers appropriate (reg 40/34 in Scotland). Further guidance on what is 'appropriate' indicates that a relevant responder must consider: (a) whether the organization carries on activities in the area relevant to contingency planning; (b) the nature and extent of those activities and whether they contribute to the prevention of an emergency occurring, the reduction, control, or mitigation of the effects of an emergency, the taking of action in connection with an emergency, social welfare, the number of staff employed by the organization, the turnover of the organization, and the nature of the organization, and in particular whether the nature of the organization is such that the advice and assistance provided by the relevant responder is likely to improve the ability of the organization to continue to carry on its activities in the event of an emergency.

It was reckoned that 70 per cent of businesses (80 per cent in the financial and retail sectors) did have plans in the survey conducted for the Business Continuity Institute's Awareness Week in March 2005.[198] The absence of plans is likely to be higher in the Small and Medium Size Enterprise (SME) sector. Promotion of BCM may be offered through an authority's economic development unit, emergency planning unit, or any other appropriate vehicle and may be delivered by materials, including booklets and videos, websites, and seminars and meetings. The exercise of the duty under s 4(1) requires account to be taken of the Community Risk Register maintained under reg 15 (reg 12 in Scotland) of the 2005 Regulations (reg 38/32 in Scotland). **4.72**

Activities likely to be involved in promotion of the statutory duty include web pages, advertising, press releases, mail shots, seminars, and standing forums.[199] **4.73**

---

[195] HL Debs vol 666 col 798 9 November 2004, Lord Bassam.

[196] HC Debs vol 421 col 1360 24 May 2004, Hazel Blears. See also HL Debs vol 664 col 1281 15 September 2004, Lord Bassam.

[197] Under the draft regulations issued in January 2004, further to alleviate the potential workload of the responder, there was a transitional provision in reg 26 by which for two years, the responder may confine its advice to local persons with a profile which include a balance sheet total (with the meaning of s 247(5) of the Companies Act 1985) in the latest financial year exceeding [6] million euros or who employs more than [250] employees.

[198] Business Continuity Institute, *Business Continuity Research* (http://www.thebci.org/BCIResearchReport.pdf, 2005). Compare: Defence Committee, Draft Civil Contingencies Bill (2002–03 HC 557) para 48.

[199] See Cabinet Office, *Emergency Preparedness: Guidance on Part 1*, para 8.66. Perhaps the most developed network is in the City of London: see Coaffee, J, 'Recasting the "Ring of Steel" in Graham, S (ed), *Cities, War and Terrorism* (Blackwell, Oxford, 2004).

The Government has promised to publish materials for local authorities to adopt so as to achieve economies of scale.[200] Because of the time being taken to develop such materials, which might take the form of a business web portal within the UK resilience website, s 4(1) is the final part of the Act to be enforced. The interim period might also be used by local authorities to consider the extent to which they wish to take on themselves this new role or whether delegation through service level agreements will be a more sensible option. Certainly, many are likely to want to be able to identify reputable commercial providers to whom detailed inquirers can be referred.[201] In the delivery of these duties, a lead responder may be identified (regs 41–43/35 and 37 in Scotland).

4.74　The broad duty in s 4(1) is followed, according to the usual pattern of the Act, by s 4(2) (and (3) in Scotland) which allows for Ministerial regulations to be issued about the extent of the duty and of the manner in which it is to be performed. Parliamentary scrutiny is limited to the negative procedure. Local authorities may under s 4(4) impose charges for advice or assistance (subject to limits specified in s 4(5)) as well as covering issues (a) to (i) and (o) to (s) as specified under s 2(5). Pursuant to s 4(4), reg 44 (reg 38 in Scotland)[202] provides that charges must not exceed the aggregate of '(a) the direct costs of providing the advice or assistance; and (b) a reasonable share of any costs indirectly related to the provision of the advice or assistance.'

4.75　Corresponding to s 3 is s 4(6) and (7) which provide for the issuance of guidance. Its impact, specified under s 4(8), corresponds with s 3(3).

4.76　One can detect several shortcomings with this scheme. First, the duty is placed on localities whereas many of the pivotal private sector enterprises in a given locality are parts of national chains. There may still be room for a useful local input even in these cases 'because of their understanding of the locality and the work that the emergency planning chief officer and his staff will have done with local businesses and so on . . .',[203] but disavowal of the bigger picture is not helpful.

4.77　Secondly, despite the importance of the vitality of private enterprises, the avoidance of regulatory burdens conduced against compulsory BCM or enforced compliance with the advice given under s 4. BCM has not yet become compulsory in public law in the same way as health and safety, though it is

---

[200] Cabinet Office, *The Government's Response to the Report of the Joint Committee*, para 45.
[201] See Cabinet Office, *Emergency Preparedness: Guidance on Part 1*, paras 8.15, 8.41.
[202] See HM Treasury, *Guide to the Establishment and Operation of Trading Funds* (2004); Scottish Executive, *Scottish Public Finance Manual* (Edinburgh, 2003). Charges should not be made for promotional materials or awareness-raising materials but can be made for attendance at local authority organized events, membership of business continuity forums, and the provision of specific information or advice: Cabinet Office, *Emergency Preparedness: Guidance on Part 1*, para 8.33.　　[203] HL Debs vol 665 col 446 14 October 2004, Lord Bassam.

increasingly required as a part of corporate governance standards by financial auditors. One might predict that the duty to provide set out in s 4 will eventually be matched by a duty to take up and action. Just as health and safety regulations are applied for the good of the enterprise, its employees, and the local community, so risk assessment can be treated as an equivalent social good.

Next, continuity advice to the wider public is not covered by s 4, though there **4.78** remains the duty in s 2(1)(g) to maintain arrangements to warn the public, and to provide information and advice to the public, if an emergency is likely to occur or has occurred. Probably the most practical warnings about emergencies are the flood alerts given by the Environment Agency and the Storm Tide Forecasting Service provided by the Meteorological Office, which also offers the National Severe Weather Warning Service. Aside from these, the Government has its own news service in the form of the Government News Network (GNN) which is the regional arm of Government communications.[204] Through the News Distribution Service (NDS), it issues official news releases. GNN offices, working with the Government Offices and through the RRFs have set up and provide support for a Regional Media Emergency Forum in each English region.[205] There is also a national Media Emergency Forum,[206] a Wales Media Emergency Forum, and moves are also afoot to establish a Scottish Media Emergency Forum.[207] These advisory groups seek to set best practice and protocol guidance for how the media might work with public sector agencies to respond to a major emergency including warning and informing the public. A 'Media protocol on public warnings during a catastrophe or major national emergency' has been agreed.[208] The media (who retain editorial control) undertake to convey the Government's announcements, instructions, and advice as to public safety information, and advice without delay. The Government undertakes only to disseminate factual and relevant information. It is not revealed whether the Government has the contingency of its own transmitting equipment, though there is a governmental Emergency Communications Network operating as an alternative to the public telephone systems.[209] So, the first recourse remains the dissemination of

---

[204] See http://www.gnn.gov.uk/aboutGNN/default.asp?page = 9. See further the attempt to set up an 'Emergency Broadcasting System' by Private Member's Bill (2002–03 HC No 85).

[205] For details, see Cabinet Office, *Emergency Preparedness: Guidance on Part 1*, Box 7.8.

[206] Http://www.ukresilience.info/mef/index.htm. See Cabinet Office, *Dealing with Disaster* (rev 3rd ed, Cabinet Office, London, 2003) para 5.34; Media Emergency Forum Joint Working Party Report, *9/11: Implications for Communications* (http://www.ukresilience.info/mef/mefreport.pdf, 2002); Harrison, S, *Disasters and the Media* (Macmillan, Basingstoke, 1999).

[207] Letter from Scottish Executive Justice Department, 30 June 2005, http://www. scotland.gov.uk/Resource/Doc/55971/0015631.pdf.

[208] Civil Contingencies Secretariat, *The Lead Government Department*, Ch 10. Note also the UK Press Card Authority: http://www.newspapersoc.org.uk/default.asp?cid = 300.

[209] House of Commons Defence Committee, *Defence and Security in the UK* (2001–02 HC 518) para 197.

information from the NDS and departments via independent media, including the BBC.[210] The GNN can also supply public bodies at the time of emergency with press officers.[211]

4.79 Beyond the GNN, the UK Resilience website would be a direct source of information on domestic issues, with the 'Preparing for Emergencies' website offering a more user-friendly version.[212] The Foreign and Commonwealth Office website is surprisingly forthright about threats abroad,[213] while the National Travel Health Network and Centre, funded by the Department of Health, informs about travel and health.[214] The MI5 website also contains advice and details of threats, including the Joint Terrorism Analysis Centre's (JTAC's) threat descriptor in relation to terrorism.[215] There is no single national system to indicate the current general level of threat. Of course, different sectors of security threat may require different assessments and warnings, so it could be misleading to have one consolidated warning. A month prior to the suicide bomb attacks in London on 7 July 2005, JTAC had calibrated downwards the threat level from 'severe general' to 'substantial' since it possessed no intelligence of a specific plot. Following the attacks, the threat assessment was raised to 'severe general', meaning further attack was anticipated, albeit that the information was not specific.[216]

4.80 In truth, there is an underlying policy here—the reluctance to reveal to the public threat levels, in line with the general view that the public should follow instructions, rather than be given raw information:[217] 'we should only give when we are clear what advice we accompany a warning with so that the public can help put themselves out of the way of danger.'[218] By comparison, in the United States, there is a five-level Threat Level System which is communicated to the public and with corresponding general advice for each stage.[219] The exclusion of the public in this and other respects represents the outmoded Cold War approach of conflict as 'an insider's affair';[220] low-level late modern conflict demands a new approach.

---

[210] HC Debs Standing Committee F col 200 3 February 2004, Fiona Mactaggart.

[211] Cabinet Office, *Dealing with Disaster* (rev 3rd ed, Cabinet Office, London, 2003) para 5.9.

[212] See www.pfe.gov.uk. Note also the guidance at http://www.ukresilience.info/councils.pdf and http://www.ukresilience.info/responders.pdf. [213] Http://www.fco.gov.uk/travel.

[214] Http://www.nathnac.org. [215] Http://www.mi5.gov.uk/output/Page4.html.

[216] See *The Times* 8 July 2005 at 11, 22 August 2005 at 6.

[217] In Northern Ireland, there is mention of the 'myth of panic' and the importance of communication: Central Emergency Planning Unit (CEPU), *A Guide to Emergency Planning Arrangements in Northern Ireland* (Belfast, 2004) para 7.2.

[218] House of Commons Home Affairs Committee, *Homeland Security* (2003–04 HC 417-I) Q75 2 March 2004, David Ormand.

[219] Http://www.dhs.gov/dhspublic/display?theme = 29. See Aguirre, BE, 'Homeland Security Warnings: Lessons Learned and Unlearned' (2004) 22.2 *International Journal of Mass Emergencies and Disasters* 103. [220] Hennessy, P, *The Secret State* (Allen Lane, London, 2002) 2.

(b) Orders to exercise conferred functions

Broader than s 4 are the powers to require intervention by Category 1   **4.81**
responders under s 5[221]—rather unhelpfully entitled 'General measures'. A
Minister of the Crown (s 5(1) relates to responders operating in England and
Wales or the whole of the United Kingdom) or the Scottish Ministers (s 5(2))
can by order require a responder to perform a function for the purpose of (a)
preventing the occurrence of an emergency, (b) reducing, controlling, or mitigating the effects of an emergency, or (c) taking other action in connection with
an emergency. Illustrative regulations are listed under s 5(4) (s 5(5) in Scotland).
An order may: (a) require consultation; (b) permit, require, or prohibit collaboration; (c) permit, require, or prohibit delegation; (d) permit or require
cooperation or (e) the provision of information; (f) confer a function on a
Minister of the Crown, on the Scottish Ministers, on the National Assembly for
Wales, on a Northern Ireland department, or on any other specified person or
body. In all cases, (g) the provision may apply generally or only to a specified
person or body or (h) may make different provision for different persons or
bodies or for different circumstances. The function specified must be 'a function
of that person or body', meaning that it must be already specified by law in the
powers of that body and not a new function. Orders under s 5 are subject to
the affirmative procedure (s 17(2)).

(c) Disclosure of information

As well as having powers to demand the provision of information regarding   **4.82**
functions under the Act itself under s 2(5)(i), s 6 allows Ministerial regulations
requiring the disclosure of information by a Category 1 responder to a Category
1 or 2 responder regarding obligations in relation to emergencies. Guidance
may be issued alongside the regulations (s 6(4), (5)). Section 6(3) allows this
power to be used in connection with, and only with, a function relating to
emergencies, whether under the 2004 Act or some other legislation, thereby
preventing furtive amendments to legislation such as the Data Protection
Act 1998.[222]

Different responders have different levels of access to classified information and   **4.83**
intelligence. Information-sharing and resilience will not make any inroads into
the current classification regime.[223] The Regulations make clear that the

---

[221] Clause 14(4) of the June 2003 draft Bill provided that provisions of s 101 of the Local
Government Act 1972 (relating to the discharge by one local authority of functions on behalf of
another and joint arrangements for discharge of functions by local authorities) were applied to
duties under cll 2(1), 4(1) and 5 of the Bill. This clause was dropped from the final version, so the
Act is a comprehensive code in itself.

[222] This is confirmed at HL Debs vol 665 col 453 14 October 2004, Lord Bassam.

[223] See Ch 8.

responder shall not comply with an inappropriate request and cannot exercise discretion.[224]

4.84 Section 12 relates to the contents of regulations under s 6 (or, indeed, any other power addressing the provision or disclosure of information). The regulations may make provision about (a) timing; (b) form; (c) end use; (d) storage; and (e) disposal.

4.85 The provision of information in connection with the performance of a duty under s 2(1)(a) to (d) or under s 4(1) is duly dealt with by Part VIII of the Civil Contingencies Act 2004 (Contingency Planning) Regulations 2005. It introduces the concept of 'sensitive information' (reg 45/39 in Scotland), which means: information the disclosure of which to the public would, or would be likely to, adversely affect national security or public safety or would, or would be likely to, prejudice the commercial interests of any person; or information which is personal data, within the meaning of s 1(1) of the Data Protection Act 1998 and disclosure would contravene any of the data protection principles or s 10 of that Act (the right to prevent processing likely to cause damage or distress) or if it is data in a category which is exempt from s 7(1)(c) of that Act (data subject's right of access to personal data), such as national security data.[225] On that point, so as to avoid any doubt or argument, reg 46 (reg 40 in Scotland) provides that a certificate signed by a Minister of the Crown specifying that disclosure of information to the public would, or would be likely to, adversely affect national security is conclusive evidence of that fact.[226] The certificate may only be issued by a Minister who is a member of the Cabinet or by the Attorney-General, the Advocate General for Scotland or the Attorney-General for Northern Ireland.

4.86 An inter-responder request for information should only be made where necessary—provided it is reasonably required in connection with the performance of functions under ss 2(1)(a) to (d), 4(1), or other emergency function and where it is not reasonable to seek to obtain the information by other means (such as through published information or informal dialogue),[227] subject to exceptions for sensitive information (reg 47/41 in Scotland). The request should be in writing, should state the name and address of the requesting responder, should describe the information requested, and should state the reason for the demand (reg 48/42 in Scotland). The receiving responder must then comply,

---

[224] HC Debs Standing Committee F col 112 29 January 2004, Douglas Alexander.
[225] Data Protection Act 1998, s 28.
[226] But 'conclusive' certificates are still subject to review where rights are issue: compare *Balfour v Foreign and Commonwealth Office* [1994] 1 WLR 681; *Johnston v Chief Constable of the Royal Ulster Constabulary*, Case 222/84 [1986] ECR 1663; *Tinnelly & Sons Ltd and others and McElduff and others v United Kingdom*, App Nos 20390/92; 21322/93, Reports 1998-IV.
[227] Cabinet Office, *Emergency Preparedness: Guidance on Part 1*, paras 2.10–2.11.

subject to considerations of sensitivity on grounds of national security or confidentiality (reg 49/43 in Scotland), and the response must be given before the end of such reasonable period as may be specified by the requesting responder (reg 50/44 in Scotland). The cooperative mode of working required of responders can be interpreted to override any restraints on collaboration imposed by the Competition Act 1998, and there is express override of s 28(2) and (7) of the Health and Safety at Work etc Act 1974[228] (reg 54), but the transferred information remains subject to laws about confidentiality,[229] so that the recipient must not pass information on to others unless forced by the 2004 Act to do so. If sensitive information is received, then reg 51 (reg 45 in Scotland) restricts further publication or disclosure by the receiving responder, and reg 52 (reg 46 in Scotland) prohibits its use save for the purpose of performing the function for which, or in connection with which, the information was requested. It is also a requirement that the received sensitive information should be kept securely (reg 53/47 in Scotland).

(d) Urgency directions

Some emergencies may by definition arise in unexpected circumstances and **4.87** without warning—'a sudden heightening of a terrorist alert or immediately following a severe attack... [or] when local responders are faced with a new scale or type of risk or threat.'[230] Accordingly, s 7 (s 8 in Scotland) allows for Ministerial intervention by 'direction' with respect to powers under ss 5 and 6 without the nicety of orders or regulations. The reference to ss 5 and 6 is narrower than the tabled Bill, which would also have applied to ss 2 to 4 and followed criticism from the House of Lords Delegated Powers and Regulatory Reform Committee that the need for urgency was not apparent in the business of planning (rather than responding) and where the negative procedure applied (unlike under s 5 but not s 6).[231] The former argument was accepted but not the latter, since it was contended that:

> ...even an hour's delay could be dangerous... For example, in Japan in 1995, following the terrorist attack on the Tokyo underground, the inability of government to put other health bodies immediately in touch with medical scientists with direct hospital experience of sarin and its effects, is claimed to have inhibited

---

[228] By s 28(2) 'Subject to the following subsection, no relevant information shall be disclosed without the consent of the person by whom it was furnished.' By s 28(7) 'A person shall not disclose any information obtained by him as a result of the exercise of any power conferred by section 14(4)(a) or 20 (including, in particular, any information with respect to any trade secret obtained by him in any premises entered by him by virtue of any such power) except—(a) for the purposes of his functions...'

[229] HC Debs Standing Committee F col 310 10 February 2004, Douglas Alexander.

[230] HL Debs vol 665 col 464 14 October 2004, Lord Bassam.

[231] *Twenty Fifth Report* (2003–04 HL 144) para 16. See HL Debs vol 665 col 468 14 October 2004, Lord Bassam.

diagnosis and treatment of casualties. If time permitted, regulations would be made under Clause 6 to require responders to share certain information. But, if there was insufficient time to draft and make regulations, it may be appropriate to require, by way of direction, a local responder to provide information to another responder.[232]

4.88 Any direction must be in writing under s 7(3). This procedural requirement represents a change from the Bill tabled in Parliament which allowed oral directions which could be later confirmed in writing.[233] There must be 'an urgent need' to make provision of a kind that falls within ss 5(1) or 6(1) while at the same time there is insufficient time for the order or regulations to be made. The provision is worded objectively, an unusual feature in this Act.[234] A direction will be binding in the same way as an order or regulation (s 7(5)). There is no requirement to inform Parliament, even though a normal regulation would be subject to the negative procedure.

4.89 Once made, a direction can be revoked or varied by a further direction and in any event must be revoked 'as soon as is reasonably practicable' (s 7(4)), though it can then be replaced with a new order or regulation under ss 5 or 6 'if or in so far as [the Minister] thinks it desirable'. This apparently subjective wording will be mitigated by the usual requirements of administrative law[235] and the less subjective wordings of ss 5 and 6, but it is thus made clear under s 7(4) that perfect knowledge is not required even if that is not actually required by administrative law in the first place.[236] Thus:

> The 'satisfied' test is a higher threshold than the 'thinks' test... If a Minister is informed of a credible threat of extensive life-threatening flooding in East Anglia, but there is a suggestion that the wind may change and the rain will fall elsewhere, he would have to balance and evaluate the evidence available. A Minister might meet the 'thinks' test if he or she reasonably considered that the flood was likely to happen. The Minister might have some doubts, but if the flood were more likely to happen than not, and he or she had reached that conclusion reasonably on the basis of the available evidence, the 'thinks' test would be met. However, to be satisfied that an emergency was about to happen, the Minister would have to be more certain. He or she would have to pretty sure that the flood was likely to occur and have no substantial doubts about the point.[237]

---

[232] HL Debs vol 665 col 472 14 October 2004, Lord Bassam.

[233] 2003–04 HC No 14 cl 7(5). HL Debs vol 666 col 816 9 November 2004, Lord Bassam. The facility for directions was doubted by the House of Lords Select Committee on Delegated Powers and Regulatory Reform, *Twenty-Fifth Report* (2003–04 HL 144) paras 14–16. The Government agreed that circumstances where directions would be necessary would be 'unusual' but did not give any examples: House of Lords Delegated Powers and Regulatory Reform Committee, *Thirtieth Report* (2003–04 HL 175) Annex 3.

[234] HL Debs vol 666 col 804 9 November 2004, Lord Lloyd. In addition, the Minister must act rationally in administrative law: HC Debs Standing Committee F col 114 29 January 2004, Douglas Alexander. [235] HL Debs vol 665 col 465 14 October 2004, Lord Bassam.

[236] HL Debs vol 666 col 811 9 November 2004, Baroness Scotland. This argument at one stage failed to convince the House of Lords: ibid, col 813; HC Debs vol 426 col 1408 17 November 2004. [237] HC Debs vol 426 cols 1411–12 17 November 2004, Ruth Kelly.

A direction automatically expires after 21 days (without prejudice to the power **4.90**
to give a new direction to the same effect). Given the period of grace under the
first direction, 'it is difficult to conceive of circumstances in which it would not
be possible to prepare and make legislation before the first direction elapsed',
save perhaps during an adjournment or recess of Parliament.[238] This form of
expiration was not replicated in the original draft Bill of June 2003 but was
conceded as appropriate in January 2004.[239]

F. Monitoring and Enforcement

As has been made clear in describing the duty holders under the Act, there **4.91**
is an unevenness between the duties are imposed on localities and the escape
of central government from statutory imposition. It follows that the main
role of central government is to be empowered to monitor and enforce compliance by local underlings. These powers respond to concerns raised in
2001 that guidance at that time was uneven, ad hoc and from a variety of
sources.[240]

By s 9(1), a Minister may, in order to ensure compliance in individual cases and **4.92**
consistency between responders, require a responder (a) to provide information
about action taken or (b) to explain why the person or body has not taken action
for the purpose of complying with a duty under Part I of the Act. The
explanations under (b) were not part of the draft Bill of June 2003. Any
'requirement' may set parameters as to deadline and form of response (s 9(3)).
The duty to comply with a lawful requirement is absolute (s 9(4)). The
information so gathered may be the basis for enforcement action against
individual persons or bodies under ss 10 or 11 or may, on reflection, trigger
general secondary legislation under ss 3 to 6. There are corresponding monitoring powers for the Scottish Executive (s 9(2)) but not for the Welsh
Assembly since the making of regulations is not devolved.[241]

These measures do not reflect some of the more radical ideas floated in the **4.93**
Cabinet Office consultation paper of 2001. It suggested that the production of
guidance could in part be transferred to individual professional bodies, such as
the Local Government Association or the Emergency Planning College.[242]

---

[238] HC Debs Standing Committee F col 120 29 January 2004, Douglas Alexander.

[239] Cabinet Office, *The Government's Response to the Report of the Joint Committee*, para 35.

[240] Cabinet Office, *Emergency Planning Review: The Future of Emergency Planning in England and Wales* (London, 2001) paras 5.6, 5.7.

[241] See further Local Government and Public Services Committee, *Report on the Civil Contingencies Bill* (National Assembly for Wales, Cardiff, 2004) para 5.

[242] Cabinet Office, *Emergency Planning Review: The Future of Emergency Planning in England and Wales* (London, 2001) para 5.9.

Equally, the paper called for regular audit, possibly by a permanent Inspectorate.[243] As shall be explained further in Chapters 8 and 9, that concept has been allowed to rest, with no emergent agency to provide joined-up and continuous leadership and a lukewarm response all round to these ideas.[244]

4.94 All of the responder duties 'are potentially susceptible to judicial review proceedings'.[245] Nevertheless, preliminary arguments could arise about the bounds of public functions as exercised by responders which are private bodies.[246] Therefore, there is a more direct power under s 10 (s 11 in Scotland) which allows for proceedings in the High Court or Court of Session in respect of failures under ss 2(1), 3(3), 4(1) or (8), 5(3), 6(6), 9(4), or 15(7). The action may be maintained by a Minister of the Crown or by one responder against another responder. The remedies are left at large: the court may grant any relief or make any order that it thinks appropriate. Legal enforcement beyond the 2004 Act will be considered further in Chapter 6.

G. Devolved Territories and London

4.95 This segment will outline the variant legislative impact of Part I in those parts of the United Kingdom subject to devolution, plus the arrangements for London.

(a) Scotland

4.96 In the original drafts of the 2004 Act, the legislation was confined to England and Wales. Mirror legislation from the Scottish Parliament was presumably expected. But after consultation, the Scottish Executive recommended that a UK-wide civil protection framework would be preferable.[247] The Government therefore extended Part 1 of the Bill to Scotland.[248] Niceties were observed, under the terms of the Memorandum of Understanding with the Devolved Administrations[249] and Devolution Guidance Note 10,[250] by passing a 'Sewel' motion in the Scottish Parliament,[251] ceding the legislative initiative on this

[243] Ibid, para 5.10.
[244] Cabinet Office, *Emergency Planning Review: The Future of Emergency Planning in England and Wales: Results of the Consultation* (London, 2002) paras 19, 20, 23.
[245] HC Debs Standing Committee F col 73 27 January 2004, Douglas Alexander.
[246] See Ch 6.
[247] Scottish Executive, *Civil Contingencies Bill: Scottish Consultation Report* (Justice Department, Edinburgh, 2004) para 14.
[248] Civil Contingencies Secretariat, *The Government's Response*, 39.
[249] Paragraph 13 (http://www.scotland.gov.uk/library2/memorandum/default.htm, 2002).
[250] http://www.dca.gov.uk/constitution/devolution/guidance/dgn10.pdf.
[251] See Scottish Parliament Justice 1 Committee col 543 25 February 2004; Debates col 6372 4 March 2004.

matter to Westminster for the sake of 'consistency of regulations and response'.[252]

Part I therefore trespasses onto the competencies of the Scottish bodies since it **4.97** involves transferred matters under the Scotland Act 1998, whereas Part II involves reserved matters. Thus, defence (except the exercise of civil defence functions by the armed forces), emergency powers, national security, the interception of communications, official secrets and terrorism, energy matters (including under the Energy Act 1976), and transport security are 'reserved' matters.[253] Otherwise, the Scottish Executive has responsibility for local resilience.

The Scottish Executive departments are linked to potential contingencies in line **4.98** with the 'Lead Government Department' principle (described in Chapter 8). If the emergency only affects Scotland,[254] then a suitable Scottish Executive department—for example, the Health Department or the Environment and Rural Affairs Department—takes charge, but if a reserved matter is the focus, then a Westminster department may lead where the competence of the devolved administration might be stretched.[255] When an emergency affects a range of services within Scotland and where coordination cannot reasonably be provided by those with the functional responsibilities in the Scottish Executive, this role may be exercised by the Emergency Management Unit within the Civil Contingencies Division of the Scottish Executive Justice Department until such time as a lead division is designated. Overall responsibility for civil protection policy sits with the Civil Contingencies Division (CCD) of the Scottish Executive Justice Department which, 'promotes and co-ordinates civil protection efforts across the Executive in a role similar to that of the Civil Contingencies Secretariat at the UK level.'[256] The Scottish Executive Justice Department also distributes financial support and guidance to local responders.

At the next tier, as indicated previously, there are eight Strategic Coordinating **4.99** Groups, based on police force areas and led by chief constables and local authority chief executives but also including health and fire brigade representatives.[257]

Despite the policy of unity, the Act itself, as well as its regulation and order- **4.100** making powers, contain variants for Scotland, mainly to take account of local

---

[252] Scottish Parliament Justice 1 Committee 1 col 547 March 2004, Hugh Henry. See also Scottish Parliament col 6372 4 March 2004. [253] Scotland Act 1998, Sch 5.
[254] For example, during the Lockerbie disaster in December 1988, the Department of Transport was in the lead, but the Scottish Executive would have played a major role in coordinating central government action in Scotland: Scottish Executive, *Dealing with Disasters Together* (http://www.scotland.gov.uk/library5/government/dealdisasters.pdf, Edinburgh, 2003) para 8.7. [255] Cabinet Office, *Emergency Response and Recovery*, para 12.10.
[256] Cabinet Office, *Emergency Preparedness: Guidance on Part 1*, para 10.10.
[257] Ibid, para 10.15.

systems (such as the court structure). To ensure that there is no conflict or unacceptable inconsistency between jurisdictions, s 14 provides for consultation between Ministers of the Crown and the Scottish Ministers before either finalises any instrument under the Act. The measure was broadened to go beyond regulations about particular classes of responders,[258] but it still does not apply to guidance.

4.101 Moving from the top to the bottom of the administrative ladder, s 15 allows for instructions about collaboration between responders across the borders of England/Wales/Northern Ireland and Scotland. Thus, Ministers and Scottish Ministers may make regulations (and issue guidance about the regulations) permitting or requiring cooperation or the provision of information. Specific reference is made to cooperation pursuant to duties imposed under s 5(1) or (2) (s 15(5) and (6)) and such orders are given equivalent status to s 5 orders (s 15(8)). By subsection (7), responders must comply with an order or regulations under s 15 and must have regard to guidance. This provision was not in the Bill as introduced in January 2004. The Minister for Crime Reduction, Policing and Community Safety explained that it arose from discussions with the Scottish Executive.[259] The problem was that Scottish Ministers can require Scottish responders to cooperate with other Scottish responders, and UK Ministers can require English, Welsh or UK-wide responders to cooperate with each other, but there was no power to require cross-border cooperation and information sharing. This power could especially apply, for example, to the British Transport Police, the Health and Safety Executive, and the Maritime and Coastguard Agency, where a Scottish responder requires information from their English-based headquarters.

4.102 The Civil Contingencies Act 2004 (Contingency Planning) (Scotland) Regulations 2005[260] implement Part I in Scotland. The details have already been related.

(b) Wales

4.103 The Welsh devolved bodies are less empowered under the Government of Wales Act 1998 than the Scottish bodies. Only agriculture, the environment, health, highways, transport, and water and flood defence are amongst the 'transferred' matters,[261] but even these limited responsibilities enable the Welsh Assembly Government to engage in aspects of civil protection work and, as a consequence, play an important coordinating role.[262] Therefore, the National

---

[258] HC Debs Standing Committee F col 122 29 January 2004, Douglas Alexander.
[259] HC Debs vol 421 col 1323 24 May 2004, Hazel Blears.   [260] SSI No 494.
[261] Government of Wales Act 1998, Sch 2.
[262] See Cabinet Office, *Emergency Preparedness: Guidance on Part 1*, para 11.7.

Assembly for Wales remitted the Civil Contingencies Bill to its Local Government and Public Services Committee for consideration. The Committee concluded that the Bill inadequately reflected the constitutional and practical implications of devolution in Wales and did not provide the National Assembly of Wales (NAW) with sufficient powers to ensure the proper preparation and implementation of emergency plans in Wales.[263] It cited the Foot and Mouth outbreak as an example of where more local decision-making would have assisted with speed and accuracy. It was also critical of the fact that the monitoring powers (s 9) applied to the Scottish Executive but not the NAW.[264]

By s 16(1), a Minister must consult the NAW before making regulations under **4.104** s 2(3), 4(2), or 6(1), issuing guidance under ss 3(1), 4(6), or 6(4), giving an order under s 5(1), giving a direction under s 7(2),[265] bringing proceedings under s 10, making an order under s 13(1) which relate wholly or partly to Wales or persons or bodies with functions in Wales. Section 16(2) provides for a stronger Welsh input. The Minister may not, without the consent of the NAW, make regulations under ss 2(3), 4(2), or 6(1), make an order under s 5(1), issue guidance under ss 3(1), 4(6), or 6(4), give a direction under s 7, bring proceedings under s 10, or make an order under s 13 where the ministerial action relates wholly or partly to a Wales-based person or body in relation to which the NAW has functions as specified in s 16(4).[266]

The Welsh Assembly has established a Wales Resilience Partnership Team.[267] **4.105** Led by a senior official, the team is 'dedicated to supporting multi-agency co-operation in Wales and engaging with the UK Government on all issues relating to civil protection and emergency preparedness.'[268] The Team reports to the Welsh Assembly Government in the Wales Resilience Forum (WRF), chaired by the First Minister and comprising responders and Welsh Ministers, which aims to provide strategic advice on civil protection.[269] Its proposals are implemented through the Wales Resilience Programme.[270] The WRF functions

---

[263] Local Government and Public Services Committee, *Report on the Civil Contingencies Bill* (National Assembly for Wales, Cardiff, 2004) para 2. [264] Ibid, para 5.

[265] An attempt to drop this provision was rejected. See Cabinet Office, *The Government's Response to the Report of the Joint Committee*, para 35. The First Minister recognized that the NAW should not be a responder: Joint Committee on the Draft Civil Contingencies Bill, para 96.

[266] Part I regulations were approved by the National Assembly of Wales on 28 June 2005. Standing Order 26 requires a vote in plenary.

[267] Cabinet Office, *Emergency Preparedness: Guidance on Part 1*, para 11.22.

[268] Ibid, para 11.31. See also http://www.wales.gov.uk/subilocalgov/content/partnership/3march05/5pmf-e.htm.

[269] Http://www.wales.gov.uk/subilocalgov/content/partnership/3march05/5pmf-e.htm.

[270] An Emergencies Working Group set up by the NAW had previously produced the Wales National Emergency Co-ordination Arrangements: Cabinet Office, *Dealing with Disaster* (rev 3rd ed, Cabinet Office, London, 2003) Annex C.

in a similar way to the Regional Resilience Forums in England 'but with a higher level of political involvement as a consequence of devolved responsibilities.'[271] Reporting to it is the Wales Risk Assessment Group, which draws from the risk assessments at LRF level. The framework also includes a Wales Resilience Stakeholders Forum which is intended to facilitate communication and information-sharing networks with Category 1 and 2 responders.[272] It is also expected that a representative of the NAW should attend the Welsh LRFs.[273]

4.106 Three other relevant groupings are: the Joint Emergency Services Group (the emergency services and armed forces); the Wales Media Emergency Forum; and the Welsh Borders Resilience Group (cooperation and information sharing between Wales and the English border areas).[274]

4.107 The four LRFs in Wales are chaired by the respective chief constable.[275]

(c) Northern Ireland

4.108 As mentioned in relation to Scotland, the original version of Part I in June 2003 applied to England and Wales only. But, after consultation, the Government concluded that Part 1 of the Bill could be extended to Northern Ireland to a very limited extent, mainly to encompass UK-wide responders and the Police Service for Northern Ireland, the only distinctly local responder.[276] Otherwise, 'The local civil protection arrangements as set out in Part 1 of the draft Bill do not read easily across to Northern Ireland.'[277] The problem is not the devolution scheme. National security and nuclear matters are 'excepted' matters (beyond the devolved bodies' competence) while civil aviation, criminal justice, public order, civil defence (including the 'subject-matter of the [then] Emergency Powers Act (Northern Ireland) 1926') are 'reserved' matters (requiring the consent of the Secretary of State if local powers are to be used).[278] Thus, as with Scotland, Part I planning is largely a devolved matter. Most functions of Category 1 responders are delivered in Northern Ireland by, or via, Government Departments forming part of the Northern Ireland Executive, such as the Department of Health, Social Services and Public Safety. Those Departments can act as Lead Government Department for devolved matters, with the

---

[271] See Cabinet Office, *Emergency Preparedness: Guidance on Part 1*, para 11.26.
[272] Ibid, para 11.23.   [273] Ibid, para 11.16.   [274] Ibid, para 11.23–11.30.
[275] Ibid, para 11.3.
[276] Civil Contingencies Secretariat, *The Government's Response*, 42.
[277] Central Emergency Planning Unit, *Civil Protection in Northern Ireland: The Implications of the Civil Contingencies Bill* (http://cepu.nics.gov.uk/consult.pdf, Belfast, 2003) para 16. Civil Contingencies Act 2004 (Contingency Planning) Regulations 2005, regs 4, 5, 9 to 11, 15 to 17, 32 to 34, and 36 to 44 do not extend to Northern Ireland.
[278] Northern Ireland Act 1998, Schs 2 and 3.

Northern Ireland Office retaining control over excepted or reserved matters. In happier times, overall coordination of civil protection arrangements would be undertaken by the Office of the First Minister and Deputy First Minister (OFMDFM). The Northern Ireland Assembly would normally have oversight of civil contingencies arrangements for transferred functions.[279] But the crux of the problem is that the local Executive and Assembly are suspended, with no immediate prospect of resurrection.[280] As a result, Northern Ireland is subject to 'direct rule' from Westminster, and the policy of the 2004 Act is not to confer duties upon Westminster Ministers, including those who currently direct and control the Northern Ireland departments. The problem is compounded by the fact that more of the relevant powers are concentrated in these government departments (and fewer in local authorities) powers than is the case in Scotland.

Northern Ireland relies upon a central strategic committee framework for **4.109** managing Civil Protection policy and responses to emergencies: 'This framework involves the Northern Ireland departments, emergency services, District Council Chief Executives and other key response organisations.'[281] Rather than listing government departments as responders, the Government proposed that 'Arrangements covering devolved functions (carried out by central departments) will be delivered, not through the Bill, but through undertakings by Northern Ireland Ministers to commit their departments to practice the civil protection duties in line with Part 1.'[282] Guidance for Northern Ireland makes it clear that 'Departments have primary responsibility for setting policy on civil protection for themselves and their agencies, NDPBs etc.'[283] Their responsibilities include strategic action, hazard awareness, the implementation of civil protection frameworks, the provision of essential services, planning of their own responses to emergencies and Business Continuity Planning, responding directly to emergencies, and supporting the emergency services. For the limited number of organizations which deliver functions that are not transferred, such as the Police Service of Northern Ireland and the Maritime and Coastguard Agency,[284] the Northern Ireland Office will ensure that activities are coordinated with the relevant Northern Ireland departments and vice versa.[285]

A key body, located within the OFMDFM, is the Central Emergency Planning **4.110** Unit (CEPU), which promotes the development of the 'Northern Ireland Civil

---

[279] See Cabinet Office, *Emergency Preparedness: Guidance on Part 1*, para 12.6.
[280] See Northern Ireland Act 2000 (Suspension of Devolved Government) Order 2002 SI No 2574; Northern Ireland Assembly (Elections and Periods of Suspension) Act 2003.
[281] Cabinet Office, *The Draft Civil Contingencies Bill*, para 6.13.
[282] Civil Contingencies Secretariat, *The Government's Response*, 42.
[283] Central Emergency Planning Unit (CEPU), *A Guide to Emergency Planning Arrangements in Northern Ireland* (Belfast, 2004) para 5.16. [284] Ibid, para 12.1.
[285] Ibid, paras 12.4, 12.7.

Contingencies Framework'.[286] It devises a policy framework for civil protection in Northern Ireland departments and public agencies and also supports central mechanisms for the coordination of policies and emergency responses by public bodies in the Province. It also links with the Civil Contingencies Secretariat in the Cabinet Office and with the emergency planning units in the Scottish and Welsh devolved administrations.[287] Assisting the CEPU on policy formation, communication, and coordination is the Infrastructure Emergency Planning Forum which brings together government agencies, Non-Departmental Public Bodies, utilities, and communications providers.[288]

4.111 The Central Emergency Management Group (CEMG) is the top level group for coordination and the development of strategy, comprising emergency services, government departments, agencies, NDPBs, and district councils. It will be chaired by the Director of the Central Emergency Planning Unit, with Liaison Officers from Northern Ireland Departments, representatives from the Executive Information Service, the Northern Ireland Information Service and Society of Local Authority Chief Executives plus representatives of the four Emergency Services. Acting akin to an RRF, it produces the Northern Ireland Civil Contingencies Framework which sets out the principles of policy and practice.[289]

4.112 The position of Northern Ireland local authorities is distinct from those in Britain, and features such as LRFs and Community Risk Registers are not replicated. District councils[290] have not had explicit functions to improve civil protection arrangements, and it was felt beyond the scope of the 2004 Act to assign new functions to them.[291] Nevertheless, district councils have been charged since 1989 with taking a lead in preparing an integrated coordination of the local response to emergencies, as well as the provision of council services during emergency response and recovery. Protocols in 2000 and 2002 cover relations with the police.[292] Belfast has its own arrangements through the Belfast Resilience project.[293] This requirement is backed by the DOE Guidance Circular LG 30/00, which sets out in greater detail the role of the district councils in emergencies, where the responsibility is placed on the Chief Executive. Their role is, however, non-statutory, since they are not Category 1

---

[286] Ibid, para 12.15. See Central Emergency Planning Unit (CEPU), *A Guide to Emergency Planning Arrangements in Northern Ireland* (Belfast, 2004) Chs 4, 6.

[287] See Cabinet Office, *Emergency Preparedness: Guidance on Part 1*, paras 12.13–12.14.

[288] Cabinet Office, *Emergency Response and Recovery*, paras 12.17–12.23.

[289] Cabinet Office, *Emergency Preparedness: Guidance on Part 1*, para 12.15.

[290] See Local Government Act (Northern Ireland) 1972.

[291] Northern Ireland Office, *Civil Protection in Northern Ireland: The Implications of the Civil Contingencies Bill* (Belfast, 2003) q E.

[292] CEPU, *The Role of District Councils in Emergency Planning and Response: Support to other Organisations* (Belfast, 2004) Annex B.   [293] Ibid, 2.

responders under the 2004 Act. Even so, under the CEPU's Guide to Emergency Planning Arrangements in Northern Ireland, issued in 2004, the expectation is reiterated that 'Most incidents are handled at a local level by the emergency services, District councils, Health and Social Services Boards and Trusts and other locally based agencies with no direct involvement by central government.'[294]

A firmer legal footing for these arrangements is now being implemented. By the **4.113** Local Government (Miscellaneous Provisions) (Northern Ireland) Order 2005,[295] Art 29:

Powers of district councils in relation to emergencies

(1) A council may make arrangements, or enter into arrangements with other bodies or persons, for
   (a) preventing the occurrence of an emergency;
   (b) reducing, controlling or mitigating the effects of any emergency which may occur.
(2) Those arrangements may in particular include
   (a) arrangements for co-operation with other bodies or persons; and
   (b) arrangements for the co-ordination of the activities of the council with those of other bodies or persons.
(3) Where an emergency occurs
   (a) a council may give effect to any arrangements made under paragraph (1); and
   (b) take any other steps which it thinks appropriate for reducing, controlling or mitigating the effects of the emergency.
(4) In carrying out its functions under paragraphs (1) to (3), a council shall have regard to any guidance issued by the Department.
(5) Before issuing any guidance under paragraph (4) the Department shall consult councils and
   (a) such associations or bodies representative of councils;
   (b) such associations or bodies representative of officers of councils; and
   (c) such other persons or bodies;
      as appear to it to be appropriate.
(6) A Northern Ireland department may by order confer or impose on district councils other functions relating to emergencies.
(7) An order under this Article
   (a) may include supplementary, incidental, consequential or transitional provisions; and
   (b) shall not be made unless a draft of the order has been laid before, and approved by a resolution of, the Assembly.
(8) In carrying out any functions conferred or imposed on it by an order under paragraph (6), a council shall have regard to any guidance issued by the Northern Ireland department which made the order.

---

[294] Belfast, 2004, para 3.39.   [295] SI No 1968 (NI 18). The Order is not yet in force.

(9) Before issuing any guidance under paragraph (8), a Northern Ireland department shall consult councils and

(a) such associations or bodies representative of councils;

(b) such associations or bodies representative of officers of councils; and

(c) such other persons or bodies,

as appear to it to be appropriate.

(10) The Department may make grants to district councils or other persons for the purpose of, or in connection with, arrangements made or entered into under paragraph (1).

(11) A grant under paragraph (10) shall be

(a) of such amount, and

(b) made subject to such conditions,

as the Department may determine.

(12) Grants shall not be made under paragraph (10) without the consent of the Department of Finance and Personnel.

(13) In this Article

'emergency', in relation to a council, means an emergency affecting the whole or part of the district of the council or all or some of its inhabitants;

'other bodies or persons' includes other councils.

This Order does not as yet bring district councils into line with their British counterparts. These are powers rather than duties, so the main focus remains on the Northern Ireland departments. However, there is the possibility of duties being conferred under (6) in the future.

4.114 Next, the Northern Ireland Fire Brigade,[296] though omitted from Schedule 1 of the 2004 Act (unlike the Police Service of Northern Ireland), is expected to act as a primary service for fires, chemical accidents, and incidents such as road traffic accidents where specialist equipment is needed, with corresponding statements being made for the Northern Ireland Ambulance Service (NIAS), (sub-regional) Health and Social Services Boards and Trusts, and the Rivers Agency, Water Service, and Roads Service.[297]

4.115 One further unique condition applies in that Northern Ireland is the international land border. At the local level it will be the emergency services who coordinate responses to incidents at or near the border. It is asserted that, 'Other arrangements are in place for co-ordinating information and efforts in areas such as public health emergencies and the response to nuclear accidents. These

---

[296] See Fire Services Order (NI) 1984 SI No 1821 (NI 11).

[297] Central Emergency Planning Unit (CEPU), *A Guide to Emergency Planning Arrangements in Northern Ireland* (Belfast, 2004) paras 5.5, 5.8, 5.10, 5.25. On health matters, the Public Health Consultants and the Health Boards and Trusts would take the lead in responding. The (Departmental) Regional Health Command Centre (RHCC) provides leadership and advice to the Health and Personal Social Services (HPSS), the CMG/CEMG, Ministers and advice to the general public, while the Emergency Medical Assistance Rescue Team (EMART) is meant to respond to large scale emergencies and Chemical, Biological, Radiological or Nuclear (CBRN) incidents.

arrangements supplement national-level arrangements for co-ordination and co-operation through agencies such as the European Union.'[298]

Other variations in relation to Northern Ireland include the system for the **4.116** issuance of information to the public which might involve, in very large-scale emergencies, the Northern Ireland Information Management Centre, activated by the CEPU, which will handle all the media activities of Northern Ireland departments. The Northern Ireland Technical Advisory Group (NITAG), formed in 1986 after the Chernobyl fall-out advises on what steps should be taken to deal with the effects of a radiation emergency.

Because of the limited range of local responders in Northern Ireland, those **4.117** UK-wide responders which operate in Northern Ireland are required by reg 57 of the Civil Contingencies Act 2004 (Contingency Planning) Regulations 2005 to have regard to any assessments of the risk of an emergency or any arrangements to warn the public and to provide information and advice to the public which have been made by any the bodies specified in paragraph (4), even where those bodies are not themselves responders. The listed bodies are a Northern Ireland government department, a district council, the Northern Ireland Housing Executive,[299] the Fire Authority for Northern Ireland, Education and Library Boards,[300] Health and Social Services trusts,[301] Health and Social Services Boards,[302] the Northern Ireland Central Services Agency,[303] the Northern Ireland Blood Transfusion Service Agency, the Health Protection Agency, the Northern Ireland Regional Medical Physics Agency, the Food Standards Agency, the Health and Safety Executive for Northern Ireland, a harbour authority, the Northern Ireland Transport Holding Company,[304] Ulsterbus Limited, Citybus Limited, Northern Ireland Railways Company Limited, and various power supply licence holders.[305]

(d) London

London is special. First, its arrangements for rule by the Greater London **4.118** Authority, as set out in the Greater London Authority Act 1999, are unique. In addition, there is a large number of boroughs (33) with no intermediate 'county' or sub-regional structure. Secondly, it has exclusive problems and

[298] Ibid, para 12.26.   [299] Housing (Northern Ireland) Order 1981 SI No 156 (NI 3).
[300] Education and Libraries (Northern Ireland) Order 1986 SI 1986 No 594, Art 3.
[301] Health and Personal Social Services (Northern Ireland) Order 1991 SI No 194 (NI 1), Art 10.
[302] Health and Personal Social Services (Northern Ireland) Order 1972 SI No 1265 (NI 14), Art 26.
[303] Health and Social Services Health and Personal Social Services (Northern Ireland) Order 1972 SI No 1265 (NI 14), Art 26.   [304] Transport Act (Northern Ireland) 1967.
[305] Electricity (Northern Ireland) Order 1992 SI No 231 (NI 1), art 10(1)(b); Gas (Northern Ireland) Order 1996 SI No 275 (NI 2), Art 8(1).

vulnerabilities in terms of resilience. It is the capital city and heart of the state bureaucracy. It is the prime economic site. It is the home to many so-called 'postcard' targets—Parliament, Buckingham Palace, and so on. It has the most complex transport and communications systems. Its population size far exceeds any other UK city.

4.119 Local resilience areas for London are set out in the Schedule to the Civil Contingencies Act 2004 (Contingency Planning) Regulations 2005. The boroughs are corralled into just six local resilience areas (still called LRFs).[306] Given that the Metropolitan Police Service covers the whole of London, it was not appropriate to base LRFs on police areas, and boroughs were also considered too fragmented, though it is possible that boroughs may still run their own forums or services which link to the area LRF.[307]

4.120 Because of the importance and vulnerability of the capital, there are three added regional structures which provide better coordination and oversight than elsewhere in the country. First, there is an executive body, the London Resilience Team (LRT), based within the Government Office for London. It consists of civil servants based within the Government Office for London and representatives from the emergency services, NHS, local authority, transport, LFEPA, communications and the Salvation Army.[308] A Team member will sit on all London LRFs.

4.121 Secondly, the London Regional Resilience Forum (LRRF) is the body that oversees the work of the LRT.[309] It runs various sub-committees and working groups, such as the Blue Lights, Utilities, Business, Transport, and Communication Sub-Committees. It is responsible for multi-agency cooperation and assumes other strategic functions. In this way, the LRRF acts as the RRF for London, but it differs significantly from others in that it is chaired by a Government Minister rather than a regional functionary, and the Mayor of London acts as deputy chair.[310] It is composed of senior officials representing the main emergency organizations and key sectors within the partnership. The Forum reports to the Government.

4.122 Thirdly, a wider consultative body has been created in the form of London Resilience. It comprises all key public and private organizations and bodies in the capital.

---

[306] See Civil Contingencies Secretariat, *The Government's Response*, 48. Since there is only one police force, its boundaries were also considered inappropriate: Cabinet Office, *Emergency Preparedness: Guidance on Part 1*, para 9.2.

[307] In Westminster, work includes a notification service: http://www.westminster.gov.uk/policingandpublicsafety/crimeandlawenforcement/crimeprevention.cfm.

[308] See http://www.londonprepared.gov.uk/resilienceteam/.

[309] See Greater London Authority Act 1999, s 330.

[310] Cabinet Office, *Emergency Preparedness: Guidance on Part 1*, para 9.14.

LRFs are also complemented and given direction by the London Fire and **4.123** Emergency Planning Authority (LFEPA) which is responsible for the London Fire Brigade.[311] Pursuant to the Government promise (described above) that local authorities might delegate some of their functions to the local Fire and Rescue Service under s 2(3), reg 55 of the Civil Contingencies Act 2004 (Contingency Planning) Regulations 2005 specifies that in London, it shall be the function of the LFEPA to (a) take the lead responsibility for ensuring that a Community Risk Register is maintained in each local resilience area in London; (b) to take the lead responsibility for performing the duty under s 2(1)(d) (duty to maintain emergency plans) in relation to pan-London emergencies; and (c) at the request of a local Category 1 responder, to assist that responder in carrying out exercises and training staff of that Category 1 responder or other persons for the purposes of ensuring that a plan maintained by that relevant Category 1 responder by virtue of s 2(1)(c) or (d) is effective. Other Category 1 responders must give full cooperation (reg 56). The LFEPA is expected to provide the secretariat for the LRFs.[312] It may be questioned whether the LFEPA has been invested with a sufficiently strong statutory coordinating role under the Bill in relation to London borough councils.

An important sectoral group is the London Emergency Services Liaison Panel **4.124** (LESLP).[313] It was established in 1973, and its Forum, chaired by the police and consisting of representatives from the emergency services, local and port authorities, and military, began in 1996. The Panel meets every three months and seeks cooperation in planning for major incidents. The Forum meets every six months to give strategic guidance. One major product is the Procedure Manual,[314] which provides summaries of the responses and responsibilities of each of the emergency services in London and the support offered by other authorities.

The LRRF, in liaison with the LESLP, has developed the Strategic Emergency **4.125** Plan,[315] which provides an overview of London's coordinated response to a catastrophic incident, relating key plans for command and control, communications, mass fatalities, evacuations, and site clearance.

Under the 2004 Act, it is intended that the London Resilience Forum will lead **4.126** Part I work on regional capabilities, with London boroughs leading on the local capability in conjunction with LFEPA.[316]

---

[311] See Greater London Authority Act 1999, s 328.
[312] Cabinet Office, *Emergency Preparedness: Guidance on Part 1*, para 9.7. See further http://www.london-fire.gov.uk/about_us/emergency_planning.asp.
[313] Http://www.leslp.gov.uk/.
[314] 6th ed, http://www.leslp.gov.uk/LESLP_man.pdf, 2004.
[315] Http://www.ukresilience.info/londonprepared/emergplan.pdf, 2005.
[316] Cabinet Office, *The Draft Civil Contingencies Bill*, para 75.

4.127   A great deal of planning work has been undertaken in London—more so than any other location in the United Kingdom. Much of it is of good quality. However, the Act may be an opportunity missed for the setting of the various bodies and relationships in a more ordered, clear, and unified structure. One might compare the unitary structure in other large cities, such as New York where just two bodies cover the field. The New York State Emergency Management Office coordinates against natural, technological, and man-made disasters.[317] There is also a New York City Office of Emergency Management.[318]

H. Legislative Procedures and Effect

4.128   Regulations and orders under Part I must be made by statutory instrument (s 17(1)). Those orders under ss 1(4), 5(1), or 13(1) may not be made by a Minister unless a draft has been laid before and approved by resolution of each House of Parliament (s 17(2)); the equivalent for Scottish Ministers is to require a draft has been laid before and approved by resolution of the Scottish Parliament (s 17(3)). As for regulations, they are subject to annulment by resolution of either House of Parliament (or Scottish Parliament as the case may be).

4.129   The effect of Part I regulations is further considered by s 17(6). Regulations may make (a) provision which applies generally or only in specified circumstances or for a specified purpose, (b) different provision for different circumstances or purposes, and (c) incidental, consequential, or transitional provision.

I. Regulations and Guidance

4.130   The purpose behind the regulations and guidance is 'to ensure that the provisions of Part I of the Act are implemented effectively' and 'to ensure that the right balance is struck between prescription and permissiveness'.[319] The Civil Contingencies Act 2004 (Contingency Planning) Regulations 2005 have been described in substance alongside the primary materials. For Scotland, there is a parallel set of regulations.[320]

---

[317]   See http://www.nysemo.state.ny.us/. Its authority is based upon the Federal Civil Defense Act of 1950 and the New York State Defense Emergency Act of 1951 but later amendments moved the emphasis from civil defence.

[318]   Http://www.nyc.gov/html/oem/. Its directorate was based in the World Trade Center in 2001.

[319]   Cabinet Office, *Civil Contingencies Act 2004: Consultation on the Draft Regulations and Guidance* (Cm 6401, London, 2004) paras 3, 7.

[320]   Civil Contingencies Act 2004 (Contingency Planning) (Scotland) Regulations 2005 SSI No 494.

J. Other Planning Powers

If, as the Minister of State for the Cabinet Office claimed when introducing the **4.131** Civil Contingencies Bill, 'The aim of the Bill and accompanying non-legislative measures is to deliver a single framework for civil protection in the United Kingdom',[321] then the process has been a failure. There remain outwith the 2004 Act numerous other statutory provisions to engage in planning against mishaps.[322] They range from the mundane problems of schools[323] all the way to the exotic contingency of being struck by near-earth objects.[324]

Diversity in organizational structure also persists. There is no single body, aside **4.132** from perhaps the Security Service, which addresses threats to the Critical National Infrastructure (CNI). Nevertheless, that concept is seen as helpful to developing a common understanding of key sectors and functions that need to be preserved as a matter of priority. The CNI cover those assets, services, and systems that support economic, political, and social life whose importance is such that any entire or partial loss or compromise could 'cause large scale loss of life; have a serious impact on the national economy; have other grave social consequences for the community, or any substantial part of the community; or be of immediate concern to the national government.'[325] These feed into the Key Capabilities Programme described in Chapter 8. There are ten critical 'sectors', communications, emergency services, energy, finance, food, government and public service, public safety, health, transport, and water. One aspect of the CNI is tackled by a specific body, namely, the National Infrastructure Security Co-ordination Centre (NISCC), established in 1999.[326] NISCC considers the risk to the critical national infrastructure from electronic attack.

It is beyond the scope of this book to undertake more than a very brief **4.133** indication of the life outside the 2004 Act. One might first mention some of its closest relations, namely, the Control of Major Accident Hazards Regulations 1999 (COMAH)[327] which implement the Seveso

---

[321] HC Debs vol 416 col 1096 19 January 2004, Douglas Alexander.

[322] See Civil Contingencies Secretariat, *Full Regulatory Impact Assessment*, Annex D.

[323] Teachernet, *Developing a School Emergency Management Plan* (http://www.teachernet. gov.uk/emergencies/resources/index.htm. Lord Clarke recommended safety strategies: Lord Cullen, *Public Inquiry into the Shootings at Dunblane Primary School on 13 March 1996* (Cm 3386, HMSO, London, 1996) para 10.19).

[324] See the Task Force on Potentially Hazardous Near Earth Objects (http://www. nearearthobjects.co.uk). It is argued that catastrophic risks (including near earth object collision) are growing and underestimated: Posner, RA; *Catastrophe: Risk and Response* (Oxford University Press, New York, 2004) i, 92, 118.

[325] See http://www.mi5.gov.uk/output/Page76.html.

[326] Http://www.niscc.gov.uk/niscc/index-en.html.

[327] 1999 SI No 743 as amended by 2005 SI No 1088. See also Control of Major Accident Hazards Regulations (Northern Ireland) 200 SR No 93, as amended by 2005 SR No 305;

Directives[328] and also respond to the explosion at a chemical plant at Flixborough in 1974.[329] Their aim is to prevent major accidents involving dangerous substances (mainly chemicals) and limit the consequences of any mishaps at around 1,200 sites.[330] The Regulations require safety planning and measures, with notification to local authorities and the adjacent inhabitants. The Joint Committee on the Draft Civil Contingencies Bill recommended that Category 2 should include all operators of establishments subject to the COMAH Regulations.[331] The Government decided that integration was best achieved on a voluntary basis but that COMAH duties should expressly over-ride those in the 2004 Act.[332] Thus, the Civil Contingencies Act 2004 (Con-tingency Planning) Regulations 2005, reg 12 provides that Category 1 responders need not perform a duty under s 2(1) in relation to any emergency which is: a major accident, within the meaning of reg 2(1) of the Control of Major Accident Hazards Regulations 1999 (or its Northern Ireland equival-ent),[333] a major accident, within the meaning of reg 2(1) of the Pipelines Safety Regulations 1996 (or its Northern Ireland equivalent);[334] or a radiation emergency, within the meaning of reg 2 of the Radiation (Emergency Pre-paredness and Public Information) Regulations 2001 (or its Northern Ireland equivalent).[335] Alongside COMAH, there are voluntary support arrangements such as the National Arrangements for Incidents involving Radioactivity (NAIR),[336] RADSAFE,[337] and CHEMSAFE.[338]

4.134 Environmental threats more generally are within the remit of the Environment Agency. Section 40 of the Environment Act 1995 allows the Minister to give

---

Dangerous Substances (Notification and Marking of Sites) Regulations 1990 SI No 304. For commentaries, see Health and Safety Executive, *Emergency Planning for Major Accidents—Control of Major Accident Hazards Regulations* (HSE Books, Sudbury, 1999); Health and Safety *Executive, Guide to the Control of Major Accident Hazards Regulations 1999* (HSE Books, Sudbury, 1999).

[328] Dirs 82/501/EEC, 87/216/EEC, 88/610/EEC, 96/82/EC, 2003/105/EC (http://europa. eu.int/comm/environment/seveso/index.htm).

[329] See Health and Safety Executive, *The Flixborough Disaste: Report of the Court of Inquiry* (HMSO, London, 1975). [330] See also Planning (Hazardous Substances) Act 1990.

[331] (2002–03 HC 1074, HL 184) para 131.

[332] Cabinet Office, *The Government's Response to the Report of the Joint Committee*, para 27.

[333] 1999 SI No 743, amended by the Greater London Authority Act 1999, s 328 and 2002 SI No 2469; Control of Major Accident Hazards Regulations (Northern Ireland) 2000 SI No 93.

[334] 1996 SI No 825; amended by the Greater London Authority Act 1999, s 328; Pipe-lines Safety Regulations (Northern Ireland) 1997 SI No 193.

[335] 2001 SI No 2975; amended by 2002 SI Nos 2099 and 2469; Radiation (Emer-gency Preparedness and Public Information) Regulations (Northern Ireland) 2001 SI No 436.

[336] See http://www.hpa.org.uk/radiation/understand/radiation_topics/radiation_incidents/ nair.htm. See further McColl, NP and Kruse, P, *NAIR Technical Handbook* (NRPB, Didcot, 2002). On 1 April 2005, the National Radiological Protection Board, which coordinates the scheme, merged with the Health Protection Agency forming its Radiation Protection Division.

[337] See http://www.radsafe.org.uk/. [338] See http://www.the-ncec.com/chemsafe/.

directions to the Agency in an emergency. More specific measures include the Hazardous Waste (England and Wales) Regulations 2005,[339] reg 62 of which allows, in the event of an emergency or grave danger, a holder of hazardous waste to take all lawful and reasonable steps to mitigate the emergency or grave danger, while reg 66 excuses breaches of regulations in pursuance of the recovery actions.[340] Local authorities also have broad powers of enforcement under ss 79 and 80 of the Environmental Protection Act 1990, while, under ss 108 and 109 of the Environment Act 1995, enforcing authorities can cause investigations and remedial action to be carried out where there is an immediate risk of serious pollution of the environment or serious harm to human health, or where circumstances exist which are likely to endanger life or health.

More specific environmental emergencies are associated with water, especially **4.135** drought and flooding. Under the Security and Emergency Measures Direction 1998 (currently under revision), issued under s 208 of the Water Industry Act 1991, water companies are required to have plans in place to deal with any emergency and to ensure the provision of essential water and sewerage services at all times.[341]

Droughts can be handled by drought permits or drought orders. Drought **4.136** permits are issued by the Environment Agency under the Water Resources Act 1991, s 79A, and allow abstraction from new sources or increased abstraction from existing sources. The more extensive 'ordinary' or 'emergency' drought orders in England and Wales[342] were initially made by the Secretary of State under s 1 of the Drought Act 1976.[343] The 1976 Act has been replaced, and orders now appear under the authority of s 73 of the Water Resources Act 1991. The Water Industry Act 1991 seeks to avert such orders by requiring the preparation and review of water resources management plans (s 37A, as amended by the Water Act 2003), while s 39A requires the preparation and review of drought plans.[344]

Despite its prominence as a source of human misery, the threat of flooding has **4.137** produced rather more powers than duties to take action, partly in order to avoid legal liability when flooding does occur.[345] First, there are around 200 Internal

---

[339] SI No 894.   [340] See also Environmental Protection Act 1990, s 33.

[341] See further Water Industry (Scotland) Act 2002 (Directions in the Interests of National Security) Order 2002 SI No 1264.

[342] For numbers issued, see http://www.environment-agency.gov.uk/commondata/103608/drought_orders_603418.txt.

[343] See Bonner, D, *Emergency Powers in Peacetime* (Sweet & Maxwell, London, 1985) 217.

[344] See Drought Plan Regulations 2005 SI No 1905.

[345] See Howarth, W, *Flood Defence Law* (Sage, Crayford, 2002); *Ship Chandlers Limited and others v Norfolk Line Limited and others* (1990, unreported, QBD).

Drainage Boards (IDBs) in areas of special drainage need (local authorities act elsewhere). Section 14 of the Land Drainage Act 1991 allows for works to be undertaken 'so far as may be necessary for the purpose of preventing flooding or mitigating any damage caused by flooding in their area'. Wider area 'flood defence' is overseen by the Environment Agency.[346] It has explained its overall approach in its Strategy for Flood Risk Management 2003–2008.[347] The Agency must arrange under s 106 of the Water Resources Act 1991 for its flood defence functions to be carried out by Regional Flood Defence Committees acting under s 14 of the Environment Act 1995. The Coastal Protection Act 1949 covers marine erosion and encroachment. The strategy envisages that some land must be sacrificed to the sea.[348] There are 88 maritime district councils which have powers to carry out works under this legislation. The Government has further encouraged the formation of voluntary coastal defence groups, which develop Shoreline Management Plans. More specific water risks have been tackled, for example, by the Thames Barrier, authorized by the Thames Barrier and Flood Protection Act 1972, completed in 1982 and operated by the Environment Agency. Large raised reservoirs are regulated by the Reservoirs Act 1975: by s 12A, flood plans may be demanded by the Secretary of State.

4.138 Moving from water to air, the overall reduction strategy for atmospheric pollution, based on technical ability rather than proof of harm, has been set by DEFRA in its paper, *The Air Quality Strategy for England, Scotland, Wales and Northern Ireland*.[349] Complex air pollution schemes were set up under Part I of the Environmental Protection Act 1990,[350] which are being succeeded by the requirements of the European Community Integrated Pollution Prevention and Control Directive (96/91) and the Pollution Prevention and Control Act 1999.[351]

---

[346] Environment Act 1995, s 6(4); Water Resources Act 1991, s 105(1). Its powers to carry out works are in the Water Resources Act 1991, ss 165, 166.

[347] Http://www.ukresilience.info/frm_strategy.pdf, 2003.

[348] See Ministry of Agriculture, *Fisheries and Food, and Welsh Office, Strategy for Flood and Coastal Defence in England and Wales* (1993); DEFRA, *DEFRA Flood Management Division, National Appraisal of Assets at Risk of Flooding and Coastal Erosion in England and Wales* (http://www.defra.gov.uk/environ/fcd/policy/NAAR1101.pdf, 2001), *National Assessment of Defence Needs and Costs for Flood and Coastal Erosion Management* (http://www.defra.gov.uk/environ/fcd/policy/nadnac0604.pdf, 2004), *Making Space for Water* (http://www.defra.gov.uk/corporate/consult/waterspace/consultation.pdf, 2004).

[349] (Cm 4548, London, 2000), as implemented by the Air Quality (England) Regulations 2000 SI No 928. A strategy is required by the Environment Act 1995, s 80. See also *Climate Change: The UK Programme* (Cm 4748, London, 2000); National Audit Office, *Policy Development: Improving Air Quality* (2001–02 HC 232).

[350] See Hughes, D, Parpworth, N, and Upson, J, *Air Pollution* (Jordans, Bristol, 1998). Note also Road Vehicles (Construction and Use) Regulations 1986 SI No 1078 (as amended); Clean Air Act 1993.

[351] See Pollution Prevention and Control (England and Wales) Regulations 2000 SI No 1973; Farthing, J, Marshall, B, and Kellett, P, *Pollution, Prevention and Control* (LexisNexis, London, 2003).

Monitoring air pollution involves over 1,500 sites, with details at the Air Quality Archive website.[352]

Another area of planning concerns the power and fuel industries, with very **4.139** extensive requirements as to reserve capacity and resilience.[353] Starting with fuel supplies, resilience is principally dealt with by the Energy Act 1976,[354] which is designed to provide responses to energy crises once an Order in Council is passed under s 3, on the basis of (a) the implementation of international obligations or (b) an actual or threatened emergency affecting fuel or electricity supplies. The only instance of invocation occurred during the September 2000 fuel crisis.[355] Any such Order in Council must be laid before Parliament after it is made. Each lasts for 28 days, and can be renewed subject to ceasing to be in force at the end of the 12 months unless there is then a resolution by both Houses of Parliament. Sections 1 to 3 deal with a crisis by regulation or prohibition, but s 6 deals with resilience by allowing directions as to stock levels which will accord with the directions given by the International Energy Agency[356] and by the European Communities,[357] both of which call for a 90-day oil stockpile.[358] By s 172 of the Energy Act 2004, the Secretary of State must annually publish a report on both the short-term and the long-term availability of electricity and gas for meeting the reasonable demands of consumers. Non-statutory supplements are the Department of Trade and Industry's Downstream Oil Emergency Response Plan[359] and a Memorandum of Understanding dated 29 September 2000[360] for the normal supply of oil fuels.

As for power utililites, a key role is played by the Office of the Gas and Elec- **4.140** tricity Markets (Ofgem),[361] which develops the Electricity Supply Emergency

---

[352] See http://www.airquality.co.uk. See the Air Quality Framework Directive (96/62/EC, and also Dirs 1999/30/EC, 2000/69/EC, 2002/3/EC), as implemented by the Air Quality Limit Values Regulations 2003 SI No 2121.

[353] But recent incidents have included a double transformer failure resulting in the loss of power in London on 28 August 2003: London Assembly, *The Power Cut in London on the 28 August 2003* (http://www.londonprepared.gov.uk/powerfailure.htm); Joint Energy Security of Supply Working Group, *International Blackouts in August & September 2003* (http://www.dti. gov.uk/energy/jess/blackout_note.pdf).

[354] See Bonner, D, *Emergency Powers in Peacetime* (Sweet & Maxwell, London, 1985) 220.

[355] Energy Act 1976 (Reserve Powers) Order 2000 SI No 2449.

[356] International Energy Programme Agreement (http://www.iea.org/Textbase/about/iep.pdf, 1974).

[357] See European Council Dirs 68/414/EEC, 98/93/EC, 73/238/EEC and Council Decision 68/416/EEC, 77/706/EEC.

[358] See Petroleum Stocks Order 1976 SI No 2162; Petroleum Stocks (Amendment) Order 1982 SI No 968; Petroleum Stocks (Amendment) Order 1983 SI No 909. There is no equivalent for gas supplies.   [359] I http://www.og.dti.gov.uk/downstream/emergencies/down_emerge.htm.

[360] See Director General of Fair Trading on the Memorandum of Understanding regarding the supply of oil fuels in an emergency [2002] UKCLR 74.

[361] See http://www.ofgem.gov.uk/ofgem/index.jsp. See the Gas Act 1986 and the Electricity Act 1989, as amended by the Utilities Act 2000. For Northern Ireland, the Office for the

Code[362] and the Gas Supply Emergency Arrangements. More detailed regulations exist under the Gas Act 1986 and the Electricity Act 1989 and via the licenses issued under them. National Grid UK,[363] as the managers of the National Gas Transmission System, has responsibility for the safe transportation and handling of gas within the grid under the Gas Safety (Management) Regulations[364] and is the National Emergency Co-ordinator for Gas. The Gas Industry Emergency Committee was set up in 2001 and was recast as the Gas and Electricity Industry Emergency Committee in 2003.[365] Section 96 of the Electricity Act 1989 grants to the Secretary of State wide powers to mitigate the effect of civil emergency. National Grid UK is again the national emergency coordinator for electricity and has developed a contingency planning code.[366]

4.141 Commensurate to the special risks in the sector, a different regime applies to nuclear power and materials. The Department of Trade and Industry coordinates policy at national level for response to an emergency with off-site effects from a licensed civil nuclear site in England and Wales. Facilities are audited by the DTI's Nuclear Installations Inspectorate.[367] The DTI also chairs the Nuclear Emergency Planning Liaison Group (NEPLG), established in 1990,[368] and produces Civil Nuclear Guidance on planning, testing, response, and recovery.[369]

4.142 The lead role of the DTI is subject to some exceptions. The Radioactive Substances Act 1993 is the principal concern of DEFRA's Radioactive Substances Division, with implementation by the Government Decontamination Service and the Environment Agency.[370] For a radiological accident overseas which may have impact on the United Kingdom (as was the case with Chernobyl in 1986), DEFRA takes the lead. Its National Response Plan, published in 1988,[371] deals with the consequences of overseas nuclear accidents, the core of which is the Radioactive Incident Monitoring Network (RIMNET).[372]

Regulation of Electricity and Gas is within the Northern Ireland Authority for Energy Regulation which was established by the Energy (Northern Ireland) Order 2003 SI No 419 (NI 6).

[362] See Electricity Act 1989, s 96, http://www.dti.gov.uk/energy/leg_and_reg/acts/elecsupply_emer_code.pdf.   [363] See http://www.nationalgrid.com/uk/.

[364] 1996 SI No 551.

[365] HC Debs vol 337 col 109w 17 Dec 2001, http://www.gisg.org.uk/geiec/index.asp.

[366] Operating Code No 9 (http://www.nationalgridinfo.co.uk/grid_code/pdfs/OC9%20i3r5.pdf, 2005).

[367] See also Health and Safety Executive, *Arrangements for Responding to Nuclear Emergencies* (HSE Books, Sudbury, 1994).

[368] See http://www.dti.gov.uk/energy/nuclear/safety/emergency.shtml.

[369] See http://www.dti.gov.uk/energy/nuclear/safety/neplg_guide.shtml, 2004.

[370] See http://www.defra.gov.uk/environment/radioactivity/discharge/rsact/. The Act implements Council Dir 80/836/EURATOM, as amended by Dirs 84/467/EURATOM and 96/29/EURATOM, referred to as the Basic Safety Standards (BSS) Directive. See further Radioactive Substances (Basic Safety Standards) (England and Wales) Direction 2000 (under the Environment Act 1995 s 40(2)).

[371] See http://www.defra.gov.uk/environment/radioactivity/response/index.htm.

[372] See http://www.defra.gov.uk/environment/radioactivity/response/rimnet.htm.

Finally, a wider range of departments, alongside the DTI, take a role in recovery and decontamination processes, including the Home Office,[373] DEFRA,[374] and the Office of the Deputy Prime Minister.[375] The Health and Safety Executive also issues guidance on Biological/Chemical Threats by Post.[376] The transportation of nuclear defence materials is a matter for the Ministry of Defence.[377] Emergency preparedness for radiological emergencies more generally is the subject of the Radiation (Emergency Preparedness and Public Information) Regulations (REPPIR) 2001, which is reminiscent of COMAH.[378] The lead regulator is the Environment Agency, which is given a broad role under s 4 of the Environment Act 1995.

The transport sector is of relevance to contingency planners both as a source of **4.143** threat from catastrophic accident and also because of the reliance upon the sector for the viability of other essential services. Probably the most prominent recent emergencies have afflicted the railways. As a result, the independent Office of Rail Regulation (ORR) is established under the Railways and Transport Safety Act 2003. Following the White Paper, *The Future of Rail*,[379] safety regulation is being transferred to the ORR by the Railways Act 2005 in order to simplify the regulatory structure and to enable the development of expertise. The Rail Accident Investigation Branch[380] was established by the Railways and Transport Safety Act 2003.[381] As with the Air and Marine Accident Investigation Branches, the Rail Accident Investigation Branch forms part of the Department for Transport. A Train Operating Companies Emergency Planning Forum has been established nationally, which brings together planners from all the train operating companies. Overseeing the system is the

---

[373] *The Decontamination of People Exposed to Chemical, Biological, Radiological, or Nuclear (CBRN) Substances or Material: Strategic National Guidance* (http://www.ukresilience.info/cbrn/cbrn_guidance.htm, 2004).

[374] *The Decontamination of the Open Environment Exposed to Chemical, Biological, Radiological or Nuclear (CBRN) Substances or Material: Strategic National Guidance* (http://www.defra.gov.uk/environment/risk/cbrn/, 2004).

[375] *Precautions to Minimise Effects of a Chemical, Biological, Radiological or Nuclear Event on Buildings and Infrastructure* (http://www.ukresilience.info/cbrn/precautions.pdf, 2004); *The Decontamination of Buildings and Infrastructure Exposed to Chemical, Biological, Radiological or Nuclear (CBRN) Substances or Material: Strategic National Guidance* (http://www.ukresilience.info/cbrn/buildings.pdf, 2004).

[376] See http://www.ukresilience.info/package.htm, 2001.

[377] See *Defence Nuclear Materials Transport Contingency Arrangements* (4th ed, http://www.mod.uk/linked_files/laesi_ver4.pdf, 2004).

[378] 2001 SI No 2975, implementing Dir 96/29/EURATOM. See also Health and Safety Executive, *Guide to the Radiation (Emergency Preparedness and Public Information Regulations 2001* (HSE Books, Sudbury, 2001); Mobbs, S *et al*, *UK Recovery Handbook for Radiation Incidents* (Health Protection Agency, 2005); http://www.hse.gov.uk/radiation/ionising/reppir.htm.

[379] (Cm 6233, London, 2004) para 3.3.5.

[380] See http://www.raib.gov.uk/home/index.cfm; Railways (Accident Investigation and Reporting) Regulations 2005 SI No 1992.   [381] See also Railway Safety Dir 2004/49/EC.

Rail Safety and Standards Board which was convened by major industry stakeholders in April 2003.[382] The Board is to facilitate an improving health and safety performance of the railways in Great Britain, and, to that end, the Railway Strategic Safety Plan 2005 analyses the current risk profile on the railway and identifies priorities for action.[383] The Secretary of State retains powers under s 118 of the Railways Act 1993 to issue directions 'in time of hostilities, whether actual or imminent, severe international tension or great national emergency'. Directions may also be given under s 119 to demand protection against acts of violence.

4.144 In the civil aviation sector, the Secretary of State may give directions under the Airports Act 1986 to the operators of airports in the interests of national security.[384] Under the Transport Act 2000, s 93, the Secretary of State may give directions in any time of actual or imminent hostilities or of severe international tension or of great national emergency; which may require participation in planning or the direction of functions or the use of assets or services. In the same circumstances, s 94 allows the issuance of order for the possession of aerodromes, aircraft, and other assets. Inspections and monitoring are undertaken by the Transport Security Directorate (TRANSEC) in the Department of Transport.[385] Following a review by Sir John Wheeler in 2002,[386] the Department for Transport and Home Office have pursued a joint programme, MATRA (Multi Agency Threat and Risk Assessment), which encourages greater joint working between all security stakeholders to produce threat assessments and management plans.[387] Further discussion takes place in the private sector United Kingdom Flight Safety Committee[388] and the United Kingdom Airlines Emergency Planning Group.[389] The British Airways Emergency Procedures Information Centre at Heathrow acts as the central airline information coordinating point following an incident instead of the usual Police Casualty Bureau arrangements.[390]

---

[382] See Lord Cullen, *The Ladbroke Grove Rail Inquiry* (HSE Books, Norwich, 2001) Pt II paras 8.38, 9.46, 10.7; Health and Safety Executive, *Hatfield Derailment Investigation* (http://www.hse .gov.uk/railways/hatfield.htm, London, 2002).   [383] See http://www.rssb.co.uk/ssr_rgsp.asp.
[384] DfT/ODPM Circular 1/2003 provides advice to local planning authorities in England and Wales regarding the safeguarding of aerodromes, technical sites and military explosives storage areas.
[385] The compliance regime is set out at http://www.dft.gov.uk/stellent/groups/dft_transsec/ documents/page/dft_transsec_030407.hcsp
[386] A review of 'Airport Security' was undertaken by Sir John Wheeler in 2002 (http:// www.dft.gov.uk/stellent/groups/dft_transsec/documents/page/dft_transsec_503590.pdf).
[387] See   http://www.dft.gov.uk/stellent/groups/dft_transsec/documents/page/dft_transsec _031313.hcsp.   [388] See http://www.ukfsc.co.uk/.
[389] See http://www.ukaepg.org.
[390] See also the independent Airsafe site: http://www.airsafe.com/. For further discussion, see Ch 8.

Road transport of dangerous goods is overseen by the Department for Transport's **4.145**
Transport Security team, including implementation of the Carriage of Dangerous
Goods and use of Transportable Pressure Equipment Regulations 2004.[391]

Rather like transport, communications are a vital facility in an emergency. As **4.146**
mentioned earlier in this chapter, the BBC has set up contacts between all BBC
local radio stations and local emergency planners to exchange information and
to offer local radio as a channel for dissemination.[392] If negotiation should falter
at any point, then, under paragraph 8 of the Agreement Dated the 25th Day of
January 1996 Between Her Majesty's Secretary of State for National Heritage
and the British Broadcasting Corporation, the BBC can be required to
broadcast any announcement when 'an emergency has arisen or continues' and
can also be required to refrain from broadcasting any matter. Likewise, by s 132
of the Communications Act 2003, the Secretary of State can require suspension
or restriction of a commercial provider's entitlement to broadcast '(a) to protect
the public from any threat to public safety or public health, or (b) in the
interests of national security'. In addition, by s 5, the Secretary of State can give
directions to OFCOM based on the interests of national security or in the
interests of the safety of the public or of public health.

As well as broadcasters, the telecommunications sector can be subject to **4.147**
directions by s 94 of the Telecommunications Act 1984 and are also subject
to licensing requirements to plan for emergencies.[393] There is a National
Emergency Plan for the UK Telecommunications Sector,[394] including a Telecommunications Industry Emergency Planning Forum which is chaired by
Ofcom.[395] The Media Emergency Forum examined communications issues
arising from the experience of September 11, 2001 and considered how the
United Kingdom could cope in those circumstances.[396]

In an emergency, telecommunications can become overloaded either through **4.148**
damage and disruption or simply because of the amount of traffic created by the
crisis. The security forces may also wish to disable the system which might be
used for hostile communications or even to trigger explosives. Consequently,
there is a need to ensure resilience and also suitable priority for the services and

---

[391] SI No 568 as amended by 2005 SI No 1732. See further Council (RID) Dir 96/49/EC;
Council (ADR) Dir 94/55/EC; Commission Dirs 2004/89/EC, 2004/110/EC, 2004/111/EC.
See also Department for Transport, *Guidance for the Security of Dangerous Goods by Road* (http://
www.dft.gov.uk/stellent/groups/dft_transsec/documents/page/dft_ transsec_038557.pdf, London, 2005).
[392] See Connecting in a Crisis, http://www.bbc.co.uk/connectinginacrisis/index.shtml.
[393] General Conditions of Entitlement (Condition 5) under the Communications Act
2003, s 45.
[394] See http://www.ofcom.org.uk/static/archive/oftel/ind_groups/emer_plan/index.htm, 2003.
[395] Cabinet Office, Emergency *Response and Recovery*, para 4.98.
[396] See http://www.ukresilience.info/mef/mefreport.htm.

authorities that are responding to the emergency.[397] The GTPS (Government Telephone Preference Scheme) deals with landlines.[398] ACCOLC (ACCess OverLoad Control Scheme) is the procedure for regulating mobile telecommunication in the event of emergencies.[399] Beyond the public networks there lies the Emergency Communications Network (ECN), which is a private switched telephone network, managed and funded by the Cabinet Office through its Central Sponsor for Information Assurance[400] and open to any 'essential service' dealing with an emergency situation.[401] There are plans afoot to migrate the ECN to a new Contingency Telecommunications Provision (CTP), which incorporates both fixed and mobile access.[402] Concerns for the security and robustness of communications within and between emergency providers have also encouraged the development in Britain, catching up with a system already in existence in Northern Ireland, of a system called Airwave.[403] The system began with the police, pursuant to a contract signed in 2000. Ambulance and fire services are also being sold the system. It is also used by the Ministry of Defence, but is not generally available within local authorities.

4.149 The resilience of the Internet became a prominent consideration at the time of Y2K, the Millennium Bug, which prompted a successful contingency planning exercise.[404] But threats continue to arise and are monitored by the National Hi-Tech Crime Unit (NHTCU), launched in April 2001.[405] The enhanced perception of the vulnerabilities of networks to terrorist attack has also resulted in the establishment within the Cabinet Office of a Central Sponsor for Information Assurance and Resilience.[406] The unit promotes information assurance, and works with the Communications-Electronics Security Group (CESG), the national technical authority within GCHQ for information assurance.[407]

---

[397] See Cabinet Office, *Dealing with Disaster* (3rd ed, London, 2003) Ch 3.
[398] See for uses HC Debs vol 300 col 651w 13 November 1997, George Howarth and vol 427 col 112w 27 April 1999, George Howarth.
[399] See Cabinet Office, *Dealing with Disasters* (3rd ed, London, 2003) paras 3.70, 3.71. For its use during the London bombings of 7 July 2005, see *The Guardian* 2 December 2005 at 11.
[400] See http://www.cabinetoffice.gov.uk/csia; Cabinet Office, *Emergency Response and Recovery*, para 4.94.
[401] Joint Committee on the Draft Civil Contingencies Bill, Appendix 9 q 81.
[402] Cabinet Office, *Emergency Response and Recovery*, para 4.94.
[403] See http://www.airwaveservice.co.uk/; http://www.pito.org.uk/what_we_do/communications/airwave.htm.
[404] There was a special Government website: http://www.millennium-centre.gov.uk/ and a committee, Action 2000. See Cabinet Office, *Modernising Government in Action: Realising the Benefits of Y2K* (Cm 4703, London, 2000).   [405] See http://www.nhtcu.org/.
[406] See http://www.cabinetoffice.gov.uk/csia/. See Defence Committee, *Defence and Security in the UK* (2001–02 HC 518) para 125.   [407] See http://www.cesg.gov.uk/.

As for communications by post, the Secretary of State may give such directions **4.150** under the Postal Services Act 2000, s 101 to the Postal Services Commission in the interests of national security.

Next, both human and animal health have prompted substantial civil **4.151** emergencies in the recent past. As for the former, the National Health Service strategic statement is the NHS Emergency Planning Guidance 2005.[408] For each NHS organization, the Chief Executive Officer will be responsible for ensuring that their organization has a Major Incident Plan. The Department of Health has also promulgated a national operational doctrine, 'Planning for Major Incidents: the NHS Guidance'.[409] It calls for mutual aid across organizational boundaries, coordinated by the Strategic Health Authority (SHA) or the Primary Care Trust leading the response, plus central support from the Department of Health and the Health Protection Agency. Regional Public Health Groups led by Regional Directors of Public Health will ensure a 24-hour capability to support the SHAs. The Health Protection Authority (HPA) Local and Regional Services Division local health protection teams will be involved in the development of local emergency plans and also provides a service through its Emergency Response Division. Its organization also includes a Centre for Infections, a Centre for Emergency Preparedness and Response, and a Centre for Radiation, Chemical and Environmental Hazards. Plans to deal with more specific crises are also being developed, including heatwaves, smallpox, and West Nile virus.[410]

Turning to animal health, following the February 2001 outbreak of Foot and **4.152** Mouth Disease, a number of inquiries were held.[411] The Government response in 2002[412] recognized the need for a stronger framework for emergency preparedness (especially at regional level), as well as revisions to animal health legislation. Subsequent contingency planning within the Department for Environment, Food and Rural Affairs has included lengthy documents about the Exotic Animal Disease, Generic Contingency Plan (covering Foot and

---

[408] Department of Health Emergency Preparedness Division, *NHS Emergency Planning Guidance 2005* (Department of Health, http://www.dh.gov.uk/assetRoot/04/12/12/36/ 04121236.pdf, London, 2005). See also Scottish Executive Health Department, *National Health Service in Scotland Manual of Guidance: Responding to Emergencies* (http://www.show.scot.nhs.uk/ emergencyplanning/guidance.htm, Edinburgh, 2004). [409] HSC 1998/197.

[410] *Heatwave: Plan for England—Protecting Health and Reducing Harm from Extreme Heat and Heatwaves* (2005); *Smallpox Mass Vaccination: An Operational Planning Framework* (2005); *West Nile Virus* (2004).

[411] See *Foot and Mouth Disease 2001: The Lessons to be Learned Inquiry by Dr Iain Anderson* (2001–02 HC 888); Royal Society, *Infectious Diseases in Livestock* (Policy Document 19/02, London); National Audit Office, *The 2001 Outbreak of Foot and Mouth Disease* (2001–02 HC 939).

[412] *Response to the Reports of the Foot and Mouth Disease Inquiries* (Cm 5637, London, 2002).

Mouth Disease, Avian Influenza, Newcastle Disease, and Classical Swine Fever).[413] There is also a Rabies Contingency Plan under revision.[414] As for legislation, the Animal Health Act 2002 quickly appeared. Consequently, there is now a substantial code of measures, consolidated within the Animal Health Act 1981, to deal with biosecurity.[415]

4.153 A more insidious crisis concerned the emergence of BSE (bovine spongiform encephalopathy—'Mad cow disease') and new variant vCJD (Creutzfeldt-Jakob disease), which were the subject of the BSE Inquiry chaired by Lord Phillips.[416] The *BSE Report* unearthed difficulties in handling risk on a precautionary basis, poor communications between different departments, and potentials for conflicts of interest arising from the Agriculture Ministry's dual responsibility for protecting public health and for sponsoring agriculture and the food industry. The latter contributed to the establishment of an independent Food Standards Agency under the Food Standards Act 1999.[417]

4.154 Food emergencies are otherwise the province of the Food and Environment Protection Act 1985, Part I,[418] which authorizes the making of emergency orders to prevent the consumption of food rendered unsuitable for human consumption in consequence of an escape of substances. The Act was first invoked to deal with the radiological fall-out from the Chernobyl nuclear plant in 1986, which led to restrictions on the slaughter and movement of sheep from specified areas.[419] Other orders, too numerous to specify fully, have related to amnesic, diarrhetic, and paralytic shellfish poisoning, dioxins, oil and chemical pollution of fish and plants, metal poisoning, and also fall-out from the Dounreay Nuclear Establishment in 1997.[420] There are also powers by ss 13(1) and 48(1) of the Food Safety Act

---

[413] See http://www.defra.gov.uk/animalh/diseases/notifiable/disease/newcastle/pdf/genericcp-final.pdf; Avian Influenza and Newcastle Disease (Contingency Planning) (England) Order 2003 SI No 2036.

[414] See http://www.defra.gov.uk/animalh/diseases/control/contingency/rabies_contingency. pdf.

[415] By contrast, in the *Interim Response to the BSE Inquiry* (DEFRA, London, 2001) para 8.4, it is stated that 'The Government does not believe there are any serious gaps in its powers to take proportionate emergency action against hazards to human or animal health in relation to animals and animal products and food.'

[416] BSE Inquiry, *Report* (1999–00 HC 887) and Government Response to the report on the BSE inquiry (Cm 5263, London, 2001). See Granot, H, 'Facing Catastrophe' (1999) 17.2 *International Journal of Mass Emergencies and Disasters* 161.

[417] See *The Food Standards Agency: A Force for Change* (Cm 3830, London, 1997).

[418] See Hawke, N, 'Contemporary Issues in Environmental Policy Implementation' [1987] *Journal of Planning & Environment Law* 241.

[419] See 1986 SI Nos 1027, 1029; Reid, CT, 'Food and Fall-out' 1986 *Scottish Law Times* 261. See further the regulations still existing in Scotland under Council Reg (EURATOM) Nos 3954/ 87, 944/89, 2218/89, 2219/89; Council Reg No 770/90, Dirs 80/836, 84/467, 96/29 (EURATOM); Council Reg (EEC) 737/90, 616/2000, 727/97.

[420] See *The Guardian* 8 December 1997 at 6.

1990[421] and under European Communities law[422] to place bans or restrictions in the interests of food safety.

Reflecting its importance to the national economy, special attention has been **4.155** accorded to UK Financial Sector Continuity.[423] The Treasury Green Paper on the Financial System and Major Operation Disruption, published in 2003,[424] questioned whether new statutory powers (such as to suspend financial obligations;[425] to direct financial markets;[426] to prohibit financial transactions;[427] to declare a bank holiday;[428] and to waive statutory requirements during a crisis) were needed. The Green Paper further emphasized that the Government concluded that any statutory powers (including presumably the Civil Contingencies legislation) should never be used in a purely financial crisis and would only be used in extreme circumstances and with support from the private sector.[429] As befits the cradle of capitalism, the Task Force on Major Operational Disruption in the Financial System,[430] convened in response to the Green Paper, concluded forcefully against statutory regulation. It reiterated that it could not readily identify ways in which the Civil Contingencies legislation might be used constructively to deal with the legal issues surrounding major operational disruption and argued that its powers should not be applied in this sector,[431] preferring non-binding recommendations as to practice. The Government has accepted this approach.[432] Progress on the recommendations has since been reported by the Tripartite Standing Committee on Financial Stability and includes the Financial Services Authority's Resilience Benchmarking Project and the establishment of the Cross Market Business Continuity Group.[433]

Contingency planning must extend offshore, given the possibilities of disaster **4.156** associated with the shipping and the oil and gas industries.[434] The impact of marine pollution is generally within the remit of the Environment Agency. It takes the lead in responding to pollution arising from land-based activities. A specific crisis arising from an incident at sea is the responsibility of the Maritime

---

[421] See Howells, GG, and Bradgate, JR, 'Food Safety—An Appraisal of the New Law' [1991] *Journal of Business Law* 320.　　[422] See the orders under the Council Dir 93/43/EEC.

[423] See its special website at http://www.fsc.gov.uk/.　　[424] Cm 5751.

[425] See Financial Services and Markets Act 2000, s 296.

[426] See Financial Services and Markets Act 2000, ss 42, 43, 138, 157, 410.

[427] See Terrorism (United Nations Measures) Order 2001 SI No 3365, Art 4; Anti-Terrorism, Crime and Security Act 2001, s 4.

[428] See Banking and Financial Dealings Act 1971, s 1.　　[429] Cm 5751, para 4.16.

[430] See http://www.fsc.gov.uk/upload/public/attachments/5/tfreportwholereport.pdf.

[431] Ibid, paras 3.39, Annex 4 para 5.7.　　[432] Ibid, Foreword.

[433] Financial Sector Business Continuity *Annual Report 2005* (http://www.fsc.gov.uk/upload/public/Files/5/64791_1.pdfannualrep.pdf).

[434] The most notable example was the Piper Alpha disaster. See Lord Cullen, *Public Inquiry into the Piper Alpha Disaster* (Cm 1310, London, 1990); Offshore Safety Act 1992.

and Coastguard Agency (MCA).[435] The MCA developed a National Contingency Plan in 2000 which recognizes that in the case of salvage activities ultimate control over all operations is the responsibility of a single designated Secretary of State's Representative for purposes of maritime salvage and intervention (SOSREP).[436] This office reflects Lord Donaldson's *Review of Salvage, Intervention and their Command and Control.*[437] Implementation of the Review has taken place through administrative arrangements, backed by legislation.[438] Subject to the powers invested in the SOSREP, the Dangerous Vessels Act 1985[439] allows a local harbour master to give directions to protect harbours, while the Offshore Safety Act 1992 extends the general purposes of the Health and Safety at Work etc Act 1974 to off-shore installations. Detailed regulations made under the former Mineral Workings (Offshore Installations) Act 1971 also remain in force. Finally, the International Convention on Oil Pollution Preparedness, Response and Cooperation 1990 (OPRC '90), as enacted into English law by the Merchant Shipping Act 1995, places a requirement on authorities and operators in charge of sea ports and oil handling facilities to prepare contingency plans. This survey demonstrates that there is substantial power to take action against off-shore civil contingencies. Indeed, one wonders whether there are any lessons to be learnt from the office of SOSREP for non-marine emergencies and whether there might be an on-shore 'National Nominated Co-ordinator' to oversee regional and local officers and thus provide a clear join to the centre. The national figure could be a departmental appointment, and thereby there would be overall political accountability for the operation in a way which is less likely to be achieved by the emergence of a localized Gold Commander.

4.157   Finally, emergency services can become even more effective if they can aid each other. The provision of mutual aid between agencies, especially amongst the 'blue lamp' services—police, fire, and ambulance—has become increasingly the subject of formal arrangements.[440] A police National Reporting Centre

---

[435] See Maritime & Coastguard Agency, *Search and Rescue Framework for the United Kingdom of Great Britain and Northern Ireland* (Southampton, 2002) Pt II.

[436] See http://www.mcga.gov.uk/c4mca/mcga-dops_environmental/mcga-dops_cp_environmental-counter-pollution/mcga-dops_cp_sosrep_role/mcga-dops_cp_ncp_ + _uk_response_to_salvage/mcga-dops_cp_national_contingency_plan.htm.

[437] (Cmnd 4193, London, 1999). See Rose, F, *Kennedy and Rose on the Law of Salvage* (Sweet & Maxwell, London, 2001).

[438] Dangerous Vessels Act 1985, Merchant Shipping Act 1995, Merchant Shipping and Maritime Security Act 1997, Pollution Prevention and Control Act 1999, and Marine Safety Act 2003, Offshore Installations (Emergency Pollution Control) Regulations 2002 SI No 1861.

[439] See also the Dangerous Vessels (Northern Ireland) Order 1991 SI No 1219; Department for Transport, *A Guide to Good Practice on Port Marine Operations* (2000) Ch 5.

[440] See National Health Service Act 1978, s 22; (see also NHS (Scotland) Act 1977, s 13; Police Act 1996, s 24; Fire and Rescue Services Act 2004, s 13; Association of Train Operating Companies, *Joint Industry Provision of Customer Care following a Major Passenger Rail Accident.*

has operated during crises in 1972 and 1984 to smooth the operation of mutual aid.[441]

K. Operational Powers

The bulk of Part I concentrates upon planning rather than operations. Indeed, **4.158** some of the former executive powers under the Civil Defence Act 1948 are discontinued, most notably the compulsory acquisition of land for civil defence purposes. There is no equivalent in Part I of the 2004 Act since its focus on planning and preparations make it redundant.[442] Beyond the Act, probably the most wide-ranging powers in the hands of central government—to acquire, produce, process, or transport articles required for the public service—are set out in the Supply Powers Act 1975.[443]

As for local government, the most pertinent power is s 138 of the Local **4.159** Government Act 1972:

(1) Where an emergency or disaster involving destruction of or danger to life or property occurs or is imminent or there is reasonable ground for apprehending such an emergency or disaster, and a principal council are of opinion that it is likely to affect the whole or part of their area or all or some of its inhabitants, the council may
   (a) incur such expenditure as they consider necessary in taking action themselves (either alone or jointly with any other person or body and either in their area or elsewhere in or outside the United Kingdom) which is calculated to avert, alleviate or eradicate in their area or among its inhabitants the effects or potential effects of the event; and
   (b) make grants or loans to other persons or bodies on conditions determined by the council in respect of any such action taken by those persons or bodies.

(1A) If a principal council are of the opinion that it is appropriate to undertake contingency planning to deal with a possible emergency or disaster which, if it occurred,
   (a) would involve destruction of or danger to life or property, and
   (b) would be likely to affect the whole or part of their area,
   they may incur such expenditure as they consider necessary on that planning (whether relating to a specific kind of such possible emergency or disaster or generally in relation to possible emergencies or disasters falling within paragraphs (a) and (b) above).

---

[441] Home Office Circular 134/1973; Loveday, B, 'Central Coordination, Police Activities and the Miners' Strike (1986) 57 *Political Quarterly* 60.
[442] Joint Committee on the Draft Civil Contingencies Bill, Appendix 9 q 56.
[443] See Law Commission, *Supply Powers Bill* (Cmnd 5850, London, 1975).

(3) Nothing in this section authorises a local authority to execute
    (a) any drainage or other works in any part of a main river..., or
    (b) any works which local authorities have power to execute under...the Land Drainage Act 1991,

but subject to those limitations, the powers conferred by subsections (1) and (1A) above are in addition to, and not in derogation of, any power conferred on a local authority by or under any other enactment, including any enactment contained in this Act.

   ...

(5) With the consent of the Secretary of State, a metropolitan county fire and civil defence authority and the London Fire and Emergency Planning Authority may incur expenditure in co-ordinating planning by principal councils in connection with their functions under subsection (1) above.

(6) In this section 'contingency planning' means the making, keeping under review and revising of plans and the carrying out of training associated with the plans.[444]

Section 138 is further amended by the Civil Contingencies Act, Schedule 2, paragraph 7. It repeals subsection (1A), which is clearly overtaken by the broader provisions in the 2004 Act. Section 138 has been interpreted as allowing expenditure not only on actions when an emergency or disaster had occurred or (possibly) was imminent but also on precautionary measures taken well in advance; but no action for breach of statutory duty is sustained where the council does not exercise its powers and damage results to third parties.[445] Section 138 does not authorize a local authority to charge for these emergency services or to enter into a contract to provide them.[446]

L. Conclusions

4.160 This chapter has exposed a vast amount of work, undertaken within government, public agencies, and major private industries and services and aimed at planning for (and against) emergencies and instilling greater resilience in the event of their occurrence. The Cabinet Office and other departments are to be commended for energizing a fervour of activity. They are as yet less assured at ensuring good standards or harmonization in action, and the approaches to governance and the instilling of a civil contingency culture still have some way to go. This industriousness is in line with what one might expect within a society obsessed with risk, where 'Every public and private place is now assessed from a safety perspective...' and where 'Risk avoidance has become an

---

[444] See also Local Government (Scotland) Act 1973, s 84.
[445] *Ship Chandlers Limited and others v Norfolk Line Limited and others* (1990, unreported, QBD).
[446] *Anglian Water Services Ltd v Crawshaw Robbins & Co Ltd* [2001] BLR 173 at para 174.

important theme in political debate and social action.'[447] Assuming society is at risk from the host of real and imagined threats now the subject of the emergency planners, is Part I of the Act worthwhile or it simply a futile reminder of the enfeeblement of human agency in late modernity? A stern test for those hypotheses is awaited. The day of reckoning, an equivalent of Hurricane Katrina,[448] has yet to dawn. But it is submitted that much of the effort and expenditure is to be supported for two reasons. First, as a matter of principle, a rationalist response to risk is surely preferable to passivity or mistrust of human invention and intervention. The second is that one ardently hopes that some of the planning will pay off—that in response to flood or a terrorist attack, the authorities will handle the situation better than they would otherwise have managed. Certainly, it was suggested that the plans worked well during the London terrorist bombings of 7 July 2005.[449] At the same time, the aftermath of those events equally demonstrate the danger of losing confidence in one's resilience, which triggered the urge to pander to fears and pass ever more restrictive legislation in the name of security,[450] so that society becomes less at ease with itself. The lessons of history in Chapter 2 offer also an important reminder not to lose sight of what counts as 'victory'—and it is not a society without risk. Perhaps, the other lesson is the vast amount of work needed to instil resilience. Part I of the Civil Contingencies Act represents an important new phase in coming to terms with risk in society, but the project is far from completion.

---

[447] Furedi, F, *Culture of Fear: Risk-taking and the Morality of Low Expectation* (rev ed, Continuum, London, 2002) 1, 2. [448] See Ch 9.
[449] Statement from the London Resilience Forum: 'The response showed the benefits of the well-prepared and well practiced plans that were in place to help the Capital respond to such an incident.' (http://www.londonprepared.gov.uk/resilienceteam/, 14 July 2005).
[450] See Terrorism Act 2006.

5

EMERGENCY POWERS: PART II
OF THE ACT

A. The Making of Emergency Regulations

Part II confers a power on Her Majesty to make regulations if an 'emergency' **5.01**
has occurred or is about to occur. In this way, Part II moves the focus of the Act
towards the action phase and away from the planning phase under Part I. The
new laws undoubtedly advance the old Emergency Powers Acts by offering
'greater flexibility, proportionality, deployability and robustness' as well as
'definitions and procedures more suited to the nature of the possible risks and
threats that the UK may face in the twenty-first century'.[1]

The regulation-making powers are of awesome scope, but the Government **5.02**
claims that sensible limits are imposed by reference to the concept of a 'Triple
Lock'—that restraints will be imposed on the triggering definitions by reference
to seriousness, necessity, and geographical proportionality. The effectiveness of
the application of these concepts, and whether they are as 'transparent and
robust'[2] as claimed will be considered further in this chapter. Some shortcomings have already been indicated in Chapter 3, namely, that the Triple Lock

---

[1] Cabinet Office, *The Draft Civil Contingencies Bill: Explanatory Notes etc*, para 4.
[2] Cabinet Office, *The Government's Response to the Report of the Joint Committee*, para 3.

demand of 'seriousness' is not adequately reflected in s 19, and the various legs of the Triple Lock are not drawn together so that they are readily visible and emphasized. Instead, they are found principally in s 19 (seriousness) and ss 21 and 23 (necessity and geographic proportionality); the Government claimed that to consolidate would 'distort the drafting'.[3] Further general problems[4] are that there is no express requirement of objectivity in any of the tests—the Minister is allowed to use powers on the basis of satisfaction without the qualification of reasonableness. The condition of necessity is left unexplained, except in s 21(5) and (6) where emergency regulations overlap with 'normal' powers or, perhaps more significantly, with more focused powers which deal with crisis, such as terrorism legislation. Next, proportionality is not sufficiently explained aside from in geographical terms. The term is baldly stated when an emergency is declared (s 20(5)(b)) and when the regulations are issued (s 23(1)(b)), but there is no requirement when the regulations are applied.[5]

(a) Invocation of powers

5.03 The actual power to make emergencies regulations is set out in s 20(1). Emergency regulations may be made by the process of Order in Council, if the Government (usually in the shape of the Home Secretary, as the Cabinet member with responsibility for domestic security and resilience) is 'satisfied' the conditions in s 21 are met. This formula is thus framed subjectively. While review by the courts and Parliament is not thereby debarred, it is likely that a less rigorous standard of proof is demanded (discussed further in Chapter 6). The Government argued there was no difference: 'We straightforwardly concluded that reasonableness is an absolute expectation of the actions of Ministers and that to import the notion specifically might query whether Ministers act reasonably in other respects.'[6] If so, why not voice an unqualified expectation as to objectivity on the face of the Act in this and every future case?

5.04 The grand device of Order in Council (which has the status of statutory instrument: s 30(1)) represented a compromise between a more commonplace regulation-making power in the hands of a Minister and the even grander device of a Royal Proclamation (as appeared in the Emergency Powers Acts). The Minister of State for the Cabinet Office was at pains to emphasize that it did not betray any incipient republican tendency: 'That improved process was determined in discussion with the Palace and reflects the centrality of the Queen's role and the need for practical arrangements in such circumstances.'[7]

---

[3] Ibid.   [4] See Joint Committee on the Draft Civil Contingencies Bill, para 38.
[5] Compare: Anti-Terrorism, Crime and Security Act 2001, ss 17 and 19; Regulation of Investigatory Powers Act 2000, ss 5, 22, 23, 28, 29, 32, 49, 51, 55, and 73–5.
[6] HC Debs vol 416 col 1178 19 January 2004, Fiona Mactaggart.
[7] Ibid, col 1109, Douglas Alexander.

In an emergency, s 20(2) allows a 'senior' Minister of the Crown to issue emer- **5.05**
gency regulations by the process of Order in Council under subsection (1) provided
not only the conditions in s 21 are satisfied, but also that it would not be possible to
proceed under (1) without serious delay.[8] Imagined scenarios are said to include
'situations in which, perhaps because of the emergency at hand, the whereabouts of
the Queen are uncertain or communication is not possible.'[9] The definition of
'senior' Minister of the Crown means (a) the First Lord of the Treasury (the Prime
Minister), (b) any of Her Majesty's Principal Secretaries of State,[10] and (c) the
Commissioners of Her Majesty's Treasury[11] (s 21(3)). The latter are a curious
choice. While in ancient constitutional theory, the Treasury is governed by a Board
of seven Lords Commissioners, including the First Lord (the Prime Minister, who
is mentioned separately in the Act to ensure he can act alone)[12] and the Second
Lord (the Chancellor of the Exchequer), the remaining five Lords (the Junior
Lords) are mere Government Whips in the House of Commons. The Government
Chief Whip's formal title is the Parliamentary Secretary to the Treasury. But
Government Whips in the House of Commons would not usually be viewed as
'senior' members of Government and do not even sit in the Cabinet. Since they are
notable for their commitment to the Government rather than to Parliament:

> ... it beggars belief that a Government could say in the Bill that the people to
> make regulations shall be Whips. That is the apotheosis of a Whips Office dream,
> and a nightmare for the rest of us.[13]

The Government lamely blamed Parliamentary counsel for their inclusion as
reflecting 'common practice in describing the Treasury in legislation'.[14]

The definition of 'serious delay' in s 20(4) requires that the delay might cause **5.06**
serious damage or seriously obstruct the prevention, control, or mitigation of
serious damage. The test is to applied objectively—it is not enough that a senior
Minister has a reasonable suspicion or belief.

Next, 'Let us suppose that Al'Qaeda chose to support [the anniversary of 9/11], **5.07**
and succeeded, with a nice barge full of explosives on the Thames in the middle of
Prime Minister's Question Time.'[15] The resilience of the 'Crown' is vital to the
operation of Part II of the Civil Contingencies Act, for it affects not only the
making of Orders in Council (s 20) but also the meeting of Parliament (s 28) and

---

[8] An attempt to demand the consent of not less than four Privy Councillors from a standing
panel appointed by the Prime Minister was rejected: HL Debs vol 666 col 856 9 November 2004,
Lord Phillips.　　　　　　　　[9] Civil Contingencies Secretariat, *The Government's Response*, 34.
[10] By the Interpretation Act 1978, Sch 1, ' "Secretary of State" means one of Her Majesty's Principal
Secretaries of State.' Under-Secretaries are therefore not authorized, nor does the term include the
Prime Minister or the Chancellor of the Exchequer.
[11] Under the Treasury Instruments (Signature) Act 1849, action must be taken by at least two
Commissioners. The 2004 Act must comply as it does not specify that 'any' commissioner can act,
unlike under (b).　　　　　　[12] HL Debs vol 665 col 758 14 October 2004, Baroness Scotland.
[13] HC Debs vol 416 col 1147 19 January 2004, Richard Shepherd.
[14] Ibid, Douglas Alexander.　　　　[15] HL Debs vol 665 col 755 14 October 2004, Lord Lucas.

the appointment of Secretaries of State and Ministers of the Crown (*passim*). It follows that it is important to ensure that in any emergency, the existence of the Crown in person is assured and that the exercise of Crown powers remains feasible. There would appear to be two responses.[16]

5.08 The first is that some contingencies for its absence or failure are built into the 2004 Act, such as the power of a senior Minister to make regulation under s 20. If all Government ministers are wiped out, but not Her Majesty, it might be possible to revert to a s 20(1) process, for while, 'in the usual way, it is the senior Ministers in Her Majesty's current Government who make up the members of the Council and who act with her. *In extremis*, of course, failing there being any, ... it would always be open to Her Majesty to invite such members of her Council who may be surviving to attend her and assist her in that regard.'[17] There are, at any one time, over 500 available Privy Counsellors, and a significant number are not resident in the United Kingdom nor even British citizens.[18]

5.09 The second point is that the Crown is in fact very robust. As regards the existence of the Crown, succession is settled by the Bill of Rights 1689, as amended by the Act of Settlement 1700 and the His Majesty's Declaration of Abdication Act 1936. According to the Royal website,[19] there are nearly 40 people in line of succession, but one could presumably go further down the line if necessary. The only circumstance which could give rise to difficulty is where a minor (under 18 years) succeeds to the Crown (s 1) or where the office holder is incapacitated by infirmity of mind or body or (under s 2) is 'for some definite cause not available' (being held captive by the enemies of the Crown might be an example). These events trigger the Regency Act 1937. A declaration as to incapacity can be made by any three or more of the Sovereign's spouse, the Lord Chancellor, the Speaker of the House of Commons, the Lord Chief Justice, and the Master of the Rolls. In addition, under the Regency Act 1937, s 6 (as amended in 1943 and 1953), the Sovereign may appoint Councillors of State when suffering from a lesser degree of incapacity or intends to be absent from the realm. The Councillors of State can then exercise specified royal functions (except dissolving Parliament or granting titles).

5.10 A measure which aids both legal and democratic accountability is s 20(5). Section 20(5) demands that regulations under this section must be prefaced by a statement which (a) specifies the nature of the emergency in respect of which the regulations are made, and (b) declaring that the person making the regulations (i) is satisfied that the conditions in s 21 are met, (ii) is satisfied that the regulations contain only provision which is appropriate for the purpose of

---

[16] See Joint Committee on the Draft Civil Contingencies Bill, Appendix 6.
[17] HL Debs vol 665 col 755 14 October 2004, Baroness Scotland.
[18] See http://www.privy-council.org.uk/output/Page76.asp.
[19] See http://www.royal.gov.uk/output/Page389.asp.

preventing, controlling, or mitigating an aspect or effect of the emergency in respect of which the regulations are made, (iii) is satisfied that the effect of the regulations is in due proportion to that aspect or effect of the emergency, (iv) is satisfied that the regulations are compatible with the Convention rights (within the meaning of the Human Rights Act 1998), and (v) in the case of regulations made under subsection (2), is satisfied as to the matter specified in subsection (2)(b). The requirement in (iv) represents an assertion of the importance of human rights which was not part of the draft Bill of January 2004.

This statement is the closest approach in Part II of the 2004 Act to the formal **5.11** declaration or proclamation of emergency under s 1 of the Emergency Powers Act 1920. It also represents a departure from a more formal, two-stage process proposed under the original draft Bill of June 2003, wherein the process started under cl 18 with a Royal Proclamation by which Her Majesty could by proclamation declare herself satisfied that (a) an emergency has occurred, is occurring, or is about to occur, and (b) it is necessary to make emergency regulations. A proclamation would have had to state (a) the nature of the emergency, and (b) the affected parts of the United Kingdom or regions. This formality reflected the view, expressed in its 2003 Consultation Paper, that, 'The formal process for declaring that special legislative measures are necessary and making emergency regulations should continue to rest with The Queen as Head of State, acting on the advice of Ministers.'[20] Only in cases of impracticality could the regulations founded upon a declaration by the Secretary of State under cl 19 (roughly equivalent to s 20(2)).[21] Those provisions were dropped from the Bill of January 2004.

Some commentators depicted the promulgation stage as an anachronism: **5.12**

> There may, of course, be good reasons for such a declaration in some legal systems, particularly under a federal constitution where a region may have access to particular powers or particular state funds after an emergency has been declared. In the UK, however, it is far from clear that the need to declare a state of emergency offers any particular safeguards.[22]

There is some support for this view in that it was unclear as to whether the failure to observe the promulgation process had any bearing on the legality of subsequent regulations.[23] This doubt aside, the proclamation served an important symbolic function, not only stressing to the public and to Parliament the seriousness of the situation, but also serving functional purposes of triggering parliamentary review. The process in Parliament involved, first, the

---

[20] Cabinet Office, *The Draft Civil Contingencies Bill*, para 5.25.   [21] Ibid, para 5.27.
[22] House of Commons Library, *Civil Contingencies Bill* (Research Paper 04/07, House of Commons, London) 18.
[23] See Morris, GS, 'The Emergency Powers Act 1920' [1979] PL 317 at 331. But the case of *Teh Cheng Poh v Public Prosecutor, Malaysia* [1980] AC 458 does suggest that proclamations of emergency have legal effect.

communication of the Proclamation by a Message achieved by the Home Secretary or Prime Minister moving a motion that it be taken into consideration, and, secondly, a Humble Address, moved again by the Home Secretary or Prime Minister and expressing thanks for the Message. The latter was the motion which was to be debated and was in principle distinct from the further debate to approve the emergency regulations or their continuance. Not all proclamations were so handled, and the Government claimed there was no legal obligation when there was a proclamation to continue an existing emergency, as occurred, for example, in 1921 and 1926.[24] Nonetheless, most were dramatic and important preludes to Parliamentary scrutiny.[25]

5.13 A more pressing explanation for the dropping of the proclamation process concerned the dangerous delay which might be incurred by the requirement to make the proclamation in the presence of the Queen in person. But the alternative of a Secretary of State's proclamation could be viewed as a dangerous usurpation of powers. It was pointed out by the Joint Committee that the Crown is itself a very resilient institution, as has already been explored, which remains vital to the operation of Part II under s 20 (the making of Orders in Council) and s 28 (requiring the meeting of Parliament).[26]

5.14 On balance, the sanction of the Privy Council is purely formal and real power resides with the Government of the day. Thus, there is an argument for not involving the Queen in person, though that argument still does not address the contentions made earlier in favour of a formal declaration which is debated by Parliament (as shall be argued in Chapter 6). It is also better to rely on the Triple Lock, and its requirements are more pertinently focused on substance rather than grandiose process. That stance is sensible but does place great reliance on both the bite of the Triple Lock tests and also the bite of the courts and Parliament.

5.15 Next, it may be commented that the requirement of mere satisfaction under s 20(1) or (2) is not a sufficiently demanding threshold to deliver on the Triple Lock. The device is also inadequate as it does not require the disclosure of evidence or intelligence.

(b) Preconditions

5.16 More explicit preconditions to s 20(1) about which Her Majesty in Council must be satisfied are set out in s 21. The first condition (s 21(2)) is that an

---

[24] HC Debs vol 198 col 2510 30 July 1926, Sir W Joynson-Hicks. See further Ch 2.
[25] For the final example in Westminster, see HC Debs vol 397 col 257, 13 November 1973. For the effectiveness of Parliamentary scrutiny, see Bonner, D, *Emergency Powers in Peacetime* (Sweet & Maxwell, London, 1985); Ewing, KD and Gearty, CA, *The Struggle for Civil Liberties: Political Freedom and the Rule of Law in Britain 1914–1945* (Oxford University Press, Oxford, 2000) Ch 4.
[26] Joint Committee on the Draft Civil Contingencies Bill, paras 204, 205. See further Ch 6.

'emergency' has occurred, is occurring, or is about to occur. The emergency is as defined by s 19, which includes the concept of seriousness as part of the Triple Lock. There is no temporal cut off point for either dealing with the aftermath of emergencies which have occurred or anticipating emergencies which are in the future. As for the future, the Government was keen not to tie down Part II to imminent emergencies but argued for a wider ambit:

> The Government believes that it is important that where possible risks should be addressed before they become emergencies. That is why the Government has arrangements to monitor and manage risk, and to scan the horizon for future threats. Preventative action is often taken—for example flood defences or inoculation programmes. This principle of pre-emption is also true in situations where an emergency becomes not just possible but probable or imminent. In those circumstances, the Government will seek to take action as soon as possible to prevent the emergency happening or to reduce its impact.[27]

In any event, whilst it might appear that a word such as 'imminent' might offer a stricter test than the Government's interpretation of 'about to occur', lessons from the invasion of Iraq in 2003, supposedly triggered by an imminent threat of missile attack,[28] conduce to scepticism about the significance of this particular phraseology.

The second condition (s 21(3)) is that the regulations are 'necessary' to make **5.17** provision for the purpose of preventing, controlling, or mitigating an aspect or effect of the emergency. This condition, which is part of the Triple Lock mechanism, previously appeared in cl 18(1)(b) of the draft Bill of June 2003. It is unlikely that, whatever the emergency, there exists a total legislative vacuum. Subsections (5) and (6) therefore make clear that where a proposed regulation covers the same ground as existing legislation or could be issued under existing legislation, it can still be shown to be 'necessary' where (a) the existing legislation cannot be relied upon without the risk of serious delay, or (b) it is not possible without the risk of serious delay to ascertain whether the existing legislation can be relied upon, or (c) the existing legislation might be insufficiently effective. Accordingly, there is a presumption against emergency regulations where existing legislation is available, but the existence of possibly pertinent legislation does not wholly bar Part II:

> It is possible that legislation that had been designed for one purpose could properly be used in an emergency situation, but it would not have been used in those circumstances before, because emergencies are emergencies, so there would not be experience of such a use.

> The law is often complex and relevant provisions might be in relatively obscure legislation. The meaning and intent might be subject to differences of interpretation

---

[27] Ibid, Appendix 9 q 9.
[28] See (Butler) Committee of Privy Counsellors, *Review of Intelligence on Weapons of Mass Destruction* (2003–04 HC 898).

and it might be necessary to seek counsel's opinion to see whether a particular power is sufficient and appropriate to be used in the emergency. In such situations, when lives might be at risk and it is not possible to ascertain with complete certainty whether a piece of legislation will be effective, and when there is not time to seek the views of a higher legal authority, the measure represents an assurance that it is not *ultra vires* to make a provision that due consideration reveals could have been made under existing legislation. Such measures would not be taken because a Minister was lazy, but because circumstances were extraordinary, and the legislative power in question perhaps had not been used in that way before.[29]

5.18 The third condition (s 21(4)) is that the need for regulations is urgent. This requirement is additional to the version in the draft Bill of June 2003 and can be seen as a strengthening of the Triple Lock requirement of necessity. The impact is to rule out regulations about, for example, global warming or the loss of biodiversity; some might count each as an 'emergency' but their alleged development and impact are hardly happening overnight and catching Ministers unawares.[30]

5.19 In addition to the three conditions, it would have been very helpful to impose statutory duties to disclose the evidence and intelligence which convinced the Minister to intervene. Otherwise:[31]

> The Government have so many powers and sources of information that they are capable of creating the illusion of a serious threat to this country. Indeed, one does not have to look back many months to see them do exactly that. We have just been to war [in Iraq] as a result of an illusion created by this Government.

There is no such duty in the Act beyond the assertive statement in s 20(5).

5.20 An unresolved issue is whether the conditions in s 21 apply as a trigger to the totality of regulation-making powers in s 22 (set out below) or whether the conditions must be satisfied for every exercise of each regulation-making power. Practice in the past was to issue regulations *en bloc*, utilizing the full panoply of powers. Thus, speaking in 1973, the Home Secretary revealed that 'It has been the practice of successive Governments—I am sure that it is a wise one—to make a complete set of emergency regulations at the outset.'[32] It is interesting that s 20(1), which contains the 'core power' does indeed refer to 'regulations' in the plural. On the other hand, s 21(3), concerning overlap with existing legislation, refers to each 'provision'. The difference will be immaterial in many cases—a crisis creates all manner of troubles, and it will be easy to find reasons to be safe rather than sorry. An example of inventive reasons may be illustrated by the response from the Cabinet Office to the question, why there is no

---

[29] HC Debs Standing Committee F cols 193 3 February 2004, Fiona Mactaggart.
[30] See HL Debs vol 665 cols 511, 730 14 October 2004, Baroness Scotland.
[31] HL Debs vol 665 col 514 14 October 2004, Lord Lucas.
[32] HC Debs vol 397 cols 680–3 15 November 1973.

distinction between the powers appropriate to a flooded river and those appropriate to sea flooding:

> The necessities would not be driven by the type of flooding, but by the specifics of the incident. Nevertheless, the Government would not agree that the only action necessary in the event of serious flooding would be evacuation (whether of people or animals). Serious flooding can disrupt water supplies, power, transport and food distribution. Action might be necessary to reduce, control or mitigate the effect of the flooding, for example the diversion of flood waters or the erection of temporary flood defences. In the aftermath, special arrangements might be necessary to support recovery. Although the expectation would be that most of these could be achieved without recourse to emergency powers, the most extreme flooding situations might require, for example, emergency powers to requisition property or restrict movement. This is equally true of both coastal and fluvial flooding. Coastal flooding has traditionally had the potential to cause greater damage, but fluvial flooding threatens more areas. Both can and have caused loss of life and dislocation of essential services.[33]

However, it is submitted that the logic of the Triple Lock, especially the precept of necessity, seems to demand that the tests in s 21 should be applied to individual regulations. The effect would be to constrain the Government from dusting off and issuing whichever powers were previously used or have been taken from a civil servant's filing cabinet. Each and every regulation in turn should be necessary.

B. The Scope and Limits of Emergency Regulations

(a) Scope

Part II both confers and delimits the ways in which emergency regulations may be **5.21** utilized. Its contents are not radically different from the 1920 Act. Thus, s 22(1) confers an almost boundless power—emergency regulations may make any provision which the issuing authority is satisfied is appropriate for the purpose of preventing, controlling, or mitigating an aspect or effect of the emergency in respect of which the regulations are made. The subjective wording of the power, combined with the broad catalogue of circumstances, leaves limited room for accountability, save for the administrative law duties to avoid irrationality or illegality. To capture the spirit, one might reiterate the alarm expressed in the popular press:

> The *Evening Standard* painted a lurid picture of what might happen: 'The army and police will be given virtually unlimited powers in the event of a terrorist attack under Government plans unveiled today. Officers would be allowed to seal off any area the Government suspects could be the focus of an attack. Exclusion zones could be enforced by shooting to kill. Permission would be given for the summary arrest of terrorist suspects and detention without charge. Authorities would also be able to order mass evacuations and take over trains, buses and airports... The

---

[33] Joint Committee on the Draft Civil Contingencies Bill, Appendix 9 q 5.

Civil Contingencies Bill is one of the most sweeping pieces of legislation ever framed in Britain and will give ministers more power than ever before.'[34]

5.22   There follows in s 22(2) a non-exhaustive list of 12 categories of purposes for which regulations may be used, including:

(a) protecting human life, health, or safety;

(b) treating human illness or injury. It is possible that this power might be used to enforce vaccinations or a quarantine,[35] though the Minister of State envisaged some limits:

> An adult of sound mind can refuse medical treatment, even if in so doing he or she is endangering their own life. Thus there is no question of a person of sound mind being compulsorily treated to save their life. When the failure of a person to accept treatment would have an effect on the health and life of others, for example if they are infectious—that circumstance could clearly be anticipated—it would be permissible to take action in relation to that person to protect the life and health of others. Such action would have to be proportionate. It would be possible to quarantine them or otherwise prevent them from infecting others.[36]

(c) protecting or restoring property;

(d) protecting or restoring a supply of money, food, water, energy, or fuel. Wider concerns about the financial system, other than the supply of money, are covered in (h) below;

(e) protecting or restoring a system of communication;

(f) protecting or restoring facilities for transport;

(g) protecting or restoring the provision of services relating to health;

(h) protecting or restoring the activities of banks or other financial institutions.[37] This leg could tie in with the need to preserve the supply of money under (e):

> 'For example, if a terrorist attack on the City of London constituted a threat of serious damage to human welfare, and it was necessary to make emergency regulations, the Government considers that it should be possible (where it was necessary to do so and in due proportion) to declare a bank holiday so as to protect the interests of financial institutions. If the Bank of England was affected, it might

---

[34] *Evening Standard* 7 January 2004 at 9, quoted by the House of Commons Library, *Civil Contingencies Bill* (Research Paper 04/07, House of Commons, London) 37.

[35] For doubts about the viability of compulsory quarantine, see Gostin, L, *Public Health Law* (University of California Press, Berkeley, 2000).

[36] HC Debs Standing Committee F col 270 10 February 2004, Douglas Alexander.

[37] The meaning is reckoned to include 'insurance companies, building societies, friendly societies, credit unions and other institutions of that type. The term is commonly used in legislation without further definition. Supermarkets, for example, would not be financial institutions, but . . . financial markets such as LIFFE might be.' HC Debs Standing Committee F col 209 5 February 2004, Fiona Mactaggart.

be appropriate to move the Government's account from the Bank so as to preserve the functioning of Her Majesty's Government.'.[38]

(i) preventing, containing, or reducing the contamination of land, water, or air;

(j) preventing, reducing, or mitigating the effects of disruption or destruction of plant life or animal life;

(k) protecting or restoring activities of Parliament, of the Scottish Parliament, of the Northern Ireland Assembly, or of the National Assembly for Wales.[39] These activities might also include 'the day-to-day activities of Government such as accounting for their actions to Parliament'.[40] Where Parliament in Westminster is disabled because of an emergency, the provision allows special measures to be used to arrange for alternative venues and arrangements;[41]

(l) protecting or restoring the performance of 'public functions'. The latter are defined in s 31(1) to mean:[42] (a) functions conferred or imposed by or by virtue of an enactment whether on a Minister or any other public body (benefits systems might be a function not mentioned elsewhere);[43] (b) functions of Ministers of the Crown (or their departments), which would include not only statutory functions but the wide and vague range of functions assumed under common law, whether of a private nature such as contract, or of a public nature such as under the prerogative especially in the fields of defence, foreign affairs, and public order; (c) functions of persons holding office under the Crown, such as police constables, whether under statute or common law; (d) functions of the Scottish Ministers; (e) functions of the Northern Ireland Ministers or of the Northern Ireland departments; and (f) functions of the National Assembly for Wales.[44] Though the legislature and the executive are mentioned in the last two legs, there is no equivalent for the courts,[45] but s 22(5) (see below) offers some assurance about court business under the Act itself.

The list in s 22(2) is non-exhaustive—these are 'in particular' illustrations which **5.23** are 'designed not to put a boundary around something but to inform and to ensure that the way in which the power can be used and the sort of regulations that can be made will go with the grain.'[46] As a whole, the list tentatively reflects

---

[38] Cabinet Office, *The Government's Response to the Report of the Joint Committee*, para 8.

[39] These were added as a distinct clause in the January 2004 draft rather than to the definition of 'public functions' which was felt to be exclusively executive in nature: Cabinet Office, *The Government's Response to the Report of the Joint Committee*, para 16.

[40] HC Debs Standing Committee F col 210 5 February 2004, Fiona Mactaggart.

[41] HL Debs vol 666 col 854 9 November 2004, Baroness Scotland.

[42] Joint Committee on the Draft Civil Contingencies Bill, Appendix 9 q 35.

[43] HC Debs Standing Committee F col 210 5 February 2004, Fiona Mactaggart.

[44] The mention of a legislature here reflects the fact that it also undertakes executive functions: Cabinet Office, *The Government's Response to the Report of the Joint Committee*, para 16.

[45] Compare: HL Debs vol 665 col 932 21 October 2004, Lord Lucas; HL Debs vol 666 col 852 9 November 2004, Lord Lucas.

[46] HC Debs Standing Committee F col 208 5 February 2004, Fiona Mactaggart.

the core meaning of 'emergency' in s 19(1), beginning with health issues (a to b), moving to environmental matters (c to d), and ending with core accoutrements of the state and political economy. It is more confined than the corresponding list in cl 21(2) of the draft Bill of June 2003, which included the protection of any 'essential commodity' and '(k) protecting or restoring activities of Her Majesty's Government' (which was in addition to the 'public functions' in (l)).[47]

5.24 An even longer list of types of possible exercise of power (17 categories) follows in s 22(3) which expresses that Part II regulations may be used even for measures which could be made by Act of Parliament or by the exercise of the Royal Prerogative (aspects of which were described in Chapter 2).

5.25 The following are specified in s 22(3) as a 'particular' (non-exhaustive) list of powers in order to confirm the power or, within each category of power, to apply the canon of *expressio unius est exclusio alterius* and thereby to exclude other powers:[48]

(a) confer a function on a Minister of the Crown, on the Scottish Ministers, on the National Assembly for Wales, on a Northern Ireland department, on a coordinator appointed under s 24, or on any other specified person. Does the latter possibility allow 'for any tinpot individual who might be directed, under these forms of emergency, to take up the reins of power'?[49] They may or may not fit that epithet, but those in mind are a chief constable, a local authority chief executive, or an Army officer;[50]

(b) provide for or enable the requisition or confiscation of property (with or without compensation). It may be noted that this power more than compensates for the loss of the powers of compulsory purchase formerly existing under the Civil Defence Act 1948, s 4;

(c) provide for or enable the destruction of property, animal life or plant life (with or without compensation). There is no corresponding reference to the destruction of human life,[51] which may place limits on the modes of disease eradication in the event of, for example, a smallpox outbreak;

(d) prohibit, or enable the prohibition of, movement to or from a specified place (such as following a bombing).[52] In this way, a cordon can be imposed around an area affected by, say, a 'dirty' bomb. Prior to the 2004 Act, statutory

---

[47] The latter still appeared in the draft Bill of January 2004 (2003–04 HC No 14 cl 21(2)(l)). The current drafting reflects the recommendations of the Joint Committee on the Draft Civil Contingencies Bill, para 76.

[48] An example might be category (o): HC Debs Standing Committee F col 221 5 February 2004, Fiona Mactaggart.

[49] HC Debs Standing Committee F col 216 5 February 2004, Patrick Mercer.

[50] Ibid, col 222, Fiona Mactaggart.

[51] House of Commons Library, *Civil Contingencies Bill* (Research Paper 04/07, House of Commons, London) 38.

[52] Ibid, col 232, Fiona Mactaggart. Preferably, domestic pets should be included in any human movements: Cabinet Office, *Dealing with Disaster* (rev 3rd ed, Cabinet Office, London, 2003) para 4.67.

cordoning powers only applied in cases of terrorism,[53] though wider common law powers exist to control crime scenes and to maintain the peace.[54] A suggested exception to (d), for the movement of Members of Parliament to the House so that it can 'fulfil its role as the cockpit of the nation at a time of crisis'[55] was not accepted. It was felt that s 22(5) (explained below) adequately protects their interests, and the Government would in any event need Parliament to be in effective session to approve the continuance of regulations;

(e) require, or enable the requirement of, movement to or from a specified place (such as an evacuation following the release of toxic gas);[56]

(f) prohibit, or enable the prohibition of, assemblies of specified kinds, at specified places or at specified times. Examples of the circumstances in which the Government might need to prohibit large gatherings of people include where there is a contagious epidemic, a risk of a mass-casualty terrorist attack, or severe disruption to transportation;[57]

(g) prohibit, or enable the prohibition of, travel at specified times;

(h) prohibit, or enable the prohibition of, other specified activities 'If a situation led to a severe water shortage, it might be necessary to impose a regulation that people could not have a bath more than once a week. If a major contaminant affected produce, it might be necessary to prevent someone from selling contaminated vegetables on a market stall';[58]

(i) create an offence[59] of (i) failing to comply with a provision of the regulations, (ii) failing to comply with a direction or order given or made under the regulations, or (iii) obstructing a person in the performance of a function under or by virtue of the regulations. These powers will usually be exercised to create offences which require intent to carry out the action which breaches the regulations, but knowledge of the regulation is not an element of *mens rea*.[60] However, the Home Office Minister expressly reserved the option of constructing offences of strict liability (such as in connection with price controls).[61] There is no specific

---

[53] See Terrorism Act 2000, s 33.

[54] *DPP v Morrison* [2003] EWHC Admin 683; *Austin v The Commissioner of Police of the Metropolis* [2005] EWHC 480 (QB).

[55] HC Debs Standing Committee F col 233 5 February 2004, Oliver Heald.

[56] Ibid, col 232, Fiona Mactaggart.

[57] Joint Committee on the Draft Civil Contingencies Bill, Appendix 9 q 26.

[58] HC Debs Standing Committee F col 224 5 February 2004, Fiona Mactaggart.

[59] This power to create offences will be construed restrictively, and it is notable that it does not allow the imposition of retrospective criminal liability; see *R (Haw) v Secretary of State for the Home Department* [2005] EWHC 2061.

[60] But the Statutory Instruments Act 1946, s 3(2) provides a defence that the instrument had not been published by the Stationery Office unless it is proved that reasonable steps had been taken for notifying the public of its purport.

[61] HC Debs Standing Committee F col 240 5 February 2004, Fiona Mactaggart. Regulatory offences under the Emergency Powers Act 1920 also sometimes involved strict liability: Morris, GS, 'The Emergency Powers Act 1920' [1979] PL 317 at 329.

power of arrest and no specific authority to use force beyond the normal formula in the Criminal Law Act 1967, s 3[62] when dealing with regulatory offences, but these can be supplied by 'normal' powers especially since the definition of offences subject to a constable's summary arrest power has been extended to all infractions of the law by s 110 of the Serious Organised Crime and Police Act 2005. However, given the protection for the non-derogable right to life under art 2 of the Human Rights Act 1998, it would be difficult to devise a lawful regulation which allowed summary shootings (such as of looters) or fire zones. That is not to say that the security forces do not already operate shoot-to-kill policies;[63]

(j) disapply or modify an enactment or a provision made under or by virtue of an enactment. This provision could afford legal immunity for soldiers or other government agents, though only from statutory provisions and not, therefore, most torts or from such offences as murder. It could arguably be used to change parts of the constitution, the implications of which are explored in Chapter 6. Less controversially, it might be necessary to disapply, for example, the Health and Safety at Work Act etc 1974 so as to allow officials or others to operate requisitioned digging equipment.[64] It may also be used to reduce or avoid statutory notice periods;[65]

(k) require a person or body to act in performance of a function (whether the function is conferred by the regulations or otherwise and whether or not the regulations also make provision for remuneration or compensation). An example might be to require private ambulance services to provide a more public service.[66] The example also illustrates a further point—that the regulation may apply in an area not affected by the emergency provided it assists with relief in the emergency region or part thereof;

(l) enable the Defence Council to authorize the deployment of Her Majesty's armed forces. This formula goes beyond that in the Emergency Powers Act 1964, s 2, which relates to agricultural work or other such urgent work of national importance, though conversely, that Act does not require a declared emergency;

(m) make provision (which may include conferring powers in relation to property) for facilitating any deployment of Her Majesty's armed forces. An

---

[62] See also Criminal Law Act (Northern Ireland) 1967, s 3; *Reference under s 48A Criminal Appeal (Northern Ireland) Act 1968 (No 1 of 1975)* [1976] 2 All ER 937; *R v Clegg* [1995] 2 WLR 80; Association of Chief Police Officers, *Manual of Guidance on Police Use of Firearms* (2001).

[63] See *McCann v United Kingdom*, App No 18984/91, Ser A vol 324 (1995); *Jordan v United Kingdom*, App No 24746/94, 04/05/2001; *McKerr v United Kingdom*, App No 28883/95, 2001-III; *Kelly v United Kingdom*, App No 30054/96, 04/05/2001; *Shanaghan v United Kingdom*, App No 37715/97, 04/05/2001; *Finucane v United Kingdom*, App No 29178/95, 2003-VIII; *Bubbins v United Kingdom*, App No 50196/99, 17/03/2005. There is pending an investigation into the shooting of John Charles de Menezes which revealed special rules ('Operation Kratos') relating to suicide bombers: *The Times* 23 July 2005 at 4.

[64] HC Debs Standing Committee F col 254 10 February 2004, Fiona Mactaggart.

[65] HL Debs vol 665 col 1007 21 October 2004, Baroness Scotland.

[66] HC Debs Standing Committee F col 232 5 February 2004, Fiona Mactaggart.

example, covering both (l) and (m), might be the use of soldiers in flood relief operations which involve the commandeering of boats and equipment;

(n) confer jurisdiction on a court or tribunal (which may include a tribunal established by the regulations). There is no qualification as to the nature of the courts or the qualifications of their personnel. As quipped by one MP, 'There is a danger that kangaroo courts will be created. I hope that the Minister can give us some assurance that we will not be appointing too many marsupials'.[67] There are indeed safeguards in ss 23(4) and 25 (below);

(o) make provision which has effect in relation to, or to anything done in (i) an area of the territorial sea (which means the territorial sea adjacent to, or to any Part of, the United Kingdom, construed in accordance with s 1 of the Territorial Sea Act 1987, (ii) an area within British fishery limits (as defined by the Fishery Limits Act 1976), or (iii) an area of the continental shelf (meaning any area designated by Order in Council under s 1(7) of the Continental Shelf Act 1964);

(p) make provision which applies generally or only in specified circumstances or for a specified purpose. This measure (and that in (q)) is not free-standing but is 'designed to allow regulations [under other subsections] to be tailored to the precise circumstances of the emergency'.[68] The example given by the Minister concerned the restriction of travel, which might apply to those forms that could facilitate the spread of an infectious disease, such as air travel, leaving other transportation unaffected;

(q) make different provision for different circumstances or purposes. The Minister's example during the passage of the Bill related to animal health—'rather than providing for the vaccination of all animals, the regulations could provide for the vaccination of animals that show symptoms of a particular disease or if they are susceptible to it.'[69]

The list was criticized by the House of Lords Select Committee on Delegated **5.26** Powers and Regulatory Reform:

> [The clause] . . . is a very wide power indeed. . . . If this were not a draft bill to make emergency provision, we would strongly question the appropriateness of a number of aspects of the power (such as, for example, sub-delegation by directions or orders whether written or oral) . . . , confiscation of property without compensation . . . destruction of property, etc. without compensation . . . prohibition of movement . . . and prohibition of assembly.[70]

Even so, there is no express power of detention without trial, as was equally the position under the Emergency Powers Act 1920. In part, the omission may relate to the fact that such powers already existed under Part IV of the

---

[67] HC Debs Standing Committee F col 261 10 February 2004, Alistair Carmichael.
[68] Ibid, col 266, Fiona Mactaggart.   [69] Ibid.
[70] Joint Committee on the Draft Civil Contingencies Bill, Appendix 3.

Anti-Terrorism, Crime and Security Act 2001. Part IV has now been replaced with control orders under the Prevention of Terrorism Act 2005, though their impact is little short of detention without trial. The 2004 Act echoes that idea in allowing regulations by which movement can be restricted, as well as a range of other curtailments imposed. The contention that detention without trial can be authorized under civil emergency legislation has never been confirmed by the courts[71] nor denied by academic commentators.[72] The Bill's sponsors refused to rule out detention without trial,[73] and, in terms of interpretation, there is a difference between creating an offence punishable without trial (expressly disallowed under s 23(4)(b)) and a power of detention without trial which can then be enforced through disciplinary offences tried in the usual way. One can envisage its use, for example, in connection with terrorist emergencies or during pandemics. The importance of carefully sequenced reviews and oversight was demonstrated by experience with Part IV of the Anti-Terrorism, Crime and Security Act 2001.[74] Such a serious incursion into liberty should have been expressly regulated on the face of the Act if it is to remain a live possibility. Nor does s 22(3) rule out powers of summary arrest, such of all persons in the vicinity of a terrorism incident or in an area affected by disease or a chemical, biological, radiological, or nuclear (CBRN) incident.

(b) Limits

5.27 Moving from empowerment to restraint, the first limit appears in s 22(5), a provision not contained in the draft Bill of January 2004 but added as a concession at Report Stage in the House of Lords.[75] A person making emergency regulations must have regard to the importance of ensuring that Parliament, the High Court, and the Court of Session are able to conduct proceedings in connection with (a) the regulations, or (b) action taken under the regulation.

5.28 More generic restraints are contained in s 23. First (subsection (1)), the regulator must be 'satisfied' (a) that the provision is appropriate for the purpose of preventing, controlling, or mitigating an aspect or effect of the emergency in

---

[71] Compare *R v Halliday, ex p Zadig* [1917] AC 260, where the House of Lords accepted that the Defence of the Realm Act 1914 could be taken to have authorized detention without trial (see Simpson, AWB, *In the Highest Degree Odious* (Clarendon Press, Oxford, 1992) Ch 2). It is arguable that wartime powers might be distinguishable from 'civil' powers (Morris, GS, 'The Emergency Powers Act 1920' [1979] PL 317 at 324–5), but a power of detention without trial in peacetime (though not its exercise) was accepted without question in *Attorney-General of St. Christopher, Nevis and Anguilla v Reynolds* [1980] AC 637.

[72] See Bonner, D, *Emergency Powers in Peacetime* (Sweet & Maxwell, London, 1985) 270.

[73] Letter from Douglas Alexander to the Joint Committee on the Draft Civil Contingencies Bill, 31 October 2003.

[74] See Walker, C, 'Prisoners of "War All the Time"' [2005] *European Human Rights Law Review* 50.   [75] HL Debs vol 666 col 854 9 November 2004, Baroness Scotland.

respect of which the regulations are made, and (b) that the effect of the provision is in due proportion to that aspect or effect of the emergency. The absence of a requirement of 'reasonableness' in the formula will not prevent judicial review[76] but may constrain its degree of intrusion.[77] Nor does s 23 restrain expressly (though one could argue that the general requirement of proportionality under s 23(1) bites here), unlike cl 21(4)(e) of the draft 2003 Bill, a provision of a kind which the person making the regulations believes is made by, or could be made by virtue of, a subsisting legislative provision, unless he also believes that use of the subsisting provision (i) would be insufficient for the purpose or (ii) would occasion a serious delay.

Next, the regulator must bear in mind geographical restraint (subsection (2)). **5.29**
Emergency regulations must specify the Parts of the United Kingdom or regions within England in relation to which the regulations are to have effect. It would be possible to specify all Parts, but the expectation is that the impact of emergency responses will be as limited as possible. Under the Emergency Powers Acts, regulations had to apply to the whole of Great Britain or to the whole of Northern Ireland. Accordingly, the 2004 Act better allows for 'response efforts to be targeted and proportionate'.[78] Some have argued this principle should be taken further and that it would be both useful and appropriate to apply special legislative measures on a sub-regional basis.[79] A precedent is the Public Order Act 1986, ss 12 to 14, by which police orders dealing with processions and meetings are designed by reference to the epicentre of the event and special measures are exerted within a set radius of it. However, too small a focus could tempt over-reliance on the emergency powers. In any event, the exercise of many emergency powers will be subject to the legal requirement of proportionality, requiring application only to affected areas within the region, and also to practical restraints:

> . . . organisations and their plans are flexible enough to deal with emergencies within and across geographical boundaries, and there is long experience of working in this way . . . In practice many emergency powers would be used on a discretionary basis and only apply within the specific area in which they are needed allowing them to be targeted on a more specific 'geographical' basis as required.[80]

---

[76] HL Debs vol 665 col 947 21 October 2004, Baroness Scotland, citing *Congreve v Home Office* [1976] QB 629.

[77] Therefore, it was criticized by the House of Lords Select Committee on Delegated Powers and Regulatory Reform, *Twenty-Fifth Report* (2003–04 HL 144) para 29. According to Lord Wilberforce in *Secretary of State for Education v Tameside MBC* [1997] AC 1014 at 1047, the court can inquire into the factual basis for the application of a subjectively worded power, but the 'evaluation of those facts is for the Secretary of State alone'. The difference between *Wednesbury* review and where the court is the primary decision-maker as to reasonableness may not be large in contemporary practice, especially where basic rights are involved: *R v Ministry of Defence, ex p Smith and others* [1996] 1 All ER 257; *Youssef v Home Office* [2004] EWHC 1884 (QB) at para 63.

[78] Cabinet Office, *The Draft Civil Contingencies Bill: Explanatory Notes etc*, para 43.

[79] Civil Contingencies Secretariat, *The Government's Response*, 29.

[80] Cabinet Office, *The Government's Response to the Report of the Joint Committee*, para 50.

5.30 There follow three subsections to s 23 in which substantive restraints are imposed. The lists are somewhat more curt than those in s 22. By s 23(3), which reflects the corresponding statements in the Emergency Powers Acts, regulations may not (a) require a person, or enable a person to be required, to provide military service, or (b) prohibit or enable the prohibition of participation in, or any activity in connection with, a strike or other industrial action.

5.31 As for military service, forced labour may also contravene Art 4 of the European Convention on Human Rights[81] and the Geneva Convention.[82] Nevertheless, the Association of Chief Police Officers complained during passage that it was an unwise restraint, as it might hamper defence or relief works:

> ... one could requisition property in that way but not necessarily require individuals to drive them on your behalf. ... in terms of heavy lifting gear, for example, if one wanted to requisition that for a rescue I would guess that the skill needed actually to operate that would be very specialised and would be very difficult to find outside of that industry, and yet there is no power in the regulations to require the individual to operate that machinery on your behalf.[83]

In response, it may be argued that ACPO's worry is met by the fact that 'work in the nature of a "function" may be required under [s 22(3)(k)], and one may interpret "function" to mean limited specific work rather than continual

---

[81] '1. No one shall be held in slavery or servitude.

2. No one shall be required to perform forced or compulsory labour.

3. For the purpose of this article the term "forced or compulsory labour" shall not include: any work required to be done in the ordinary course of detention imposed according to the provisions of Article 5 of this Convention or during conditional release from such detention; any service of a military character or, in case of conscientious objectors in countries where they are recognised, service exacted instead of compulsory military service; any service exacted in case of an emergency or calamity threatening the life or well-being of the community; any work or service which forms part of normal civic obligations.'

[82] By the Fourth Geneva Convention relative to the Protection of Civilian Persons in Time of War 1949, Art 51: 'The Occupying Power may not compel protected persons to serve in its armed or auxiliary forces. No pressure or propaganda which aims at securing voluntary enlistment is permitted. The Occupying Power may not compel protected persons to work unless they are over eighteen years of age, and then only on work which is necessary either for the needs of the army of occupation, or for the public utility services, or for the feeding, sheltering, clothing, transportation or health of the population of the occupied country. Protected persons may not be compelled to undertake any work which would involve them in the obligation of taking part in military operations. The Occupying Power may not compel protected persons to employ forcible means to ensure the security of the installations where they are performing compulsory labour. The work shall be carried out only in the occupied territory where the persons whose services have been requisitioned are. Every such person shall, so far as possible, be kept in his usual place of employment. Workers shall be paid a fair wage and the work shall be proportionate to their physical and intellectual capacities. The legislation in force in the occupied country concerning working conditions, and safeguards as regards, in particular, such matters as wages, hours of work, equipment, preliminary training and compensation for occupational accidents and diseases, shall be applicable to the protected persons assigned to the work referred to in this Article. In no case shall requisition of labour lead to a mobilization of workers in an organization of a military or semi-military character.'

[83] Joint Committee on the Draft Civil Contingencies Bill, para 158.

employment as in a "service".'[84] Furthermore, the wording of Art 4 is elaborated upon in Art 4(3) as expressly not including, '(c) any service exacted in case of an emergency or calamity threatening the life or well-being of the community'. 'Emergency' is interpreted widely, and certainly covers situations other than what normally would be viewed as a 'disaster'. For example, in *Iversen v Norway*,[85] service in northern Norway could be imposed upon dentists because of a shortage of volunteers. Likewise, a requirement that those holding shooting rights should participate in the gassing of foxholes to control rabies was upheld in *S v Germany*.[86]

5.32 Turning to the protection for industrial action, the original draft in June 2003 seemed to be more narrowly drawn than the equivalent provision in the 1920 Act, s 2(1) of which prevented regulations being issued to make it an offence 'to take part in a strike, or peacefully to persuade any other person or persons to take part in a strike.' The difference relates to whether peaceful picketing or other forms of pressure[87] can for sure be defined as 'industrial action'. Peaceful picketing at one's own place of work is included in s 220 of the Trade Union and Labour Relations (Consolidation) Act 1992 as a form of activity under the Part V heading of 'Industrial Action' but forms of 'secondary action' are not immune from suit under s 224.[88] Next, there was concern in the draft Bill that cl 21(3)(k) (requiring the performance of a function) could be invoked to impose new forms of statutory duty which could then be breached by industrial action, liability for which would fall outside the normal scope of immunities for industrial action.[89] Subsequently, the words 'or any activity in connection with' were added to s 23(3)(b) which may remove this danger. The 2004 Act is now arguably more favourable to strikers than the 1920 Act[90] since it covers industrial action other than a strike (which should be interpreted as involving the taking part in it, as well as the calling of it)[91] and covers prohibitions of a civil as well as criminal nature. But the Government declined to clarify the position regarding peaceful picketing by adding a specific regulation-making power under s 22(3) and removing picketing from the ambit of s 22(3)(f) so that the matter could be specifically signalled.

5.33 Though strikes and industrial action are generally permitted by the 2004 Act— in the eyes of some, a 'gold-plated' right offered as 'a sop to the trade unions'

---

[84] Ibid, para 159.   [85] App No 1468/62, 6 YBEC 278 (1963).
[86] App No 9686/82, 39 DR 90 (1984).
[87] See *Knowles and another v Fire Brigades Union* [1996] 4 All ER 653.
[88] Compare the view of the Joint Committee which thought the difference was that it could be made unlawful under the Bill for *specific groups* to take part in industrial action (Joint Committee on the Draft Civil Contingencies Bill, para 160). Lord Norton's note (ibid, 79) more accurately captures the problem.   [89] Joint Committee on the Draft Civil Contingencies Bill, para 161.
[90] But Emergency Regulations were altered in late 1973 to adopt the wider forms of protection: see Morris, GS, 'The Emergency Powers Act 1920' [1979] PL 317 at 324.
[91] *Smith v Wood* (1927) 43 TLR 179.

despite the disruption they can cause[92]—they may be constrained by local union agreements[93] and by laws other than in the 2004 Act. These include the prohibitions on trade union activity in relation to the police, armed forces, and seafarers,[94] no-strike agreements[95] and also by the Trade Union and Labour Relations (Consolidation) Act 1992, s 240, by which:

> A person commits an offence who wilfully and maliciously breaks a contract of service . . . knowing or having reasonable cause to believe that the probable consequences of his so doing . . . will be (a) to endanger human life or cause serious bodily injury, or (b) to expose valuable property, whether real or personal, to destruction or serious injury.[96]

Of course, the right to free association under Art 11 of the European Convention on Human Rights will provide some protection for collection action, though on a more balanced basis than the absolute grant in s 22(3)(b),[97] which is arguably a more sensible approach for an emergency.[98]

5.34 The next restraint relates to criminal justice. By s 23(4), regulations may not (a) create an offence other than one of the kind described in s 22(3)(i) (in other words, failing to comply with a provision of the regulations or directions or orders under a regulation or obstructing a person carrying out a function under the regulations), or (b) create an offence other than one which is triable only before a magistrates' court or, in Scotland, before a sheriff under summary procedure, or (c) create an offence which is punishable (i) with imprisonment for a period exceeding three months, or (ii) with a fine exceeding level 5 on the standard scale, or (d) alter procedure in relation to criminal proceedings.[99] The latter is arguably a more valuable restraint than the formula in the Bill's cl 21(4)(d)(iii), by which regulations could not create an offence which is

---

[92] HL Debs vol 665 cols 1014 and 1018 21 October 2004, Baroness Balcombe and the Earl of Onslow. [93] See Ch 2.

[94] Army Act 1955, ss 34, 193; Police Act 1996, ss 64, 91; Incitement to Disaffection Act 1934; Merchant Shipping Act 1995, ss 58, 59; Postal Services Act 2000, ss 83, 84.

[95] See Department of Employment, *Trade Union Immunities* (Cmnd 8128, London, 1981) para 330; Morris, GS, *Strikes in Essential Services* (Mansell, London, 1986); Welsh, R, 'Re-establishing Trade Union Rights' in McColgan, A (ed), *The Future of Labour Law* (Cassell, London, 1996). They were rejected in regard to the NHS (*Royal Commission on the National Health Service* (Cmnd 7615, London, 1979)), the civil service (*Inquiry into Civil Service Pay* (Cmnd 8590, London, 1982)), prisons (*Committee Inquiry into the Prison Service* (Cmnd 7676, London, 1979)), and fire-fighters (Independent Review of the Fire Service, *Report* (http://www.irfs.org.uk/, 2002) para 10.33; Office of the Deputy Prime Minister, *Our Fire and Rescue Service* (Cm 5808, London, 2003) para 7.25.

[96] See further *Gouriet v Union of Post Office Workers* [1978] AC 435. Section 4 of the Conspiracy and Protection of Property Act 1875 (which relates to breaches of contract by persons employed in the supply of gas or water) was repealed by the Industrial Relations Act 1971.

[97] See, for example, *Council of Civil Service Unions v United Kingdom*, App No 11603/85, (1987) Decisions and Reports 50, 228. In 1944, 3,714,000 days were lost to strike action, in 1974 14,750,000 days, and in 1984 27,135,000 days: HL Debs vol 666 col 1337 16 November 2004, Baroness Scotland. [98] HL Debs vol 666 col 1332 16 November 2004, Lord Lester.

[99] Compare Emergency Provisions Act 1920, s 2(3).

punishable without trial before a magistrates' court, the Crown Court, a district court, or the sheriff.[100] It was not present in the draft Bill of June 2003 but followed pressure from the Joint Committee that regulations should not alter any existing procedure in criminal cases in any way which is inconsistent with Art 6 of the European Convention.[101] As already considered, since s 23(4) is confined to offences, trial, and punishment, it does not rule out the making of a regulation providing for internment or detention without trial.

Next, by s 23(5), emergency regulations may not amend Part II of the Act or **5.35** any part of the Human Rights Act 1998. The treatment of constitutional laws and Human Rights Act are considered further in Chapters 6 and 7. Protection for the Human Rights Act represents a *volte face* during the legislative process. The original intention was explicitly to allow for amendments, but fierce opposition forced a withdrawal of the offending cl 25. The Government then sought to keep the issue off the face of the Bill,[102] but Parliamentary opposition continued, and s 23(5)(b) was eventually conceded.[103]

By contrast, the opposition was not so fierce as to coerce the Government into **5.36** listing as sacrosanct a host of other 'constitutional' legislation. However, it is reported that 'Parliamentary Counsel have advised that the effect of the normal principles of the construction of delegated powers is that substantive amendments could not be made by emergency regulations to provisions of an enactment which are of constitutional significance.'[104] Though this reassurance is welcome, it is less certain, for reasons explained in Chapter 6, that it will be an infallible guide to the outcome of future litigation.

A further limit appears in s 26. Emergency regulations must lapse at the end of **5.37** the period of 30 days beginning with the date on which they are made, or at such earlier time specified in the regulations. Even within the specified time, Ministers have promised to keep the situation under review, as is required by the considerations of necessity and proportionality.[105] This provision echoes the limit in the Emergency Powers Act 1920. Clause 23(4) of the draft Bill of June 2003 sought to allow the extension of existing regulations under a new emergency order in council, but this possibility was dropped after criticism that it obviated a fresh look at the regulations.[106] However, this restraint does not rule out the making of new regulations in identical form. Nor does it affect the

---

[100] This clause envisaged the possibility of trial by indictment for a regulatory offence, not a possibility under the Emergency Provisions Act 1920, s 2(3).

[101] Cabinet Office, *The Government's Response to the Report of the Joint Committee*, para 30. The draft Bill of June 2003 had allowed for trial in the Crown Court and trial on indictment (See Joint Committee on the Draft Civil Contingencies Bill, para 177), thus departing from the 1920 Act.

[102] Cabinet Office, *The Government's Response to the Report of the Joint Committee*, para 32.

[103] HL Debs vol 666 col 876 9 November 2004.

[104] *Explanatory Notes to Civil Contingencies Act 2004* (Stationery Office, London, 2004) para 53.

[105] HL Debs vol 665 col 1032 21 October 2004, Baroness Scotland.

[106] Joint Committee on the Draft Civil Contingencies Bill, para 202.

validity of 'anything done' by virtue of the regulations before they lapse, so that any rights or liabilities can persist after that time. Further consideration of the duration of emergencies is given in Chapter 6.

C. Regional Emergency Coordinators

5.38 Part II envisages that responses to emergencies will be organized at regional levels. Some details of this tier have been given in Chapter 4, and it is confirmed in guidance that the Government Offices are often the first place that 'government departments turn to for a situation report on non-terrorist incidents' and, in the event of a UK-wide emergency, 'When the national emergency co-ordination arrangements are activated, GOs will provide situation reports for the Central Situation cell, copied to the Lead Government Department and any other government department with a significant interest.'[107] Chapter 4 further described how escalating emergencies would be handled at three levels. As calamities mount, a strong coordinator is needed.

5.39 In the absence of any democratic device at regional level, s 24 allocates the power of appointment of coordinators to Ministers. By subsection (1), whenever emergency regulations are invoked, a senior Minister must appoint (a) for each Part of the United Kingdom other than England (in other words, Northern Ireland, Scotland, or Wales), in relation to which the regulations under Part II have come into effect, a person to be known as the 'Emergency Coordinator'. For each region of England in relation to which the regulations come into effect, a person to be known as the 'Regional Nominated Coordinator' (RNC) is to be appointed for that region. These officers have been explained to some extent in Chapter 4 and represent a statutory recognition of a post which has long antecedents, as explained in Chapter 2.

5.40 The RNCs are appointed in cases of severest emergencies, when Part II is triggered, as the executive heads of the Regional Civil Contingencies Committees (RCCC) (which are not mentioned in s 24). Their terms and conditions of appointment, along with job specifications, will also be specified as the need arises by emergency regulation. But by subsection (5), a coordinator shall not be regarded as the servant or agent of the Crown or as enjoying any status, immunity, or privilege of the Crown. This provision was not reflected in the draft Bill of June 2003 but does helpfully ensure certainty by ruling out resort to prerogative powers. It is also envisaged that the appointments will be covered by a concordat between the UK Government and each devolved Administration who will be consulted on the appointment except when the urgency of the situation prevents it.[108]

---

[107] Cabinet Office, *Emergency Response and Recovery*, paras 7.1–7.10.
[108] HC Debs Standing Committee F col 222 5 February 2004, Fiona Mactaggart.

By subsection (3), the principal purpose of the appointment shall be to facilitate **5.41**
coordination of activities under the emergency regulations (whether only in the
Part or region for which the appointment is made or partly there and partly
elsewhere, such as in neighbouring regions). In this way, the RNCs must add
value through the strategic coordination of the regional response[109] rather than
through direct powers of direction or operational input.[110] Their functions may
be further regulated under subsection (4) by directions and guidance issued by the
senior Minister. This power makes it clear that, once appointed, the RNCs will in
turn be answerable to the senior Minister. Presumably, such directions
and guidance will make clear that the RNCs can override the wishes of elected
representatives of the district or county or, even more controversially, the
local police chief.[111] These relationships are not, unfortunately, delineated in
the Act and will not be revealed in public until the emergency arises and
depending on the nature of the emergency. In the meantime, it should be
'business as usual':

> The Government believes that, wherever possible, usual lines of command and
> control should not be altered in times of emergency. The purpose of this is to avoid
> confusion and disruption of established procedures and relationships. This principle
> is widely supported across the civil protection community. Nevertheless, in the most
> serious emergencies, in might be necessary for the Government to be more directive.
> This is particularly true in situations where local areas do not have a full appreciation
> of the national picture. That is why the draft Bill enables, where appropriate,
> regional and emergency co-ordinators to be empowered to give strategic direction to
> local agencies in terms of setting priorities and facilitating co-ordination.[112]

Potential RNCs will be pre-nominated by the organizations represented on the **5.42**
Regional Resilience Forums, and each nominee will invariably be:

> ...a highly trained crisis-management expert with considerable personal
> authority. They will also need a degree of specialist knowledge regarding the
> area(s) affected and the various organizations involved, their procedures and
> capabilities. The RNC's relationship with existing organisations and multi-agency
> groups will be one of strategic overview and co-ordination.[113]

Even in early 2004, work was underway to develop a comprehensive list of
technical experts covering areas of responsibility, such as public health or animal
health, a list that will be added to with people skilled, trained, and experienced
in crisis management.[114] A suitable candidate for the RNC will then be selected

---

[109] Cabinet Office, *The Draft Civil Contingencies Bill*, paras 4.10, 4.13.
[110] Cabinet Office, *Emergency Response and Recovery*, para 13.15.
[111] There could also be direction of the Secretary of State's Representative (SOSREP) (see Ch 4).
[112] Joint Committee on the Draft Civil Contingencies Bill, Appendix 9 q 71.
[113] Civil Contingencies Secretariat, *The Government's Response*, 28. See also HL Debs vol 665
col 1027 21 October 2004, Lord Bassam.
[114] HC Debs Standing Committee F col 281 10 February 2004, Douglas Alexander.

from the pool of pre-nominees with sets of different skills and experience as and when the crisis arises and according to the nature of the emergency.[115] There is no set rule as to how many pre-nominations must be made nor how often the list should be reviewed.[116] It is by no means certain that senior police officers will always fulfil the role,[117] to the dismay of some who believe that they are best placed to exude authority and also to operate around the clock and in all conditions. But given that Chief Constables will almost certainly be heavily involved in police work, they may not be best placed to assume the additional burden of the office of RNC.

5.43 The Government envisages appointees who can show 'specialist subject knowledge, strong leadership and effective emergency management',[118] and it is suggested that there are three broad areas of expertise in which regional and emergency coordinators might specialize:[119]

> —Subject matter specialisation. Some Regional Nominated Co-ordinators (RNCs) might be appointed on the basis of expertise in particular types of emergency. For example, a senior health professional might be nominated to take on the role in the case of an epidemic in humans.
> —Crisis management. Some RNCs might be appointed on the basis of strong track records in leadership positions during emergencies. An example of this might be a senior police officer.
> —Regional knowledge. Some RNCs might be appointed on the basis of experience in a particular region, and experience of working closely with key players. An example of this would be the Director of a Government Office.

There could then be selection dependent on the nature of the incident and which Lead Government Department (described in Chapter 8) is to assume responsibility:

> For flooding, it might be an official from DEFRA. In the case of a flu pandemic, it would be likely to be the Regional Director of Public Health. In terrorist incidents, it would most likely be a Chief Police Officer. The person selected as RNC might work within the region already (for example, the Regional Director of Public Health) but could be attached from central government, with appropriate regional expertise provided in support.[120]

---

[115] Cabinet Office, *The Government's Response to the Report of the Joint Committee*, Annex A para 6.
[116] Joint Committee on the Draft Civil Contingencies Bill, Appendix 9 q 62.
[117] Police chiefs have acted as regional coordinators, such as in the Yorkshire and Humber region, whereas others have had civil servants as the chair (such as East Midlands). The previous regime under the Cabinet Office, *Dealing with Disaster* (rev 3rd ed, Cabinet Office, London, 2003) para 82 assumed police leadership.
[118] Joint Committee on the Draft Civil Contingencies Bill, Appendix 9 q 60.
[119] Ibid, Appendix 9 q 61. Compare Joint Committee on the Draft Civil Contingencies Bill, para 267: 'A proven crisis manager is preferable to a specialist in the discipline most closely connected to it.'   [120] Cabinet Office, *The Draft Civil Contingencies Bill*, para 4.18.

Pending a full-blown Part II emergency, it is envisaged that they might cut their teeth as 'RNC-designates', appointed to deal with any serious wide area emergency.[121]

The post will normally expire whenever the regulations issued under Part II cease **5.44** to have effect. Curiously, there is no express power of dismissal—one would have expected to see some requirements of efficiency and effectiveness along the lines applied to Chief Constables.[122] But these possibilities of termination could presumably be imported by regulation or by the contract of employment.

The creation (or, rather, recognition) of a distinct regional response mechanism **5.45** is unique to the Act which otherwise builds on existing structures. The result is that Parts I and II appear to adopt inconsistent approaches while simultaneously promising to ensure integration. A more comprehensive blueprint for regional and national structures, especially under Part I, would have avoided this problem. The reason for the contrast between Part I and Part II is, of course, that Part I involves a 'bottom up' local response to planning and crisis, where local authorities and others are left to take the lead, whereas Part II involves a 'top down' central response, with civil servants in the lead whether in the regional office or Whitehall. Yet the full scheme is not revealed in Part II, and the RNCs alone are mentioned without reference to the RRFs or RCCCs described in Chapter 4. There is, as a result, a lack of clarity as well as a total absence of accountability to the localities affected. Accountability to Parliament is also weakened and should have been enhanced by requiring under s 24(4) periodic reports on the work of the regional tier, designate or real.

D. Emergency Tribunals

Section 22(3)(n) envisages that regulations may either confer jurisdiction on a **5.46** tribunal or establish a new tribunal '[a] tribunal that is set up to approve quarantining is one example; in the event of appeals against quarantining being made during a major infectious disease outbreak or another such civil contingency'.[123] New tribunals will not be required for criminal cases, since the creation of new offences, their trial other than before existing summary courts and the alteration of criminal process are all circumscribed by s 23(4). So, it would not appear possible to assume the jurisdiction exercised by existing criminal courts. If the criminal court system were to collapse in an emergency, then further legislation would be needed, the pattern which was followed in the Second World War.[124] It

---

[121] Cabinet Office, *The Government's Response to the Report of the Joint Committee*, Annex A para 6. [122] Police Act 1996, s 11(2).

[123] HC Debs Standing Committee F col 287 10 February 2004, Douglas Alexander.

[124] See Emergency Powers (Defence) (No 2) Act 1940 and the Defence (War Zone Courts) Regulations 1940 SR & O 1444; Jennings, WI, 'The Rule of Law in Total War' (1941) 50 *Yale*

follows that the s 22(3)(n) powers will deal with civil matters and could arise in two ways. One would be where it is decided to change jurisdiction over an existing issue—to take it from one tribunal and give it to another either for the sake of convenience or because of necessity where the emergency prevents the tribunal from sitting. The second situation would be where it is decided to set up a new tribunal under the 2004 Act to deal with issues arising under it, such as the issuance of licenses or the assessment of compensation.[125]

5.47 Special care must be taken when forms of judicial process are instituted as part of the emergency regulations.[126] Therefore, by s 25(1), which was not reflected in the 2003 draft Bill but followed discussions with the Council on Tribunals, emergency regulations which establish a tribunal may not be made unless a senior Minister of the Crown has consulted the Council on Tribunals.[127] Exceptions arise under subsection (2) on the basis of urgency (but the duty to consult then applies retrospectively under subsection (6)) or where the Council on Tribunals have already consented to the new body. By subsection (3), the Council must make a report to the Minister, and the Minister shall not make the regulations until the report is received—unless, once again, the issue is urgent. But a failure to satisfy subsections (1) or (3) shall not affect the validity of the relevant regulations. Where a report is received, it shall be laid before Parliament together with a statement of the extent to which the regulations give effect to any recommendations in the report, and an explanation for any departure from recommendations in the report (subsection (5)).

5.48 In all, s 25 may turn out to be 'just a bit of window dressing',[128] given that the Council cannot at any stage override the will of the Minister and that the events may have transpired before it can react.

E. Oversight

5.49 Parliamentary scrutiny is designed to bite at a number of stages. First, by s 27(1), at the time when emergency regulations are made, a senior Minister of the Crown shall as soon as is reasonably practicable lay the regulations before Parliament, and the regulations shall lapse at the end of the period of seven days beginning with the date of laying unless during that period each House of Parliament passes a resolution approving them. Once regulations are in force,

Law Journal 365. That broad unspecified powers can remove access to the courts is doubted in *Chester v Bateson* [1920] 1 KB 820.

[125] See Joint Committee on the Draft Civil Contingencies Bill, Appendix 9 q 27.
[126] Ibid, para 166.
[127] Cabinet Office, *The Government's Response to the Report of the Joint Committee*, para 31. The Council would have preferred that all tribunals be established by primary legislation.
[128] HC Debs Standing Committee F col 285 10 February 2004, Oliver Heald.

then by subsection (2), a resolution of each House can cause them to cease to have effect at a specified time (or the day after in default).[129] Alternatively, under subsection (3), a resolution can specify amendments to the emergency regulations. But none of these forms of oversight affect the powers of Ministers to make new regulations or affect anything done by virtue of regulations before they lapse, cease to have effect or are amended (subsection (4)).

The power to amend in s 27(3) is a notable exception to the general rule that **5.50** statutory instruments cannot be amended during their Parliamentary passage. The facility was applied to the Bill introduced in January 2004 after lobbying by the Joint Committee.[130] This procedure is known as the 'super-affirmative procedure' and has been utilized under an array of recent legislation, including the Northern Ireland Act 1998, the Local Government Act 2000, and the Regulatory Reform Act 2001. However, it is not entirely a recent innovation for, under s 2(4) of the Emergency Powers Act 1920, emergency regulations 'shall have effect as if enacted in this Act, but may be added to, altered, or revoked by resolution of both Houses of Parliament'. The procedure allows for deeper analysis and scrutiny of the regulations, as compared to the 'take it or leave it' stance under the affirmative procedure, which would almost certainly result in acceptance in the heat of a crisis.[131]

Special rules under s 28 cater for the difficulties which might be caused by **5.51** emergencies which arise during times when Parliament is not sitting. Even though Parliament may not be sitting or even if it has been dissolved pending a General Election, the Privy Council still exists and Ministers still retain office and so can take action under Part II. By s 28(1), if Parliament stands prorogued to a day after the end of the period of five days beginning with the date on which the regulations are made, there must be issued a proclamation under the Meeting of Parliament Act 1797 to require Parliament to meet on a specified day within that period. If Parliament has been dissolved, then no recall or reinstatement is possible. But once a new Parliament has been elected, it will be required to meet quickly, but in the meantime, any regulations would still carry legal force for 30 days.[132] Alternatively, under subsection (2), if the House of Commons stands adjourned to a day after the end of the period of five days beginning with the date on which the regulations are made, the Speaker (or alternate as specified by House Standing Orders) must arrange for the House to meet on a day during that period. A corresponding duty is imposed upon the

---

[129] See Cabinet Office, *The Government's Response to the Report of the Joint Committee*, para 37.
[130] Joint Committee on the Draft Civil Contingencies Bill, para 198; Cabinet Office, *The Government's Response to the Report of the Joint Committee*, para 37.
[131] See Bonner, D, *Emergency Powers in Peacetime* (Sweet & Maxwell, London, 1985) 42–6.
[132] Joint Committee on the Draft Civil Contingencies Bill, Appendix 9 q 18; HL Debs vol 665 col 1041 21 October 2004, Baroness Scotland.

Lord Chancellor[133] (or alternate as specified by House Standing Orders) in the case of House Lords adjournments (subsection (3)).

5.52 By s 30, all emergency regulations shall be made by statutory instrument. This designation even applies to those made under s 20 by Order in Council. Section 30 goes on to state, in pursuance to the Government's change of heart after initial consultation and described further in Chapter 7, that the regulations must be treated under the Human Right Act 1998 as subordinate legislation and not primary legislation.[134] Secondary legislative instruments can be declared to be invalid under administrative law to the extent of their incompatibility with their parent Act.[135]

F. Devolved Territories and London

5.53 As mentioned earlier, the devolution arrangements produce uneven complexity which is to be smoothed by administrative arrangements.

5.54 Scrutiny by the devolved administrations is dealt with by s 29. The relevant Minister of the Crown must consult before making regulations the Scottish Ministers, the (Northern Ireland) First Minister and Deputy First Minister, or the National Assembly for Wales, as the case may be. The inclusion of Northern Ireland is in contrast to the position under Part I of the Act. The process of consultation is subject to overriding needs of urgency, and, in any event, a failure to satisfy a requirement to consult shall not affect the validity of regulations.

5.55 The Consultation Paper of 2003 mooted, in the interests of flexibility and subsidiarity, allowing each of the devolved administrations to make such a declaration of emergency under Part II and to invoke measures so far as they are within their competence.[136] During consultation, there was said to be general agreement, with especial enthusiasm in Scotland.[137] The idea was quickly dropped as wiser counsel prevailed in Westminster.[138] A change in the rules from previous practice (except for Northern Ireland) was a risk, it would have been messy in administrative terms because the competence of the devolved administrations is patchy, which means that the centre would still become embroiled. As for Northern Ireland, there is also the unfortunate historical

---

[133] The office is to be replaced by the Speaker of the House of Lords: Constitutional Reform Act 2005, s 18.

[134] These seem to be the prime reasons for s 30(2): HC Debs Standing Committee F col 304 10 February 2004, Douglas Alexander.

[135] See, for example, *Secretary of State for Education v Tameside MBC* [1977] AC 1014; *Bromley LBC v Greater London Council* [1982] 1 All ER 129.

[136] Cabinet Office, *The Draft Civil Contingencies Bill*, para 6.18.

[137] Civil Contingencies Secretariat, *The Government's Response*, 44, 45.   [138] Ibid, 46.

experience of the abuse of special emergency powers, resonances which might be better avoided.[139]

Concordats are being devised between Westminster and the devolved admin-   **5.56**
istrations covering arrangements for the use of special legislative measures; the
target publication date is January 2006.[140] Indications have been given that they
will cover 'the use and nature of regulations and the identity and functions of
the emergency coordinator'.[141] During any suspension of devolved bodies in
Northern Ireland, there will be a trilateral agreement between the Head of the
Northern Ireland Civil Service, the Northern Ireland Office, and the Cabinet
Office—in other words, the Government will agree to talk with itself!

Turning to specific Parts of the United Kingdom, in Northern Ireland, national   **5.57**
security and nuclear matters are 'excepted' matters (beyond the devolved bodies'
competence) while civil aviation, criminal justice, public order, civil defence
(including the 'subject-matter of the [then] Emergency Powers Act (Northern
Ireland) 1926') are 'reserved' matters (requiring the consent of the Secretary of
State if local powers are to be used).[142] Section 29(2) specifies that the Northern
Ireland First Minister and Deputy First Minister must be consulted, where
possible, in advance of use affecting Northern Ireland.

When a major emergency occurs in Northern Ireland, the Head of the Northern   **5.58**
Ireland Civil Service in consultation with Permanent Secretaries or the nominated departmental emergency planning liaison officer will convene the Central
Emergency Management Group (CEMG) of the Central Emergency Planning
Unit (CEPU).[143] The CEMG also meets at the request of the Lead Department
or others in actual emergencies where a strategic multi-agency involvement is
anticipated or required.[144] The Crisis Management Group (CMG) will be
convened, along with the CEMG, to deal with serious live emergencies.[145] It
will only be called in the event of a major disaster or crisis where existing
departmental arrangements are insufficient to provide and coordinate the
resources necessary to deal with the situation. Crisis management response will
be coordinated through the activation of the Northern Ireland Office Briefing
Rooms (NIOBR) which will be chaired by the Secretary of State for Northern
Ireland, a designated Minister, or senior official.[146] In addition, the Northern
Ireland Information Management Centre (NIIMC) acts in an emergency, or

---

[139] See Hogan, G and Walker, CP, *Political Violence and the Law in Ireland* (Manchester
University Press, Manchester, 1989).
[140] Civil Contingencies Secretariat, *The Government's Response*, 39 (Scotland), 40 (Wales).
[141] Joint Committee on the Draft Civil Contingencies Bill, Appendix 9 q 25. See further
Cabinet Office, *Emergency Preparedness: Guidance on Part 1*, para 11.4.
[142] Northern Ireland Act 1998, Schs 2 and 3.   [143] See http://cepu.nics.gov.uk.
[144] Cabinet Office, *Emergency Response and Recovery*, para 11.13.
[145] Ibid, para 11.16.   [146] Cabinet Office, *Central Government Arrangements*, Annex B.

potential emergency, focusing on inter-organizational communications or operations, and in support of the CEPU or specifically the CMG or CEMG. There is also an Infrastructure Emergency Planning Forum.

5.59 In Scotland, defence (but not the exercise of civil defence functions by the armed forces), emergency powers, national security, the interception of communications, official secrets and terrorism, energy matters (including under the Energy Act 1976) and transport security are 'reserved' matters.[147] The Scottish Executive must be consulted, where possible, in advance of the use of special legislative measures under Part II which affect Scotland (s 29(1)).

5.60 The Emergency Management Unit within the Civil Contingencies Division of the Scottish Executive Justice Department is responsible for the Scottish Executive Emergency Room (which will be activated to deal with inter-departmental coordination, to facilitate communication, to contact UK Government departments and COBR, to contact Scottish Executive departments, and to inform the Scottish public and media).[148] The Scottish Executive Justice Department Secretary chairs the Scottish Emergencies Coordinating Committee (akin to an English RCCC), available for the coordination of an emergency response led by any Scottish Executive division and whose work mirrors the Civil Contingencies Secretariat (CCS).[149] This 'ensures that steps are taken to respond to the changing risk environment and determines the national strategy for the development of civil protection.'[150] Its membership includes Executive Departments, emergency services, local authorities, and the military.

5.61 In Wales, only agriculture, the environment, health, highways, transport, and water and flood defence are amongst the 'transferred' matters.[151] The Welsh Assembly must be consulted, where possible, in advance of the use of special legislative measures under Part II affecting Wales. But, in common with other devolved administrations, it is not granted any power to promulgate emergencies, contrary to the wishes of the Welsh Assembly itself.[152] An Emergency Co-ordination Centre at the Welsh Assembly Government will operate during an emergency. The Wales Civil Contingencies Committee can act akin to the RCCCs in England.[153]

---

[147] Scotland Act 1998, Sch 5.
[148] Cabinet Office, *Emergency Response and Recovery*, para 9.14. It will receive information from the Scottish Police Information and Co-ordination Centre.
[149] See Ch 8 and http://www.civilcontingenciesscotland.gov.uk/.
[150] Cabinet Office, *Emergency Preparedness: Guidance on Part 1*, para 10.13.
[151] Government of Wales Act 1998, Sch 2.
[152] Local Government and Public Services Committee, *Report on the Civil Contingencies Bill* (National Assembly for Wales, Cardiff, 2004) para 7.
[153] Cabinet Office, *Emergency Response and Recovery*, para 10.2.

As for London, it should be noted that during response and recovery phases, 'the **5.62** London Resilience Team would provide the [Government Office] representation to either the SCG or the RCCC.'[154] LFEPA will support pan-London local authority strategic arrangements, it will deliver an effective control centre to support strategic coordination groups, it will provide training to support pan-London arrangements and will provide means of exercising pan-London arrangements.[155]

G. Regulations and Guidance

In contrast to Part I, the Government made no firm promises to publish Part II **5.63** regulations in draft and in fact has not done so. Its only concession was to pass onto the Joint Committee a confidential copy of the existing draft emergency regulations. Those draft regulations resembled those issued under the Emergency Powers Act 1920 in the 1970s. The Cabinet Office indicated in 2003 that a review was being undertaken and that it intended 'to identify an indicative list of draft regulations that would be possible under [the Act]'.[156] Later signals are that generic regulations have been drafted with each Lead Government Department in mind and that there will be selection within the menus according to the nature of the emergency which actually arises.

An illustration of emergency regulations might be those issued in November **5.64** 1973, the last set ever promulgated in Britain.[157] The first tranche of regulations dealt with transportation. Thus, they concerned the regulation of ports (regs 3, 4, and 5) and changes to restrictions on road vehicles (regs 6 to 15). Public services and utilities, including the supply of power were next on the agenda (regs 16 to 20), followed by its consumption and supply (regs 21 to 29). The requisitioning of chattels and the taking possession of land was allowed by regs 30 to 31. Enforcement was the subject of most of the remainder, with special offences of sabotage, trespassing and loitering, interference with soldiers, constables, or persons performing essential services, and inducing disaffection of soldiers and constables, as well as summary powers of arrest and penalties (up to three months' imprisonment) (regs 30 to 41).

Whilst often prosaic in character, past emergency regulations did contain some **5.65** measures which were very threatening to a broad range of human rights. History belies the claim of the sponsors of the 2004 Act that its design made it 'practically

---

[154] Ibid, para 8.13.
[155] Cabinet Office, *Emergency Preparedness: Guidance on Part 1*, para 9.6.
[156] Joint Committee on the Draft Civil Contingencies Bill, Appendix 9 q 22.
[157] Emergency Regulations 1973 SI No 1881. They were renewed by 1973 SI No 2089; 1974 SI No 33; 1974 SI No 175; 1974 SI No 350. For a fuller description of regulations issued under the 1920 Act, see Morris, GS, 'The Emergency Powers Act 1920' [1979] PL 317 at 326–31.

impossible for any government to abuse these powers, even if any government was likely to be disposed to do so.'[158] For example, the offence in reg 19 of the 1921 Emergency Regulations relating to inducing mutiny, sedition, or disaffection of soldiers, constables, fire officers, or of the civilian population was invoked in *Inkpin v Roll* in 1921 to convict the Secretary General of the Communist Party of Great Britain for the publication of, *inter alia*, the Theses and Statutes of the Party, thus denying the very legality of a political party.[159] During the General Strike of 1926, as well as 1,760 prosecutions for breaches of the regulations (150 relating to incitements),[160] new regulations were passed to prevent payments from abroad, a reaction to the scares about the payment of 'Moscow gold' to sympathetic labour and political bodies.[161] Regulations also existed at that time for the banning of meetings and processions[162] and were queried as being more designed 'not to preserve order but to suppress Communist propaganda'.[163]

5.66 Aside from the substance of the regulations, another controversial aspect has been the deliberate policy of overkill—the practice to issue this relatively settled collection of regulations in full on every occasion.[164] There was no attempt to provide for designer regulations *à la carte* according to the crisis being faced. But should the same pattern be replicated under the 2004 Act? The 1920 Act was in practice, if not design, wholly confined to industrial emergencies, so the fact that the same regulations appeared time and again was consistent with their use. But the 2004 Act applies to a much wider range of scenarios. The demands of the Triple Lock and especially the criterion of necessity suggests that the same set of regulations should not be drafted for every occasion. This point echoes the questions of interpretation around s 21 and whether it allows a full panoply of regulations to be issued without discernment or whether each and every regulation must meet each and every Triple Lock test. Since it was concluded that the latter was the outcome more consistent with s 21 and the philosophy of the Triple Lock, so the idea of a single set of standard draft regulations should no longer be tenable. The greater emphasis on resilience planning under Part I should also result in more selection amongst regulations rather than a 'one size fits all' approach.

---

[158] HL Debs vol 42 col 104 28 October 1920, Viscount Milner.

[159] (1922) 82 JP 61. The regulations continued to contain a disaffection offence even after the Incitement to Disaffection Act 1934.

[160] HC Debs vol 196 col 824 2 June 1926, Sir W Joynson-Hicks.

[161] 1926 SR & O 556 (reg 13A). See Thurlow, R, *The Secret State* (Blackwell, Oxford, 1994) 147; Bennett, G, *A Most Extraordinary and Mysterious Business* (Foreign & Commonwealth Office, London, 1999). Foreign payments are now regulated by the Political Parties, Elections and Referendums Acts 2000, s 54.

[162] Regulation 22 was used on 22 occasions from 1 May until 19 October 1926, and 63 from then until 18 November 1926: HC Debs vol 199 col 1847 17 November 1926; ibid, col 2090 18 November 1926; ibid, col 2107 19 November 1926.

[163] HC Debs vol 199 col 2107 19 November 1926, George Lansbury.

[164] Morris, GS, 'The Emergency Powers Act 1920' [1979] PL 317 at 325.

It might also be hoped that the Cabinet Office will eventually agree to them **5.67** being scrutinized in draft. The lack of available information on the contingencies envisaged for crises other than industrial emergency means that, if ever needed, there will an absence of considered debate by both Parliament and a lack of preparedness on the part of agencies affected.[165] If public scrutiny is a step too far for the official psyche, then perhaps there could be review in confidence by the parliamentary Intelligence Committee. Unfortunately, there is no present intention to shed further light on the draft regulations. Maintaining secrecy remains here the 'dominant characteristic of British contingency planning'.[166] Consequently, the focus of Ministerial pronouncements has been on why details cannot be given:

> There are three strong arguments against making the full set of draft regulations available. Firstly, the draft regulations are subject to frequent change. They are updated at least every two years, and more often if necessary. Secondly, 'standard' (or publication of draft) regulations would not necessarily offer a clear indication of the content of future emergency regulations. By their very nature emergencies are unpredictable and can occur in, and include, a very wide-range of circumstances and events. Emergency regulations will in large measure be tailored to the particular emergency at hand. The draft regulations are very much a starting point, and it is highly likely that any actual regulations would diverge from the standard draft. Thirdly, draft regulations will be designed to respond to emergencies including terrorism and disruptive industrial action. Wide access to draft emergency regulations could highlight both potential weaknesses or targets and likely counter-measures.[167]

These arguments are feeble.[168] As for frequent change, publication in line with **5.68** the review period of two years is surely feasible, especially if web-based. Further, if the facilities exist to issue regulations in a real emergency 'within a minimum of six hours',[169] it is hardly onerous to require the publication of drafts from time to time. Secondly, a single set of regulations would indeed be a poor offering, given the range of emergencies covered by the Act, but it would be better than nothing and it will surely be evident from the subjects covered and the ordering in Part which regulations are relevant to which types of emergencies. Thirdly, the idea that laws (as opposed to operations pursuant to them) must remain secret because they might give forewarning is unacceptable in the light of the principle of the rule of law. Furthermore, this argument could work

---

[165] See Defence Committee, *Draft Civil Contingencies Bill* (2002–03 HC 557) para 13; New Zealand Law Commission, *Report No 22, Final Report on Emergencies*, (Wellington, 1991) para 6.79.

[166] Morris, GS, *Strikes in Essential Services* (Mansell, London, 1986) 122. In the prelude to the Second World War, the planning of civil defence precautions was conducted in secret from 1924 to 1935: O'Brien, TH, *Civil Defence* (HMSO, London, 1955) 6.

[167] Joint Committee on the Draft Civil Contingencies Bill, Appendix 9 q 23. See further, Cabinet Office, *The Government's Response to the Report of the Joint Committee*, para 36.

[168] See Joint Committee on the Draft Civil Contingencies Bill, para 196.

[169] Cabinet Office, *Emergency Response and Recovery*, para 13.10.

both ways—publication in advance could allow a dialogue which ensures weaknesses are reduced, and it could ensure training and better enforcement.

5.69 Just as there are no regulations, so the Cabinet Office guidance, *Emergency Response and Recovery*, is less detailed and less binding than its Part I counterpart. It reveals that the relevant Lead Government Department will make a recommendation as to necessary regulations, but such a serious issue may well come before Cabinet.

5.70 It was mentioned at the outset of this chapter that measures could be taken if an emergency 'has occurred, is occurring or is about to occur' (s 21(2)). The practice under the Emergency Powers Act 1920 was latterly to promulgate at an early stage of the crisis or even as a precautionary measure.[170] However, one cannot be certain that this practice will continue under the 2004 Act, and the signs from recent events such as the spread of H5N1 Avian Influenza and the July 2005 London bombings are that Ministers are very reluctant to resort to the Part II powers as a precaution.

H. Other Emergency Powers

5.71 Corresponding to the position on Part I, a range of statutory powers beyond Part II grant emergency powers and will often take precedence. This ordering is a matter of law by virtue of the Triple Lock mechanism which includes a 'necessity' standard (especially through s 21(6)). In addition, it is a matter of administrative and political preference:

> Where possible, Governments have enacted legislation to deal with specific emergencies. As a consequence, a large body of sector specific emergency legislation exists. Where possible, the Government would turn to the powers available under such legislation first in the event of an emergency.[171]

5.72 The reasons for the precedence include the political imperative of not dramatizing a crisis. As described in Chapter 2, the post-1979 practice was increasingly to turn to sectoral legislation. This analysis is denied by the Cabinet Office:

> The unusually high number of States of Emergency between 1970 and 1973 seems likely to have been the consequence of a number of factors, rather than simply the attitude of the Government of the day. The early 1970s saw a combination of serious emergencies coupled with an increasingly interdependent and technologically dependent society.[172]

---

[170] See Morris, GS, 'The Emergency Powers Act 1920' [1979] PL 317 at 335–8.
[171] Joint Committee on the Draft Civil Contingencies Bill, Appendix 9 q 1.
[172] Ibid, Appendix 9 q 2.

This denial is unconvincing—society became even more interdependent and technologically dependent after 1979 and was troubled not only by major industrial strife (as in the case of the NUM strike from 1984 to 1985)[173] but also by major bouts of street riots in 1981 and 1985.[174] Yet the Emergency Powers Acts remained quiescent.

An example of a more technical interplay between the sectoral and the emergency **5.73** legislation might be given in the context of animal diseases. One obviously relevant additional power that the Civil Contingencies Act offers over and above the Animal Health Acts 1981 to 2002 would appear to be the power to slaughter animals without paying compensation. Other Part II powers might also be handy:

> ... an animal health emergency might have a collateral impact on the national infrastructure in an unexpected way that cannot be catered for by the Animal Health Acts. For example, an animal health emergency might affect the food supply, or necessitate movement bans. In such circumstances recourse might be needed to powers in the draft Bill to address the wider impact.[175]

Several examples of alternative sectoral legislation have been cited in Chapter 4, **5.74** for many straddle Parts I and II of the Act. Of relevance purely to Part II and perhaps the most extensive and prominent sectoral legislation relevant to Part II is that dealing with terrorism. The Terrorism Act 2000, the Anti-Terrorism, Crime and Security Act 2001, and the Prevention of Terrorism Act 2005 form together the code on the subject, alongside a pending Terrorism Bill 2005–06, and there is a considerable array of other terrorism legislation beyond these specialist Acts.[176] The Terrorism Act 2000 is divided into eight parts, with six reflecting its substantive themes: proscribed organizations; terrorist property; terrorist investigations; counter-terrorism powers; criminal offences; and extra measures relating to criminal process and policing powers confined to Northern Ireland, including non-jury 'Diplock' trials. The Terrorism Bill will extend the grounds for proscription, elongate police powers of detention following arrest, and create new offences of encouragement of terrorism. The Anti-Terrorism, Crime and Security Act 2001 is organized into 14 parts. The first three deal with the forfeiture of terrorist property, the seizure of terrorist cash, and the freezing of foreign property held by UK institutions.[177] Part IV addresses immigration and asylum matters, the most notable of which, concerning detention without trial, was repealed and replaced by 'control orders' under the

---

[173] See Ch 2.

[174] See Lord Scarman, *The Brixton Disorders* (Cmnd 8427, London, 1981); Home Office, *Review of Public Order Law* (Cmnd 9510, London, 1985).

[175] Joint Committee on the Draft Civil Contingencies Bill, Appendix 9 q 6.

[176] See Walker, C, *The Anti-terrorism Legislation* (Oxford University Press, Oxford, 2002); Privy Counsellor Review Committee, *Anti-Terrorism, Crime and Security Act 2001 Review: Report* (2003–04 HC 100); Home Office, *Counter Terrorism Powers* (Cm 6147, London, 2004); Walker, C, 'Terrorism and Criminal Justice' [2004] Crim LR 311.

[177] See Privy Counsellor Review Committee, *Anti-Terrorism, Crime and Security Act 2001 Review, Report* (2003–04 HC 100) paras 123, 126, 133).

Terrorism Act 2005.[178] It follows that the powers in Part II of the Civil Contingencies Act may supplement the specialist legislation by allowing for a general power of internment without trial. Returning to the 2001 Act, Parts VI to X govern dangerous substances (weapons of mass destruction (Part VI) and pathogens and toxins (Part VII)[179]) and acute vulnerabilities (such as nuclear and aviation facilities (Parts VIII and IX)). Aspects of Parts X and XI cover policing and surveillance measures. Finally, there are criminal offences, such as s 117 (withholding information about terrorism).

5.75 Finally, even the broad acres of Part II of the 2004 Act would not suffice in wartime: 'In a war, the Government would bring forward primary legislation.'[180] The vast catalogue of measures under the Emergency Powers (Defence) Act 1939[181] and related legislation points the way. The strategy would be to maintain a nucleus of central civil control but with maximum devolution to regional commissioners, all armed with 'ferocious' new powers.[182]

I. Conclusions

5.76 Part II of the Civil Contingencies Act is the most powerful and extensive peacetime legislation ever enacted. It reflects a determination to respond to extreme risk and does so by methods which involve command and control rather than a broad governance approach. Indeed, it contains within it the tools for dismantling civil society. The process of scrutiny and passage did, however, result in the strengthening of the Triple Lock safeguards. Politics and history may also conduce against its intrusion. As for politics, the Government is wary of Part II regulations and has asserted that 'The presumption is against their use.'[183] As for history, the Emergency Powers Act 1920 has lain dormant since 1974, especially as more focused alternatives to it were brought into being. One wonders whether it will be decades longer before Part II of the 2004 Act will spark into life, particularly if planning and resilience is improved under Part I. One hopes so, for its necessity would represent both a calamity and, to some extent, a failure of planning and resilience. It would also create grave dangers to the constitution and to human rights, as shall now be further explored.

---

[178] See *A and others v Secretary of State for the Home Department* [2004] UKHL 56; Walker, C, 'Prisoners of "War All the Time" ' [2005] *European Human Rights Law Review* 50.

[179] See further Walker, C, 'Biological Attack, Terrorism and the Law' (2004) 17 *Journal of Terrorism and Political Violence* 175.

[180] HL Debs vol 665 col 1024 21 October 2004, Lord Bassam. [181] See Chs 2, 6.

[182] Hennessy, P, *The Secret State* (Allen Lane, London, 2002) 154. See also ibid, 121–8, 140, 171. [183] Cabinet Office, *Emergency Response and Recovery*, para 13.14.

6

CONSTITUTIONAL ISSUES

The next three chapters will seek to draw together from different Parts of the **6.01** Act the themes of principle, policy, and implementation which make it so controversial. The first two highlight constitutional principle. As for this chapter, the core question, as raised in Chapter 2, is how emergency powers can be instituted without providing 'any exploitable opportunity to misuse emergency powers and potentially, in a worst case scenario, allow for the dismantling of democracy.'[1] The same questions can be asked of judicial review and other forms of independent scrutiny. Such weighty concerns notwithstanding, these discussions should be viewed in the light of the conclusion reached in Chapter 2 that the Act is in principle justifiable.

A. The Principles of Constitutionalism

An unrelenting policy debate adverts to the delicate conflict between the pro- **6.02** tection of rights to public safety[2] and the diminution of individual rights as well as other constitutional values such as executive accountability. To give the

---

[1] Joint Committee on the Draft Civil Contingencies Bill, para 6.
[2] See *R v Director of Public Prosecutions, ex p Kebilene* [1999] 3 WLR 972 at 1000 per Lord Hope.

complete picture,[3] full constitutional governance requires adherence to constitutionalism—principles which reflect the values of individual rights,[4] democratic and legal accountability and review and respect for other central tenets of the UK constitution's rules and principles.[5] It also demands that the overall purpose should be the restoration of normal existence without special powers and that, once achieved, special powers should be lifted. One might also criticize the very language of 'balance' and 'trade off'. Rights should not be traded in return for security but rights to security of both life and liberty should be secured,[6] as expressed two centuries earlier (and declaimed at the outset of this book) by Lord Mansfield in *R v Wilkes*:

> We must not regard political consequences; however formidable soever they may be; if rebellion was the certain consequence, we are bound to say '*fiat justitia, ruat caelum*'. ('Though the heavens fall, let justice be done').[7]

6.03 The task then is to ensure that Civil Contingencies Act powers are only triggered by serious threats, are proportionate to them, and do not disproportionately or unlawfully curtail individual rights, and will terminate when the serious threat has dissipated either through the efforts of the special measures or otherwise. Furthermore, any processes involving the Act must involve clear laws for which there is legal and legislative accountability. As contended by the Joint Committee on the Draft Bill, 'The rule of law demands that the courts and Parliament are not impotent in response to the might of executive power in an emergency.'[8]

6.04 Have these objectives been secured? This chapter will consider: whether the Act is designed overall in a way which encourages constitutionalism; whether the powers potentially affecting the constitution are sufficiently constrained; what

---

[3] Compare: *Inquiry into Legislation against Terrorism* (Cm 3420, London, 1996) para 3.1; Privy Counsellor Review Committee, *Anti-Terrorism, Crime and Security Act 2001 Review Report* (2003–04 HC 100) Pt C3; Home Office, *Counter-terrorism powers* (Cm 6147, London, 2004) Pt II para 1; Ignatieff, M, *The Lesser Evil: Political Ethics in an Age of Terror* (Princeton University Press, 2004) 23–4.

[4] See Council of Europe, *Guidelines on Human Rights and the Fight Against Terrorism* (3rd ed, Strasbourg, 2005).

[5] Walker, CP, 'Constitutional Governance and Special Powers against Terrorism' (1997) 35 *Columbia Journal of Transnational Law* 1.

[6] See Ashworth, A and Redmayne, M, *The Criminal Process* (3rd ed, Oxford University Press, Oxford, 2005) Ch 2.

[7] (1770) 98 ER 327 at 347. For the position during the emergency of the Second World War, see Berriedale Keith, A, 'The War and the Constitution' (1940) 4 MLR 1; Carr, CT, 'Crisis Legislation in Great Britain' (1940) 40 *Columbia Law Review* 1309; Kidd, R, *British Liberty in Danger* (Lawrence & Wishart, London, 1940); Jennings, WI, 'The Rule of Law in Total War' (1941) 50 *Yale Law Journal* 365; Rossiter, C, *Constitutional Dictatorship: Crisis Government in the Modern Democracies* (Princeton University Press, Princeton, 1948); Cotter, CP, 'Constitutionalizing Emergency Powers' (1953) 6 *Stanford Law Review* 382; Townshend, C, *Making the Peace* (Oxford University Press, Oxford, 1993) Ch 6.

[8] Joint Committee on the Draft Civil Contingencies Bill, para 180.

are the opportunities for democratic scrutiny; what are the opportunities for judicial scrutiny; what are the opportunities for other independent scrutiny; and is there respect for the constitutional settlement represented by devolution?

B. Overall Design

Whilst there should be civil contingencies legislation, it is less evident that **6.05** legislative effort should have focused upon one Act, albeit in skeletal form and alongside the likes of the Energy Act 1976. This 'framework' or 'skeletal' approach carries the important benefit of being able to adapt to unexpected situations. However, a less felicitous consequence is that very sweeping powers are necessitated to cover all eventualities.

One might contrast the skeletal approach with the 'sectoral' approach adopted by **6.06** the New Zealand Law Commission when it considered the design of emergency laws.[9] Its *First Report on Emergencies* of 1990 and *Final Report on Emergencies* in 1991[10] centrally recommended that when emergency powers are required, they should be conferred via 'sectoral legislation'—legislation deliberated upon and designed in advance of the emergency and tailored strictly to the needs of each particular kind of emergency.[11] The sectoral approach is not reflected in the Civil Contingencies Act though it adds to, rather than removes, the prior and extensive sectoral legislation. The sectoral approach can, through a matrix of legislative provision provide state responses which are predictable and effective but, at the same time, will be regulated whatever direction they take. A sectoral approach may also encourage inter-agency cooperation since more clarity and focus is imparted by the legislation concerning the task in hand.[12] If a 'sectoral' approach had been executed, especially by the update and amendment of the many sectoral Acts (such as those listed in Chapter 4) which do exist, then the need for the full panoply of regulation-making powers under the 2004 Act would have been reduced in practice and by the effect of s 21:

> It would be possible to have a Flooding Act to provide appropriate powers for tackling flooding, a Supply of Essential Services Act, and so on. The Animal Health Act 2002 already provides those powers considered to have been lacking in the epidemic of Foot and Mouth Disease in 2001.[13]

---

[9] New Zealand Law Commission, *Report No 12, First Report on Emergencies* (Wellington, 1990) 11. See further Ch 9.
[10] New Zealand Law Commission, *Report No 12, First Report on Emergencies* (Wellington, 1990); Report No 22, *Final Report on Emergencies* (Wellington, 1991).
[11] *Final Report on Emergencies*, x.
[12] *First Report on Emergencies*, para 8.65; *Final Report*, paras 8.76, 9.6, 9.7, 9.20.
[13] House of Commons Library, Civil Contingencies Bill (Research Paper 04/07, House of Commons, London) 7, 8.

C. Breadth of Powers Over Constitutional Matters

6.07 It ought to be a source of consternation that Ministers are ever given powers to amend any previous Acts of Parliament. However, these 'Henry VIII' clauses[14] are now almost commonplace, being especially prominent in the Deregulation and Contracting Out Act 1994 and the Regulatory Reform Act 2001 and even in the Human Rights Act 1998, s 10.

6.08 The Civil Contingencies Bill tabled in January 2004[15] was sensibly altered during passage to curtail the use of Henry VIII clauses. A very important addition by s 23(5) was that emergency regulations may not amend Part II of the Act itself (which could have been a prelude to limitless, Doomsday powers) or any part of the Human Rights Act 1998 (considered further in Chapter 7).

6.09 But there was felt to remain acute danger to the constitutional fabric from the regulation-making power in what became s 22(3)(j)—to 'disapply or modify an enactment or a provision made under or by virtue of an enactment'. In response the Joint Committee on the Draft Civil Contingencies Bill commented that:

> In the wrong hands, this could be used to remove all past legislation which makes up the statutory patchwork of the British constitution. We believe that the Bill should list a number of fundamental parts of constitutional law that should be exempt from modification or disapplication.[16]

6.10 Likewise the House of Lords Select Committee on Delegated Powers and Regulatory Reform[17] questioned the power to override constitutional rights. Amongst the statutes which might be put on the endangered species list were: Magna Carta 1297, the Bill of Rights 1688, the Crown and Parliament Recognition Act 1689, the Act of Settlement 1700, the Union with Scotland Act 1707, the Union with Ireland Act 1800, the Parliament Acts 1911–49, the Life Peerages Act 1958, the Emergency Powers Act 1964, the European Communities Act 1972, the House of Commons Disqualification Act 1975, the Ministerial and Other Salaries Act 1975, the British Nationality Act 1981, the Supreme Court Act 1981, the Representation of the People Act 1983, the Government of Wales Act 1998, the Northern Ireland Act 1998, the Scotland Act 1998, and the House of Lords Act 1999.[18]

6.11 As noted above, the Government accepted just two candidates which were also on this list—the Human Rights Act 1998 and the Civil Contingencies Act 2004. The official reason for the former is that 'unlike other constitutional

---

[14] For their 'proper' use, see *Report of the Committee on Ministers' Powers* (Cmd 4060, London, 1932) para 61.   [15] 2003–04 HC No 14.

[16] Draft Civil Contingencies Bill (2002–03 HC 1074) para 13.

[17] *Twenty-Fifth Report* (2003–04 HL 144) para 23.

[18] Joint Committee on the Draft Civil Contingencies Bill, paras 183, 184.

legislation, the Human Rights Act sets out the relationship between the individual and the state that is at the heart of the operation of the emergency powers and fundamental to concerns about their possible misuse'.[19] As for the latter, it came to be recognized that the 2004 Act itself 'should similarly be incapable of being amended by emergency powers, in order to ensure that the tests and protections it contains cannot be amended'.[20] For the rest, the Minister rejected the necessity for formal protection.

One reason was that there is no precedent for a constitutional list[21] and that **6.12** 'because of the unwritten nature of the British constitution, there are considerable difficulties in defining in such a Bill exactly what constitutional legislation is'.[22] True enough, though one might counter that the Civil Contingencies Act goes beyond any precedent, including the Emergency Powers Acts. Whilst one might not be able to agree on a full list of constitutional statutes, one could surely consider specific nominated cases, as represented by the foregoing list. Next, the Minister contended, because there is no clear category of 'constitutional' legislation, then Acts which one might see as predominantly 'constitutional' may contain a rag-bag of irrelevant provisions, some of which should be subject to alteration in an emergency. Of course, this problem could be countered by specifying just those sections or Parts which are incontrovertibly to be protected. Such a process of debate and delineation need not be an attempt 'exhaustively to list all such constitutional enactments',[23] but should focus upon those most vital to the constitution and most at risk in an emergency.

A second reason for standing firm against the guarding of the constitutional **6.13** jewels was that the Act should not be interpreted at all as affecting constitutional law: 'For the record and for the purposes of interpretation, I make it clear that, in crafting this Bill . . . It does not include any constitutional arrangements.'[24] The rationale (based on expert advice) is set out in full below:

> We have sought advice from Parliamentary Counsel as to the scope of the power to 'modify or disapply an enactment', and in particular whether it would permit regulations under Part 2 of the Bill to modify an enactment which has constitutional importance—such as the Human Rights Act 1998 or the Bill of Rights Act 1689.

> They have advised that each proposed exercise of such a power must be assessed by reference to whether or not it is within the class of action that Parliament must have contemplated when conferring the power. There are certain rebuttable presumptions as to what Parliament must have intended in conferring a power of this kind.

---

[19] HL Debs vol 666 col 867 9 November 2004, Baroness Scotland.   [20] Ibid.

[21] Cabinet Office, *The Government's Response to the Report of the Joint Committee*, para 34.

[22] HC Debs vol 416 col 1110 19 January 2004, Douglas Alexander.

[23] HL Debs vol 666 col 1351 16 November 2004, Baroness Scotland. The further argument that '*expressio unius est exclusio alterius*' could easily be countered by an 'in particular' clause.

[24] HL Debs vol 666 col 869 9 November 2004, Baroness Scotland. See also HC Debs vol 416 col 1113 19 January 2004, Douglas Alexander.

These may be presumptions of common law (for example, the presumption against the imposition of taxation) or presumptions based on statute (for example, section 2 of the European Communities Act 1972 or section 3 of the Human Rights Act 1998). These presumptions apply even where Parliament has used general language. The courts have also suggested rules in relation to provisions of particular constitutional importance, requiring statutory modification to be express.

The Bill does not contain any express provision that enables regulations under Part 2 of the Bill to modify or disapply a constitutional enactment. While the specific powers listed in the Bill are very wide ranging, they are capable of being exercised without interfering with a constitutional enactment. In particular, they are capable of being exercised in accordance with the Convention rights. Nor is the permission to do anything that an Act could do sufficiently precise to displace the general approach detailed above.

In light of this, Parliamentary Counsel have advised that, in exercising the power conferred under Part 2 of the Bill, in the unlikely event of needing to use this power, Parliament will not permit interference either with a general presumption or with a 'constitutional' enactment. However, it may be safe to assume that Parliament intended to confer the power to interfere with such a statute if the interference is trivial in so far as it concerns the substance of the presumption or the constitutional enactment.

Given the inherent limits on the scope of the power, Parliamentary Counsel have advised that if we wished to be able to modify or disapply a constitutional enactment, we should take an express power to do so. We do not propose to do this.

Without such an express power, we cannot presently envisage circumstances in which this power would lawfully enable us to make a substantive amendment to a constitutional enactment.[25]

6.14 These propositions were rested on the prime precedent[26] of *Thoburn v Sunderland City Council*.[27] The defendant, a food trader, was charged with two offences of having in his possession weighing machines which were not calibrated in metric units. The failure was deliberate since the defendant objected to European legislation[28] which overrode, by the operation of s 2(2) of the European Communities Act 1972, the use of Imperial measures under the Weights and Measures Act 1985. The Divisional Court held that that in determining whether a 'constitutional' statute had been repealed or confined, it must be shown that the legislature's actual intention, and not its imputed, constructive, or presumed intention, was to bring about the repeal or limitation. That test could only be met by express words in a later statute or by words so specific that the inference of an actual determination to effect the result was

[25] Cabinet Office, *The Government's Response to the Report of the Joint Committee*, para 34.
[26] HC Debs Standing Committee F col 249 5 February 2004, Fiona Mactaggart.
[27] [2002] EWHC 195 (Admin).
[28] Weights and Measures (Metrication Amendments) Regulations 1994 SI No 1851; Weights and Measures Act 1985 (Metrication) (Amendment) Order 1994 SI No 2866; Units of Measurement Regulations 1994 SI No 2867; Price Marking Order 1999 SI No 3042.

irresistible. According to Lord Justice Law, the ordinary rule of implied repeal of an earlier Act by a later Act[29] did not satisfy that test and therefore had no application to constitutional statutes:

> For the repeal of a constitutional Act or the abrogation of a fundamental right to be effected by statute, the court would apply this test: is it shown that the legislature's actual—not imputed, constructive or presumed—intention was to effect the repeal or abrogation? I think the test could only be met by express words in the later statute, or by words so specific that the inference of an actual determination to effect the result contended for was irresistible. The ordinary rule of implied repeal does not satisfy this test. Accordingly, it has no application to constitutional statutes. . . . A constitutional statute can only be repealed, or amended in a way which significantly affects its provisions touching fundamental rights or otherwise the relation between citizen and state, by unambiguous words on the face of the later statute.[30]

On the facts, the European Communities Act 1972 was palpably a constitutional statute. Therefore, the 1985 Act could not by implication delimit the regulation-making power under s 2(2) of the 1972 Act.

In the light of this precedent, the argument that rules of construction impugn any **6.15** constitutional change attempted under the 2004 Act seems shaky. In the first place, the Act unambiguously sends a signal that it allows such a power and a reading of Hansard will reveal that the Government consistently resisted restrictions for the sake of the listed constitutional laws (beyond the two exceptions mentioned). The message is that if the emergency is deemed sufficiently serious (presumably by Ministers), trust must be placed in the cultures of the ruling classes to observe the restraints in the Act.[31] In the second place, the 2004 Act could itself be construed to attain the status of constitutional law and so is on a par with the status of any legislation it seeks to alter, a far cry from the Weights and Measures Act 1985. Thirdly, there can be no doubting the vast override powers under the 2004 Act. The *Thoburn* case rejects implied repeal not express measures.

Finally, examples of the 'trivial' amendments which the advice from Parlia- **6.16** mentary Counsel does envisage as legitimate include:

> . . . timing elements or insubstantial matters. . . . Under the [European Communities] Act [1972], I understand that Ministers are designated via Order in Council before they can implement Community law. . . . It might not be possible to operate the Order in Council mechanism, which is the usual mechanism for implementing the provision under the European Communities Act. We may require therefore a minor modification that would, for example, give a Minister the power to fulfil the responsibility created under the Act in a particular way.[32]

---

[29] *Vauxhall Estates Ltd v Liverpool Corpn* [1932] 1 KB 733; *Ellen Street Estates Ltd v Minister of Health* [1934] 1 KB 590.   [30] [2002] EWHC 195 (Admin) at [63].

[31] Barnum, DG, Sullivan, JL, and Sunkin, M, 'Constitutional and Cultural Underpinnings of Political Freedom in Britain and the United States' (1992) 12 OJLS 362.

[32] HC Debs Standing Committee F col 249 10 February 2004, Fiona Mactaggart.

D. Democratic Scrutiny

(a) Parliament

6.17 Undeniably, 'The context of emergency powers exacerbates problems faced [by Parliament] in its oversight of executive power.'[33] Systemic obstacles arise because emergency authority, including prelogative powers, does not require authorization or confirmation or relies upon statutory instruments offered (subject to s 27(3) of the 2004 Act) on a 'take it or leave it' basis, if notified at all. Furthermore, the political context of emergency renders a greater reluctance than usual to produce detailed facts or explanations, and patriotic fervour dampens partisan cut and thrust.

General measures

6.18 To allow Parliament to be at least 'fairly effective',[34] a general prerequisite is to make debates about the legislation more principled and informed. This process could be aided by stating explicitly some of the desirable limiting principles adduced earlier. In addition, the adoption of a sectoral approach would hone for each issue covered by the Act criteria by which to judge its value or dispensability and its proportionality so that there can be a distinct and informed assessment and vote on each part.

Pre-publication review

6.19 Two triggers could have been suggested for the encouragement of pre-enforcement review. First, there should be a requirement that Parliament should see drafts of regulations before their formal laying. The requirement could be reduced or even avoided in case of urgency, though that consideration is less likely under Part I than Part II. The refusal to publish drafts under Part II has been discussed in Chapter 5. Secondly, for Part II, the device of proclamation/declaration of emergency could be turned into a formal motion which must be approved by Parliament. In this way, 'specific parliamentary endorsement for such a declaration would give democratic legitimacy to a whole range of measures which could not be examined in detail, would give confidence to those carrying them out that they were properly authorized and would assure the courts that Ministers were not acting beyond their political authority.'[35] It would also ensure there is no unaccountable use of Part II by an emergency being declared and successive sets of regulations being made for no more than seven days at a time.[36]

---

[33] Bonner, D, *Emergency Powers in Peacetime* (Sweet & Maxwell, London, 1985) 37.

[34] Eaves, J, *Emergency Powers and the Parliamentary Watchdog* (Hansard Society, London, 1957) 187. See further Franks, CES, *Parliament and Security Matters* (Ministry of Supply and Services Canada, Ottawa, 1980).

[35] Joint Committee on the Draft Civil Contingencies Bill, para 189.   [36] See s 27(1)(b).

The mechanisms for post-publication review

The Bill introduced to Parliament in January 2004 was altered in a number of **6.20**
respects during passage so as to strengthen Parliamentary review. One change
was that Ministers approved the idea of giving a human rights statement akin to
s 19 of the Human Rights Act 1998 in respect of Part II regulations.[37]

As for Part I regulations, an amendment was made to require use of the affir- **6.21**
mative process for regulations under s 1 (s 17(2) and (3)). But calls for the
affirmative process for regulations under s 2 was not accepted.[38] Under Part II,
the draft of which borrowed heavily from the Emergency Powers Act 1920,[39]
the regulations are all subject to the affirmative procedure (s 27), and, as noted
in Chapter 5, s 27 applies the 'super-affirmative procedure'.[40]

The intervention of Parliament could have been strengthened still further,[41] such **6.22**
as by a requirement to inform Parliament of the issuance of urgent directions
under s 7 (s 8 in Scotland) which take the place of orders or regulations. The
omission is all the more an egregious assault on Parliamentary accountability since
consultation of the National Assembly of Wales is required under s 16.[42]

A proposal for an 'Emergency Powers' Select Committee was made during the **6.23**
parliamentary passage[43] but to no avail. The Government preferred recourse to
collaboration with representatives of key parties on Privy Council terms 'to
build consensus across the political spectrum', as the Minister of State for the
Cabinet Office put it, rather than the less closed or managed setting of a Select
Committee.[44] One hopes that one or other select committee, such as the Joint
Committee on Human Rights or the House of Commons Home Affairs
Committee, will both review the implementation of the legislation under Part I
and also will dissect each and every invocation of Part II.

Duration and expiry

While Part II regulations lapse after 30 days, a sunset clause for the whole Act (or **6.24**
probably just Part II) could ensure a review, perhaps after five years.[45] In this

---

[37] See Ch 7.

[38] Defence Committee, *Draft Civil Contingencies Bill* (2002–03 HC 557) para 71.

[39] See Morris, GS, 'The Emergency Powers Act 1920' [1979] PL 317 at 331.

[40] See Joint Committee on the Draft Civil Contingencies Bill, para 198.

[41] Several of these ideas derive from Joint Committee on the Draft Civil Contingencies Bill,
paras 181, 190, 197.

[42] See Joint Committee on the Draft Civil Contingencies Bill, para 192.

[43] HC Debs Standing Committee F col 298 10 February 2004 and HC Debs vol 421 col
1388 24 May 2004.

[44] Ibid col 299, Douglas Alexander. See also HC Debs vol 421 col 1392 24 May 2004, Fiona
Mactaggart.

[45] See (three years) HL Debs vol 666 col 1337 16 November 2004, col 1653 18 November
2004; (five years) Joint Committee on the Draft Civil Contingencies Bill, para 201.

respect, the model would be similar to the Armed Forces Act which lapses after five years and is replaced by a new Act, passed after scrutiny by a Select Committee. Other emergency-related legislation with finite shelf-life includes the Terrorism Act 2000, Part VII, and the Prevention of Terrorism Act 2005. The design did not commend itself to the Government because it saw little doubt about the necessity for Part II in five years' time.[46] Equally, one might predict there will also be a need for an Army Act, given that one has existed since the seventeenth century—the point is not about predictability of need but the effectiveness of review. Furthermore, Cabinet Office guidelines about sunset clauses suggest they might be appropriate where there are 'measures extending the powers of the state or reducing civil liberties, reserve powers that may never be used or bodies that are set up but not immediately given any powers to do anything.'[47]

6.25 Some of the anti-terrorism legislation also requires annual renewal, but the proposed application of this device to the 2004 Act was also rejected.[48] A more modest alternative would have required a debate within one year of any use of Part II.[49] The insertion of a deadline of 90 days on the life of any Part II emergency[50] was rightly considered unduly inflexible.

(b) Local democracy

6.26 Some concern was expressed during consultation that emergency planning and resilience under Part I is becoming too much a technical matter for professionals. Not only were the public out in the cold but also elected councillors are not given a sufficiently prominent role. In response, the Government pointed to their role as part of the key authorities, even if 'it is accepted that elected members do not play a direct role in emergencies and that planning and response, for example, are a matter for their professional officers.'[51] One hopes that local authorities will reflect upon this observation and will ensure that the securitocracy is held to local democratic account.

E. Judicial Scrutiny

6.27 The application, and indeed non-application, of legal powers and duties under the Act should be subjected to judicial control so far as possible.[52] The ensuing survey

---

[46] Cabinet Office, *The Government's Response to the Report of the Joint Committee*, para 38; HL Debs vol 666 col 1361 16 November 2004, Baroness Scotland.

[47] HC Debs vol 426 col 1376 17 November 2004, Ruth Kelly.

[48] See HL Debs vol 666 cols 1633, 1653 18 November 2004.

[49] HL Debs vol 666 col 1623 18 November 2004, Baroness Buscombe.

[50] HC Debs Standing Committee F col 289 10 February 2004.

[51] Civil Contingencies Secretariat, *The Government's Response*, 50.

[52] See Joint Committee on the Draft Civil Contingencies Bill, Appendix 12.

will be confined to national jurisdictions (though human rights review is mainly covered in Chapter 7). There is no international case-law relating to the predecessors to the Civil Contingencies Act 2004, though several cases exist about anti-terrorism laws which evince, above all, a strong penchant for excusing state action under the doctrine of 'margin of appreciation'.[53] Perhaps the closest foreign-related judgment is the *'Greek'* case[54] in 1969. A military junta declared a state of emergency in 1967, following their coup, so as to repress increased Communist Party influence in a pending election. The European Commission of Human Rights rejected the regime's claim to the existence of an emergency. The wholly undemocratic nature of the assumption and exercise of power, which also involved torture, plus the fact that the imagined threats could not realistically be depicted as the heinous activity of terrorism,[55] all conduced against the usually indulgent review of state emergencies.[56]

(a) Administrative law review

Part I

The position seems straightforward under Part I. As mentioned in Chapter 4, all **6.28** of the responder duties 'are potentially susceptible to judicial review proceedings'.[57] Furthermore, while the Local and Regional Resilience Forums have no separate legal personality, this invisibility may not preclude their recognition as public bodies under administrative law.[58] Their duties set forth in the official guidance must therefore be observed as they may create legitimate expectations in the local community.[59] Yet, it is not expected that judicial review will have more than marginal impact.

---

[53] See *Ireland v UK*, App No 5310/71, Ser A No 25 (1978); *Brannigan and McBride v United Kingdom*, App Nos 14553/89, 14554/89, Ser A vol 258-B (1994); *Marshall v United Kingdom*, App No 41571/98, Judgment 10 July 2001; *Kerr v United Kingdom*, App No 40451/98, Judgment 10 July 2001.

[54] *Denmark, Norway, Sweden, and the Netherlands v Greece*, App Nos 3321/67, 3322/67, 3323/67, 3344/67, (1969) 12 *Yearbook of the European Convention on Human Rights* 1.

[55] *Klass v Germany*, App No 5029/71, Ser A 28 (1978), paras 48–9, 59; *Brogan v UK*, App Nos 11209, 11234, 11266/84, 11386/85, Judgment Ser A No 145-B (1988) para 48; *Fox, Campbell and Hartley v UK*, App Nos 12244, 12245, 12383/86, Ser A 182 (1990) para 44.

[56] See Chowdhury, SR, *Rule of Law in a State of Emergency* (Oxford University Press, Oxford, 1989); Finn, JE, *Constitutions in Crisis: Political Violence and the Rule of Law* (Oxford University Press, New York, 1991); Oraá, J, *Human Rights in States of Emergency in International Law* (Clarendon Press, Oxford, 1992); Fitzpatrick, J, *Human Rights in Crisis* (University of Philadephia Press, Philadelphia, 1994); Walker, C, 'Constitutional Governance and Special Powers against Terrorism' (1997) 35 *Columbia Journal of Transnational Law* 1; Gross, O, 'Once More into the Breach' (1998) 23 *Yale Journal of International Law* 440.

[57] HC Debs Standing Committee F col 73 27 January 2004, Douglas Alexander.

[58] See Ch 4.

[59] *A-G of HK v Ng Yuen Shiu* [1983] 2 AC 629; *R v Secretary of State for the Home Department ex p Khan* [1985] 1 All ER 40; *In re Preston* [1985] AC 835; *R v Secretary of State for the Home Department, ex p Ruddock* [1987] 1 WLR 1482; *R v Secretary of State for Health, ex p United States Tobacco International Inc* [1992] QB 353.

Part II

6.29 Deeper sensitivities arise under Part II, but the rule of law demands that the courts are not impotent in response to the might of executive power in an emergency. Lord Atkin's famous dissent in the case of *Liversidge v Anderson* may warrant repetition:

> In this country, amid the clash of arms, the laws are not silent. They may be changed, but they speak the same language in war as in peace. It has always been one of the pillars of freedom, one of the principles of liberty for which on recent authority we are now fighting, that the judges are no respector of persons and stand between the subject and any attempted encroachments on his liberty by the Executive, alert to see that any coercive action is justified in law. In this case I have listened to arguments which might have been addressed to the Court of King's Bench in the time of Charles I.[60]

The courts perform a distinct role in that they enjoy the feature of independence from government which is far from assured during the parliamentary process.

6.30 Yet, despite the indispensable role, the judges generally show restraint in reviewing the decisions of the executive in such high state issues because of the claims to the superior ability of the executive to assess national security. The only reported successful challenge to previous emergency regulations occurred in *Smith v Wood*,[61] concerning the prosecution of union officials for threatening to withdraw safety cover at a coal mine, a prosecution depending on a regulation which was unlawful as it effectively made it an offence to take part in a strike. It is true that times have moved on since the wartime emergency regulations were challenged unsuccessfully in *Liversidge v Anderson* when a wartime detention without trial could be upheld without substantive reasons being disclosed. That general approach was condemned in *Nakkuda Ali v Jayaratne*,[62] *Ridge v Baldwin*,[63] and in *IRC v Rossminster Ltd*,[64] albeit in situations which did not involve the trauma of war or terrorism. In 2004, Douglas Alexander, the Minister of State for the Cabinet Office, boldly pronounced that, 'We are satisfied that it is now accepted that the dissenting judgment of Lord Atkin is good law, and that the courts will inquire about the reasonableness of the Minister's belief.'[65] Nevertheless, the judges have continued to exhibit considerable deference in fraught circumstances, as can be demonstrated by later disputes, such as

---

[60] [1942] AC 206 at 244. But see the dismissive assessment in Simpson, AWB, *In the Highest Degree Odious* (Clarendon Press, Oxford, 1992) 363. See also Dyzenhaus, D, 'Intimations of Legality Amid the Clash of Arms' (2004) 2 *International Journal of Constitutional Law* 244.
[61] (1927) 43 TLR 178.
[62] [1951] AC 66 at 76 per Lord Radcliffe. [63] [1964] AC 40 at 73 per Lord Reid.
[64] [1980] AC 952 at 1011 per Lord Diplock.
[65] HC Debs Standing Committee F col 308 10 February 2004, Douglas Alexander.

challenges to orders proscribing terrorist groups,[66] the compulsory ballot of workers whose union intends to call a strike,[67] the broadcasting ban applied to Sinn Féin,[68] and refusal of entry or deportation on grounds of national security.[69] In summary, Lord Pearce stated truly in *Conway v Rimmer* that 'the flame of individual right and justice must burn more palely when it is ringed by the more dramatic light of bombed buildings'.[70] Thus, there will prevail 'the "Presumption of Executive Innocence"... which has continued to embody the attitude of the judiciary to executive power in such cases.'[71]

Assuming review is not halted altogether by claims to non-justiciability,[72] it will **6.31** now be outlined how far it might bite. There are various forms of error affecting jurisdiction that may trigger judicial intervention. In the *ASLEF (No 2)* case, Lord Denning suggested that '... if the Minister does not act in good faith, or if he acts on extraneous considerations which ought not to influence him, or if he plainly misdirects himself in fact or in law', then the court can interfere.[73]

As for procedural errors, the courts might ensure that an order has actually been **6.32** made and has been approved by the proper authority.[74] In addition, orders served should correspond with orders signed by the Minister.[75] Furthermore, regulations may set up procedures such as appeals and reviews which must be adhered to and must embody the basics of natural justice.[76]

Judicial review may next set aside abuses of discretion. By and large, evidential **6.33** difficulties will ensure that challenges on the basis of irrationality have little impact, subject to some exceptions. First, if, contrary to its common practice,

---

[66] *McEldowney v Forde* [1971] AC 632; *Re Williamson's Application for Judicial Review* [2000] NI 281. [67] *Secretary of State for Employment v ASLEF (No 2)* [1972] QB 455.
[68] *R v Secretary of State for the Home Department, ex p Brind* [1991] AC 696.
[69] See *R v Secretary of State for the Home Department, ex p NSH, The Times* 24 March 1988; *Chahal v Secretary of State for the Home Department* [1994] Imm AR 107.
[70] [1968] AC 910 at 982.
[71] Simpson, AWB, *In the Highest Degree Odious* (Clarendon Press, Oxford, 1992) 30.
[72] Walker, C, 'Review of the Prerogative: The Remaining Issues' [1987] PL 62; *R v Jones (Margaret) and others* [2004] EWCA Crim 1981. In cases of exclusion of terrorist suspects, the courts wavered in favour of non-justiciability but generally opted for an indulgent standard of review: Walker, C, 'Constitutional Governance and Special Powers against Terrorism' (1997) 35 *Columbia Journal of Transnational Law* 1. [73] [1972] QB 455 at 493.
[74] Ministers of State can probably approve on behalf of their Secretary of State: *Govt of Malaysia v Mahan Singh* [1975] 2 MLJ 155; *Najar Singh v Govt of Malaysia* [1976] Ch 30, *McKernan v Governor of Belfast Prison and another* [1983] NI 83. But it is less certain that a civil servant can act: *R v Secretary of State for the Home Department, ex p Oladehinde* [1991] 3 WLR 797.
[75] *R v Home Secretary, ex p Budd* [1941] 2 All ER 749, [1942] 1 All ER 373, [1943] 2 All ER 452.
[76] See *R v Secretary of State for the Home Department, ex p Gallagher* (1992) QBD, *The Times* 16 February 1994, CA. Compare: *Re McElduff* [1972] NI 1; *Re Mackey* (1972) 23 *Northern Ireland Legal Quarterly* 173. See de Smith, SA, 'Internment and Natural Justice' (1972) 23 *Northern Ireland Legal Quarterly* 331; Lowry, DR, 'Internment in Northern Ireland' (1976) 8 *Toledo Law Review* 169.

the executive reveals the information on which it acted, these reasons can be scrutinized.[77] But a failure to specify reasons beyond the formulaic may not be sufficient to disturb the decision.[78] The next forms of abuse of discretion are disregarding legally relevant, or taking into account legally irrelevant, considerations. Problems of proof are also the main hindrance to the final form of abuse of discretion, namely bad faith on the part of the decision-maker[79] or simply inattention—a large number of legal decisions being processed in a short time does not raise such an inference.[80]

6.34 Some might argue that the House of Lords decision on the legality of the detention without trial of foreign terrorist suspects in December 2004 in *A and others v Secretary of State for the Home Department*[81] evinces a fundamental change of judicial heart. Certainly, Lord Hoffman has recently produced starkly inconsistent views. He may be best remembered for his later thoughts in that case, when he trenchantly stated:

> Terrorist violence, serious as it is, does not threaten our institutions of government or our existence as a civil community. . . . The real threat to the life of the nation, in the sense of a people living in accordance with its traditional laws and political values, comes not from terrorism but from laws such as these. That is the true measure of what terrorism may achieve. It is for Parliament to decide whether to give the terrorists such a victory.[82]

Yet, the same Lord Hoffman offered these opinions just a couple of years previously in *Secretary of State for the Home Department v Rehman*:

> *Postscript.* I wrote this speech some three months before the recent events in New York and Washington. They are a reminder that in matters of national security, the cost of failure can be high. This seems to me to underline the need for the judicial arm of government to respect the decisions of ministers of the Crown on the question of whether support for terrorist activities in a foreign country constitutes a threat to national security.[83]

6.35 If not signalling a wholly unequivocal turning point, does the case *A and others v Secretary of State for the Home Department* presage any more detailed shifts in judicial policy? The litigation raised the issue of whether there was a sufficient degree of public emergency within the meaning of Art 15 of the European

---

[77] For example *Attorney-General of Saint Christopher, Nevis and Anguilla v Reynolds* [1980] AC 637.

[78] *R v Secretary of State for Home Affairs, ex p Stitt* (1987) *The Times* 3 February, 1987 (QBD); *R v Secretary of State for the Home Department, ex p Cheblak* [1991] 2 All ER 319; *R v Secretary of State for the Home Department, ex p McQuillan* [1995] 4 All ER 400.

[79] See *Sebe and others v Government of Ciskei* 1983 (4) SA 523 (CKSC).

[80] *Stuart v Anderson and Morrison* [1941] 2 All ER 665.

[81] [2004] UKHL 56. See Walker, C, 'Prisoners of "War All the Time"' [2005] *European Human Rights Law Review* 50.   [82] [2004] UKHL 56 at paras 96, 97.

[83] [2001] UKHL 47 at para 62.

Convention on Human Rights for the derogation issued in 2001 to persist in 2004 and beyond. In response, the Special Immigration Appeals Commission (SIAC) ruled that there was a sufficient emergency for derogation (but ruling also that the regime was discriminatory).[84] The Court of Appeal[85] adjudged that SIAC had correctly considered the issue of whether there existed a public emergency threatening the life of the nation and that SIAC had been entitled to conclude, rightly giving 'considerable deference' to the views of the Home Secretary.[86]

This verdict was in turn reversed by an extraordinary nine-member bench of the House of Lords in December 2004.[87] The leading judgment of Lord Bingham can be analysed in three stages. The first issue is whether there has been shown to be a public emergency sufficient to warrant the issuance and continuance of the derogation notice, for it was accepted that the detentions could not be compatible with Art 5(1)(f) of the European Convention save for the derogation. A majority of the Court found for the Government on the basis of several arguments.[88] Quite simply, it had not been proven that SIAC, the venue where the facts had been most fully ventilated, had misdirected itself in any respect.[89] Next, the jurisprudence of the European Court of Human Rights did not seem to require as a trigger the actual experience of widespread loss of life caused by an armed body dedicated to destroying the territorial integrity or other fundamental characteristics of the state.[90] This finding raises issues both about the relationship between the European and national court and also as to the appropriate tests to be applied. In regard to the relationship, Lord Bingham was content to apply the same kinds of approaches as does the European Court itself, including the recognition of a margin of appreciation for executive discretion, on the basis of s 2(1) of the Human Rights Act 1998. Other judges argued that the European Court expected a more searching review in domestic courts,[91] though it may be difficult to sustain that view in the light of the fact that no judicial review was possible until the Human Rights Act 1998 came into force[92] and yet the European Court had never condemned this legal omission in

6.36

---

[84] Appeal No SC/1–7/2002, para 35. The author thanks Shami Chakrabati, Liberty, for the supply of a copy. [85] [2002] EWCA Civ 1502.

[86] Ibid, para 40.

[87] [2004] UKHL 56, Lord Walker dissenting as to the outcome.

[88] There were dissents from Lord Hoffmann (ibid, para 96) and Lord Hope (para 119). Lord Scott entertained 'very grave doubts' (para 154).

[89] Ibid, para 27 (Lord Bingham).   [90] Ibid, para 28 (Lord Bingham).

[91] For example, Lord Hope argued for a narrower margin of appreciation when liberty rights under Art 5 were at stake (ibid, para 108) as well as arguing that stricter domestic review was expected (ibid, para 131). Lord Rodger likewise was in favour of stiffer domestic review in line with the extent of the incursion into rights (ibid, paras 176, 178).

[92] Some judges were of the view that review of derogation is only possible because of the passage of s 30 of the Anti-Terrorism, Crime and Security Act 2001: Lord Rodger, ibid, para 164.

its earlier judgments. More trenchant domestic review would also require the development of a far more sophisticated jurisprudence than has yet been devised by the European Court. As for the appropriate tests to be applied, the idea of a margin of appreciation was approved: 'great weight should be given to the judgment of the Home Secretary, his colleagues and Parliament on this question, because they were called on to exercise a pre-eminently political judgment'.[93] In addition, it was accepted that it would be sufficient that an emergency was 'imminent', and there was no contradiction of the arguments of the Attorney-General that 'an emergency could properly be regarded as imminent if an atrocity was credibly threatened by a body such as Al-Qaeda which had demonstrated its capacity and will to carry out such a threat, where the atrocity might be committed without warning at any time.'[94] Two cases cited by a number of judges as supporting the notion of a 'cold' rather than 'hot' emergency were *Lawless v Ireland (No 3)*,[95] in which detention without trial was justified on the basis of a relatively low-level of IRA activity in Ireland in 1957, and *Marshall v United Kingdom*,[96] where the 1988 derogation notice was upheld even though by the time of challenge in 1998 the actual incidence (as opposed to the threat) of terrorist violence had much diminished owing to the 'Peace Process' in Northern Ireland.

6.37 The second argument was one of proportionality—was detention without trial as currently constructed 'strictly required by the exigencies of the (emergency) situation'? Unlike the reluctance to add to the jurisprudence of the meaning of 'emergency', here the Court did strike out on its own. It made clear that in handling this more familiar standard, the Court was not hidebound by 'any doctrine of deference' and should apply a 'greater intensity of review'.[97] Furthermore, it added flesh to the bones of the concept of proportionality by adopting the interpretation given by the Privy Council in *de Freitas v Permanent Secretary of Ministry of Agriculture, Fisheries, Lands and Housing*.[98] In determining whether a limitation is arbitrary or excessive, the court must ask itself, 'whether: (i) the legislative objective is sufficiently important to justify limiting a fundamental right; (ii) the measures designed to meet the legislative objective are rationally connected to it; and (iii) the means used to impair the right or freedom are no more than is necessary to accomplish the objective.' The chief objection of the appellants was directed towards (ii) and (iii) and contended that detention without trial under Part IV of the Anti-Terrorism, Crime and Security Act 2001 was not rationally connected to the objective of public safety

---

[93] Ibid, para 29 (Lord Bingham).   [94] Ibid, para 25 (Lord Bingham).
[95] App No 332/56, Ser A 3, [1961] 1 EHRR 15. See Hogan, G and Walker, C, *Political Violence and the Law in Ireland* (Manchester University Press, Manchester 1989) Ch 9.
[96] App No 41571/98, 10 July 2001.
[97] [2004] UKHL 56 at paras 42, 44 (Lord Bingham).   [98] [1999] 1 AC 69 at 80.

against terrorism. It was further suggested that the legislative objective could have been achieved by less severe restrictions on liberty; this argument was not much explored by the House of Lords, though it was accepted[99] (and is taken up later in this Chapter). Returning to the point about rational connection, there were two main features of Part IV which were out of keeping with the objective of public safety and were ultimately sustained as disproportionate. One was that Part IV only applied to deportable aliens. While they represented the predominant threat, they were not the only problem—to ignore terrorism threatened by British citizens was wrong. The other was that the creation of a 'prison with three walls'—the absent fourth wall allowing foreign terrorists to depart the jurisdiction and plot abroad likewise made no sense.

The third stage of Lord Bingham's judgment addressed the discriminatory impact **6.38** of the detention regime, which could either be taken as a further challenge as to proportionality or could be said to be a challenge under the requirement of Art 15 that there be no inconsistency with other obligations under international law (such as Art 14). In short, the House of Lords found a breach of Art 14.

The headlines about the case may create the impression of a judicial revolt. But **6.39** one should be more cautious about claims that a new era of judicial activism has dawned. First, this case must be set against the vast majority where executive decisions have been upheld. Secondly, and crucial to the argument about the Civil Contingencies Act, the House of Lords did sustain the finding of emergency. It was the aspect of discrimination which resulted in the adverse findings. All the same, there is a suggestion that proportionality emerges as a stronger test than in previous jurisprudence and that a more searching scrutiny will now be applied. This advance does, however, beg the question as to how far the courts will be prepared to demand a fresh look at the original evidence, a problem which faced the courts in the wartime detention cases.[100] In *A and others v Secretary of State for the Home Department*, the defects were inherent in the legislative policy from the face of the Act—but what if the defects had been more personal and latent?

(b) Liability in torts

Many of the wide ranging duties in Part I of the Act are designed to protect **6.40** the public.[101] If a responder fails to act as required, especially a Category 1 responder, will an action lie in tort? An example might be the failure to give adequate warnings to the public under s 2(1)(g). Community Risk Registers may also become potential sources of liability. Separate risk assessments by

---

[99] [2004] UKHL 56 at para 35 (Lord Bingham).
[100] See Simpson, AWB, *In the Highest Degree Odious* (Clarendon Press, Oxford, 1992) 330.
[101] For liability to staff, especially for psychiatric injuries, see *White and others v Chief Constable of the South Yorkshire Police and others* [1999] 2 AC 455.

various responders may mitigate this danger of civil action, but, if there are widely variant assessments as a result, that might undermine s 2. A further difficulty will be how far local responders can make assessments about what are essentially national risks, such as terrorism. The Act's blindness to central government, which would be the more natural source of strategic national risk assessments, again comes into play. The local responders receive guidance from central government but are sworn to secrecy as to its contents—whether to the extent of litigation remains to be determined by the doctrine of public interest immunity. In reply to these musings, the 2004 Act is silent on civil liability. There is no express grant of a claim to civil damages for default or immunity therefrom[102] or consideration as to whether liability might be personal or corporate.[103]

Fault liability

6.41 Civil action for the tort of negligence does not appear a promising avenue of approach.[104] There is in s 2 of the 2004 Act no direct duty to prevent an emergency. The duties are to assess and plan. Section 5, on the other hand, can involve a duty to perform a function to prevent, reduce, control, or mitigate an emergency. But even these duties do not guarantee that there will be no emergency or losses arising from it. The local authority is not being statutorily required to provide all-risks insurance.

6.42 As for the imposition of liability in negligence, the courts are sympathetic to the need to take quick action in circumstances of great pressure and inadequate information and will not impose duties of care to the general public, even when someone has been identified as a victim in need of help either by the police, as in the case of *Brooks v Metropolitan Police Commissioner and others*,[105] or by the fire brigade as in *Capital and Counties plc v Hampshire County Council*.[106] There can usually be found 'compelling considerations rooted in the welfare of the whole community which outweigh the dictates of individualised justice'.[107] The leading case is *Hill v Chief Constable of West Yorkshire*.[108] The House of Lords

---

[102] For examples of immunities, see Restoration of Order in Ireland (Indemnity) Act 1923 and, with reference to events in New Orleans in 2005, La RS s 29:735 (2005).

[103] Compare the personal liability offences in the Environmental Protection Act 1990, s 157. For discussion relating to corporate manslaughter, see Home Office, *Corporate Manslaughter* (Cm 6497, London, 2005).

[104] See further Walker, C, 'Liability for Acts of Terrorism: United Kingdom Perspective' in *European Centre for Tort and Insurance Law Liability for Acts of Terrorism* (Springer, Vienna, 2004).

[105] [2005] UKHL 24. See also *D v East Berkshire Community Health NHS Trust; K and another v Dewsbury Healthcare NHS Trust and another; K and another v Oldham NHS Trust* [2005] UKHL 23.   [106] [1997] 2 All ER 865.

[107] *Elguzouli-Daf v Commissioner of Police of the Metropolis* [1995] QB 335 at 350 per Lord Steyn.

[108] [1989] AC 53. See further *Alexandrou v Oxford* [1993] 4 All ER 328; *Osman and another v Ferguson and another* [1993] 4 All ER 344; *Cowan v Chief Constable of Avon and Somerset Constabulary* [2001] EWCA Civ 1699; *Orange v Chief Constable of West Yorkshire* [2001] EWCA Civ 611; *Vellino v Chief Constable of Greater Manchester* [2001] EWCA Civ 1249; *Brooks v*

concluded on grounds of public policy that it would be undesirable to impose civil liability on the police for failing to prevent further murders by a serial killer. It preferred to allow them to use their professional judgement in the allocation of resources[109] and the direction of policies unworried by the threat of potentially endless civil actions by concerned or disgruntled citizens, even those within the most foreseeable circle of victims (here, young women). The European Court of Human Rights in the case of *Osman*[110] later cast considerable doubt on any absolute exclusion of liability on grounds of public policy and required on grounds of due process the possibility of a claim to be considered case by case. However, it subsequently understood, in *TP and KM v United Kingdom*,[111] that these requirements of Art 6 of the European Convention are met by the approach of English law. Surely, the same considerations of the encouragement of professional judgement and the absence of limitless resources would affect any ascription of liability in respect of actions in response to emergencies.

Numerous illustrations might be given of the liability of emergency responders.[112] **6.43** Within the constraints of the present study, the following gives a flavour. In *Thames Trains Ltd v Health and Safety Executive*,[113] the claimants, who were held responsible for the Ladbroke Grove Junction train crash in 1999, arising through faulty signalling equipment, sought a contribution from the HSE for negligence by its Railway Inspectorate. It was shown that there was knowledge of the dangerous state of the signalling system over a period of three years, and so the claim was allowed to proceed. The opposite result was sustained in *Great North Eastern Railway Ltd v Hart and others*.[114] Hart was the driver of a vehicle which, through his inadvertence, veered off a motorway and came to rest on a main rail line at Little Heck, Selby, as a result of which a train crashed into the vehicle and then another train, killing ten people in 2001. His insurers admitted liability but sought

---

Commissioner of Police for the Metropolis [2005] UKHL 24. But the possibility of liability to a suitably narrow category of persons remains: *Swinney v Chief Constable of Northumbria (No 1)* and *(No 2)* [1997] QB 464, (1999) 11 Admin LR 811 (informant at risk from attack); *Kirkham v Chief Constable of Greater Manchester* [1990] 2 QB 283 (known suicide risk); *Reeves v Metropolitan Police Commissioner* [2000] 1 AC 360 (known suicide risk).

[109] See further *Rigby and another v Chief Constable of Northamptonshire* [1985] 1 WLR 1242 (regarding the non-purchase of specialist equipment, but liability was sustained for the planning of the operation).

[110] *Osman v United Kingdom*, (1998) App No 23452/94, Reports 1998-VIII. See Giliker, P, 'Osman and Police Immunity in the English Law of Torts' (2000) 20 *Legal Studies* 372; Gearty, CA, 'Unravelling Osman' (2001) 64 MLR 159.

[111] *X v Bedfordshire County Council* [1995] 2 AC 633; *TP and KM v United Kingdom* App No 28945/95, 2001-V; *Keenan v United Kingdom* App No 27229/95, 2001-III; *Z v United Kingdom* App No 29392/95, 2001-V.

[112] See also *Lister v Hesley Hall Ltd* [2001] UKHL 22; *Weir v Bettison (sued as Chief Constable of Merseyside)* [2003] EWCA Civ 111.

[113] [2002] EWHC 1415. In turn, Thames Train Ltd was fined £2m in April 2004 on a prosecution brought by the HSE.   [114] [2003] EWHC 2450 (QB).

to counterclaim against the Secretary of State for Transport (acting through the Highways Authority) for failing to provide adequate safety barriers. It was held that there was a duty of care on the part of the Highways Authority, but it had applied a 'risk-ranking tool' to assess the site and found no justification in safety or cost terms for any additional protection measures.

6.44 In 'hotter' emergencies, liability can result, provided the emergency services have taken charge of the situation.[115] In *Knightley v Johns*,[116] a serious road accident occurred near the exit of a tunnel. The police inspector in charge at the scene of the accident failed to close the tunnel to oncoming traffic in accordance with police force standing orders for road accidents, and there was another crash. The inspector was negligent. The emergency services may thus be liable for failing to protect an identifiable person or group of persons where a responsibility for intervention has been assumed. Many cases have concerned crashes by emergency vehicles on the way to the scene of an accident or hazard and, in several instances, passing through red traffic lights. The courts have accepted the policy of non-prosecution,[117] but a high standard of care will be imposed which might not even be met by proceeding cautiously in a large red fire engine with three sets of lights flashing.[118]

6.45 It would, alternatively, be possible to claim damages against the direct tort-feasors, where an emergency is clearly the result of human agency, but such actions are not uncommon owing to the absence of resources and exclusiveness of the likes of terrorist bombers.

6.46 An occupier of land, fixed, or moveable structures (including vehicles) has a duty under the Occupier's Liability Act 1957 to take such care as is reasonable to keep visitors reasonably safe. The liability can extend to the harm or loss caused by trespassers such as terrorists, but a high degree of foreseeability then seems to be required,[119] and liability can be displaced by evidence of security checks and protective architecture. This level of foreseeability has rarely been sustained, though examples include *Ship Chandlers Ltd and others v Norfolk Line*

---

[115] The emergency services have limited duties in civil law to respond at all: *Ancell v McDermott* [1993] 4 All ER 355; *Capital & Counties Plc v Hampshire County Council* [1997] QB 1004; *OLL Ltd v Secretary of State for Transport* [1997] 3 All ER 897. But ambulance services are said to have a public law duty to provide an ambulance (*Kent v Griffiths* [2001] QB 36) and the police have duties to maintain the peace (*R v Dytham* [1979] QB 722). See Hickman, T, ' "And that's Magic!"—Making Public Bodies Liable for Failure to Confer Benefits' (2000) 59 *Cambridge Law Journal* 532.   [116] [1982] 1 WLR 349.

[117] *Buckoke and others v Greater London Council* [1971] Ch 655.

[118] *Purdue v Devon Fire and Rescue Service* [2002] EWCA Civ 1538.

[119] *Smith v Littlewoods Organization plc* [1987] AC 241. See also *P Perl (Exporters) Ltd v Camden London Borough Council* [1984] 1 QB 342; *King v Liverpool City Council* [1986] 1 WLR 890; *Topp v London Country Bus* [1993] 1 WLR 976; *Jeffrey v Commodore Cabaret Ltd* (1995) 128 DLR (4th) 535. Compare: *Hosie v Arbroath Football Club Ltd* 1978 SLT 122.

Ltd and others,[120] where a sandbag wall, inadequately constructed by a water authority and its associates, around temporary building works failed to repel the highest (but forecast) flood surge in Great Yarmouth since the Great Flood of 1953. In *Cunningham and others v Reading Football Club Ltd*, the presence of football hooligans at a football ground was dolefully predictable and made worse by inadequate crowd segregation and control and the crumbling state of the terraces which could too easily be transformed into handy concrete missiles.[121] To date, there have been no recorded instances of claims in the United Kingdom for damages against property owners or carriers arising from terrorism.[122] However, cases in the United States include[123] the *Lockerbie* case, where Pan Am was held liable for wilful misconduct for failing to check baggage at Frankfurt Airport.[124]

If liability is founded, then the emergency may also give rise to issues of con- **6.47** tributory negligence,[125] third party causation, necessity,[126] or even (as a denial of liability) Act of God.

Aside from the liability of responders and others, there may be a residual duty of **6.48** protection vested in the state. As discussed in Chapter 2, the state's recognition and protection of rights to life, liberty, and property involve its most fundamental duties. But do these broad normative considerations translate into legal duties? The answer would seem to be that protection by the state is a legitimate demand both in domestic law under the broad reach of the Royal Prerogative.[127] And one might reach similar conclusions in regard to international law. The focus here might first be turned towards the interpretation of the right to life under Art 2 of the European Convention on Human Rights and Fundamental Freedoms of 1950. The state must put in place laws, security personnel and security tactics such as can secure a reasonable amount of protection for all citizens.[128] This duty has now been reinforced domestically by the Human Rights Act 1998.

---

[120] (1990, unreported, QBD).   [121] *The Times* 22 March 1991.
[122] But there is a current attempt to sue members of the Real IRA who are alleged to be responsible for the death and injury resulting from the bombing in Omagh in August 1998: 'Writs over Omagh Delivered under Grey Irish Sky' *The Times*, 27 July 2002 at 4.
[123] See also *Sakaria v TWA* (1993) 8 F 3d 164; *Shah v Pan Am* (1999) 148 3d 84. For state sponsors, see Foreign Sovereign Immunities Act (28 USC s 1602), Alien Tort Act (28 USC s 1350).
[124] *In re Air Disaster at Lockerbie, Scotland* (1994) 37 F3d 804, cert denied, 513 US 1126 (1995). The action against the airline was followed by several actions against the Libyan Government: *Smith v Socialist People's Libyan Arab Jamahiriya* (1996) 101 F 3d 239, *cert denied*, 520 US 1204; *Rein v Socialist People's Libyan Arab Jamahiriya* (1998) 162 F 3d 748.
[125] *Purdue v Devon Fire and Rescue Service* [2002] EWCA Civ 1538.
[126] *Rigby and another v Chief Constable of Northamptonshire* [1985] 1 WLR 1242.
[127] *Calvin's case* (1609) 77 English Reports 377; *Mutasa v Attorney General* [1980] QB 114; *R (Abassi and another) v Secretary of State for the Home Department* [2002] EWCA Civ 1598.
[128] *McCann, Savage and Farrell v United Kingdom* App No 18984/91, Ser A vol 324 (1995).

Strict liability

6.49 In a recent risk assessment, the Security Service (MI5) was said to have identified 350 key terrorist targets, including nuclear plants, telecommunications infrastructure, fuel depots, and governmental and military centres.[129] Since commercial insurance coverage is often unavailable for such targets, insurance is often provided through a mixture of Pool Re[130] and other statutory schemes.

6.50 The residual common law form of legal strict liability is that set out in the rule in *Rylands v Fletcher*,[131] by which[132] 'the person who for his own purposes brings on his lands and collects and keeps there anything likely to do mischief if it escapes must keep it at his peril, and, if he does not do so, is prima facie answerable for all the damage which is the natural consequence of its escape.' Applying this form of liability to the release of 'dangerous things' as a result of disasters, whether through human agency or Act of God,[133] the property owner whence the damage arises does have the defence of an independent act by a third party, in the case of outside agency, subject to it not being reasonably possible for the defendant to have foreseen and guarded against the intervention.[134] Otherwise, the occupier can be liable for an escape even if there is no negligence in foreseeing or preventing the escape. It is also probably the case that the tort does not allow for claims for personal injury but must relate to damage to land-owning interests.[135]

6.51 As well as *Rylands v Fletcher* liability, there exists a range of statutes which relate to exceptionally dangerous substance or processes and any harm or loss arising from them. The first example relates to nuclear materials.[136] Under the Nuclear Installations Act 1965, ss 7 and 8, there is imposed a strict duty to secure that

---

[129] *The Times* 6 June 2002 at 1.

[130] See Walker, C, 'Political Violence and Commercial Risk' (2004) 56 *Current Legal Problems* 531.

[131] (1868) LR 3 HL 330. See Rogers, WVH, *Winfield and Jolowicz on Tort* (16th edn, Sweet & Maxwell, London, 2002), Ch 15.

[132] (1866) LR 1 Ex 265 at 279–80 per Blackburn J. See for its modern restatement *Cambridge Water v Eastern Counties Leather plc* [1994] 2 AC 264.

[133] *Greenock Corporation v Caledonian Railway* [1917] AC 556.

[134] *Perry v Kendricks Transport Ltd* [1956] 1 WLR 85 at 87 per Singleton LJ. See further *Box v Jubb* (1879) 4 Ex D 76; *Richards v Lothian* [1913] AC 263; *Northwestern Utilities v London Guarantee and Accident Co Ltd* [1936] AC 108; *H & N Emanual Ltd v Greater London Council and another* [1971] 2 All ER 835; *Ribee v Norrie*, The Times 22 November 2000.

[135] Compare *Hunter v Canary Wharf* [1997] AC 655.

[136] See Miller, CE, 'Radiological Risks and Civil Liability' (1989) 1(1) *Journal of Environmental Law* 10; Tromans, S and Fitzgerald, J, *The Law of Nuclear Installations and Radioactive Substances* (Butterworths, London, 1997); Radetzki, M, 'Liability of Nuclear and Other Industrial Corporations for Large Scale Accident Damage' (1997) 15(4) *Journal of Energy & Natural Resources Law* 317; Lee, M, 'Civil Liability of the Nuclear Industry' (2000) 12(3) *Journal of Environmental Law* 317; Ghiassee, NB, 'Nuclear Terrorism and the Environment' (2002) 8 *Environmental Law* 15; Horbach, LJT, 'Terrorism and Nuclear Damage Coverage' (2002) 20(3) *Journal of Energy & Natural Resources Law* 231.

there is no injury, loss of life, or damage to any property arising from the licensed use of a nuclear plant. If there is a breach of that duty, then liability arises under s 12,[137] though other forms of legal action[138] are debarred. There is a defence under s 13 where the loss or harm arises from hostile action in armed conflict.[139] To ensure that claims can be met, the licensee must obtain under s 19 insurance up to a specified amount (£140 million is the figure currently set under the Energy Act 1983, s 27); under s 18, the Government underwrites claims beyond that sum and claims after ten years.[140]

The second sector involves poisonous pollution. Should there be the deposit on land of any poisonous, noxious, or polluting waste, then an offence is committed and civil liability arises under the Environmental Protection Act 1990, s 73(6), except where the damage '(a) was due wholly to the fault of the person who suffered it; or (b) was suffered by a person who voluntarily accepted the risk of the damage being caused'. The liability attaches to any person who deposited it or knowingly permitted it to be deposited. However, there is a defence under s 33(7) for one who can prove, *inter alia*, '(a) that he took all reasonable precautions and exercised all due diligence to avoid the commission of the offence; or . . . (c) that the acts . . . were done in an emergency in order to avoid danger to the public'. Another code on pollution relates to oil spillages from ships, namely, the Merchant Shipping Act 1995, Part VI.[141] Section 153 imposes liability on the tanker owners in respect of damage caused by contamination arising from the discharge or escape of oil.[142] Owners of ships are required by s 163 to insure up to a set amount,[143] beyond which the International Oil Pollution Compensation Fund[144] will meet the claims under s 175. There are defences in s 155 which absolve the owner of liability from

6.52

---

[137] Pure economic loss was not allowed in *Merlin v British Nuclear Fuels* [1990] 2 QB 557.

[138] An action under *Rylands v Fletcher* would seem otherwise possible: *Blue Circle Industries plc v Ministry of Defence* [1999] Ch 289.

[139] Terrorism probably does not qualify: Walker, C, 'Irish Republican Prisoners—Political Detainees, Prisoners of War or Common Criminals?' [1984] 19 *Irish Jurist* 189; Rona, G, 'International Law under Fire' (2003) 27 *Fletcher Forum of World Affairs* 55.

[140] Sections 16, 18.

[141] See Abencassis, DW and Jarashow, RL, *Oil Pollution from Ships* (2nd ed, Stevens, London, 1995); Fogarty, ARM, *Merchant Shipping Legislation* (LLP, London, 1994); Phillips, N and Craig, N, *Merchant Shipping Act 1995* (LLP, London, 2001).

[142] See, for example, *Alegrete Shipping Co Inc v International Oil Pollution Compensation Fund* [2002] EWHC 1095 (Admlty). The legislation is based on the International Convention on Civil Liability for Oil Pollution Damage 1969 (Cmnd 6183, London, 1975).

[143] Their liability is limited by s 157.

[144] See http://www.iopcfund.org/; Wilkinson, D, 'Moving the Boundaries of Compensable Environmental Damage Caused by Marine Oil Spills: The Effect of Two New International Protocols' (1993) 5(1) *Journal of Environmental Law* 71; Gaskell, N, 'Oil Pollution and the IOPC Funds 1971 and 1992' (1992) 6(7) *International Maritime Law* 177. The IOPC Funds arise from the International Convention on Civil Liability for Oil Pollution Damage (Cmnd 6183, London, 1975).

contamination which, *inter alia*, 'resulted from an act of war, hostilities, civil war, insurrection or an exceptional, inevitable and irresistible natural phenomenon'.

6.53 The third sector relates to public utilities. For example, by s 209(1) of the Water Industry Act 1991,[145] 'Where an escape of water, however caused, from a pipe vested in a water undertaker causes loss or damage, the undertaker shall be liable, except as otherwise provided in this section, for the loss or damage.' In this case the liability extends to acts of third parties (unless the claim is by the person at fault for the loss or damage which has been sustained) and even Acts of God. Likewise,[146] under the Gas Act 1965, s 14, a 'gas transporter shall be absolutely liable in civil proceedings in respect of damage caused by gas in an underground gas storage, or by gas in the boreholes connected with an underground gas storage, or which is escaping from or has escaped from any underground gas storage or any such boreholes.' Again, the liability does not extend to a person who suffers the damage as the result of his own fault.

6.54 The fourth sector relates to aircraft. Aircraft are not considered 'dangerous things' and so their use in attacks cannot per se give rise to *Rylands v Fletcher* liability.[147] However, liability is imposed by the Civil Aviation Act 1982, s 76(2) for damage from aircraft while in flight, taking off, or landing, without proof of negligence. There is no defence that the loss or damage is caused by a third party, though the owner may seek an indemnity from that nefarious third party.

Legal assistance

6.55 Where there are multi-party actions, for example, arising from sudden disasters, the Law Society guidelines on Disaster Litigation[148] seek to combine claims for the sake of cost and time savings. The instructed firms should notify the Multi-Party Actions Information Service and should then arrange for a lead firm and steering committee to be appointed.

(c) Exclusion of jurisdiction

6.56 The Home Office Minister indicated during passage that 'For the purposes of the record, I am saying that the courts are not excluded and we do not propose to add anything to the Bill which would exclude them or which would be implied or deemed to exclude them. That is not our intention.'[149] Furthermore, even if a Minister did have the intent (for example, under s 22(3)(j)), the courts

---

[145] But no special civil liability is imposed by the Reservoirs Act 1975 (s 1(5)).
[146] There is no equivalent under the Electricity Act 1989.
[147] *Fosbroke-Hobbes v Airwork Ltd.* [1937] 1 All ER 108.
[148] *Disaster Litigation: Practice Guidelines for Solicitors* (Law Society, London, 2003).
[149] HL Debs vol 665 col 750 14 October 2004, Baroness Scotland.

have shown in *Chester v Bateson* hostility to the exclusion of jurisdiction where the parent Act does not expressly allow it.[150]

(d) Victims

There is no special scheme relating to the compensation of victims of disaster **6.57** within the United Kingdom, their relief being left to charity and voluntary groups.[151] There are various schemes for compensation relating to criminal personal injury and, in Northern Ireland only, criminal property damage.[152]

F. Other Independent Scrutiny

There was considerable disagreement during the Parliamentary process about **6.58** the possibilities of expert and sustained independent inquiry. The matter was settled at the last hour by a concession that:[153]

> . . . within one year of the end of the point at which the emergency regulations fall, a senior Privy Counsellor appointed by the Government will review the operation of the Act in that instance. That process would be repeated for each and every emergency during which the Act was used. That review will be published and available to Parliament—and there will be a debate on the review.

The Government assured that a single 'independent' Privy Councillor will lead the review, assisted by a review team.[154] It is regrettable that the review and the appointment of the reviewer are not based on statute and that the reviewer was afforded no statutory powers to gather evidence. The Inquiries Act 2005 now provides a ready blueprint for delivery. The model adopted derived from a mixture of two sources. First, the Terrorism Act 2000, which is subject to the extra-statutory concession that an annual report will be delivered by an independent reviewer (who turned out to be Lord Alex Carlile) reporting to the Secretary of State who is then obliged to lay the report before Parliament.[155] Secondly, by the Anti-Terrorism, Crime and Security Act 2001, s 122, a committee of Privy Councillors had to be appointed to conduct a one-off review of the Act.[156]

It has also been suggested in Chapter 5 that periodic reports to Parliament **6.59** should be forthcoming in respect of the activities of any appointed Regional Nominated Coordinator. No concession was made on this point, nor any whatsoever in relation to Part I of the Act.

---

[150] [1920] 1 KB 820. See Ch 5.   [151] But see further Ch 8.
[152] See Walker, C, 'Political Violence and Commercial Risk' (2004) 56 *Current Legal Pro-blems* 531.   [153] HL Debs vol 666 col 1655 18 November 2004, Lord Bassam.
[154] HC Debs vol 426 col 1515 18 November 2004, Ruth Kelly.
[155] HC Debs Standing Committee D col 315 8 February 2000, Charles Clarke.
[156] For the results, see Privy Counsellor Review Committee, *Anti-Terrorism, Crime and Security Act 2001 Review, Report* (2003–04 HC 100).

G. Devolution

6.60 Albeit that it was not originally conceived in this way,[157] the Civil Contingencies Act 2004 is a unifying measure in the sense that there is now only one Act, covering the whole of the United Kingdom. The separate legislation for Northern Ireland has been abolished. Nevertheless, the Act provides for geographical variation on a much greater scale than present in the previous legislation. In part, this change reflects policy—the need to act proportionally, which is a prominent consideration as part of the Triple Lock in Part II. In addition, the influence of constitutional devolution plays a role in Part II, as described in Chapter 4.

H. Conclusions

6.61 Despite the efforts of the Joint Committee and of Parliamentary debate, the Civil Contingencies Act has the potential to inflict terrible damage on the constitution of the United Kingdom. Nor can the courts be wholly relied upon, based on their past record, as a bulwark against Ministerial excesses, though the auguries for resistance have improved since the House of Lords delivered *A and others v Secretary of State for the Home Department* and, more generally, since the passage of the Human Rights Act 1998, the influence of which will be explored further in the next chapter. One might take some heart in the knowledge that the previous Emergency Powers Act 1920 was used rather sparingly and that, whenever there is an equivalent invocation under Part II, a Privy Counsel review will now be instigated. In this way, political constraints and the fear of unpopularity and the exposure of error will ultimately continue to be more influential than legal restraint. But if Part II remains a dead-letter, the price might be an extensively securitized society under Part I, with increasing focus on risk management and prevention and its attendant emanations such as pervasive surveillance and physical security, expensive equipment, remote and shadowy organizations and programmes, and the skewing of social agendas.

---

[157] Cabinet Office, *The Draft Civil Contingencies Bill*, para 6.2.

7

HUMAN RIGHTS ISSUES

This chapter is infused with the same spirit as the prior one, not surprisingly **7.01** as human rights are a key element of contemporary constitutionalism. Consequently, many relevant points have already been made in Chapter 6 and elsewhere, hence the comparative brevity of this chapter. One might, however, reiterate that the position of human rights in an emergency is precarious and that 'The true test of the viability of any legal system is its ability to respond to crisis without permanently sacrificing fundamental freedoms.'[1]

The Emergency Powers Act 1920 was passed three decades before the United **7.02** Kingdom ratified the European Convention of Human Rights and nearly eight decades before the Human Rights Act 1998 (HRA) incorporated the Convention rights into UK law. As a result of these developments, there is now heightened attention and precision afforded to rights discourse. In addition, since 1920, wartime legislation and laws against terrorism have provided solid experience of the relevance of rights instruments as well as the powers and limitations of judicial and Parliamentary protections. It follows that the level of protection of human rights against regulatory change (especially under Part II where the threats are most acute) offers great interest. It is necessary to consider how far the 2004 Act might permit, either by design or default, breaches of the HRA.

---

[1] Lee, HP, *Emergency Powers* (Law Book Co, Sydney, 1984) ix. See further Hope Report, Royal Commission on Intelligence and Security (4th Report PP 249 (Cth, Canberra, 1977); Bingham, T, 'Personal Freedom and the Dilemma of Democracies' (2003) 52 ICLQ 841.

A. Human Rights Requirements

7.03 Many substantive rights can be affected by emergencies. For the purposes of this discussion, concentration will be mainly confined to those in the European Convention on Human Rights,[2] since they are directly enforceable in domestic law through the Human Rights Act 1998.[3] Prominent amongst these will be rights to liberty (Art 5), fair process (Art 6), privacy (Art 8), free expression and association (Arts 10 and 11), plus freedom from discrimination (Art 14), and rights to enjoy private property (Art 1 of Protocol 1) and to take part in elections (Art 3 of Protocol 1).[4]

7.04 As well as the specification of substantive rights of relevance, European Convention jurisprudence has developed further values concerning their application in relation to each other and in relation to other values (at least where not absolute).[5] These include legality—that any measure affecting rights is 'prescribed by law'— that it must have a basis in domestic law, must be accessible (available and readily comprehensible) and must be sufficiently precise in terms.[6] Restraints must be 'necessary in a democratic society'—the objective pursued by the restriction is compatible with a democratic society and serves a pressing social need which outweighs the trump cards of rights. There must be proportionality. Even if new provisions are in principle necessary, were the actions taken by public authorities proportionate to the threat or crisis which they are seeking to act against? There must be no greater restriction of rights than necessary to achieve the objectives, a stricter standard than applied by English courts in judicial review under the *Wednesbury* test.[7] Finally, restraints must be non-discriminatory.[8]

7.05 Alongside the care for rights, as noted in Chapters 2 and 6, the European Convention concedes that times of emergency may demand adjustment to standard norms, and so declares in Arts 15 and 17:

> 15. (1) In time of war or other public emergency threatening the life of the nation any High Contracting Party may take measures derogating from its obligations under this Convention to the extent strictly required by the exigencies

---

[2] See especially Clayton, R and Tomlinson, H, *The Law of Human Rights* (Oxford University Press, 2000); Feldman, D, *Civil Liberties and Human Rights in England and Wales* (2nd ed, Oxford University Press, Oxford, 2001); Lester, A and Pannick, D, *Human Rights Law and Practice* (2nd ed, LexisNexis, London, 2004).

[3] For other sources, see the discussion in *A and others v Secretary of State for the Home Department* [2004] UKHL 56 and also n 63 below.

[4] See Council of Europe, *Guidelines on Human Rights and the Fight Against Terrorism* (3rd ed, Strasbourg, 2005).   [5] See Joint Committee on The Draft Civil Contingencies Bill, para 139.

[6] See *Silver v United Kingdom*, App No 5947/72, 6205/73, 7052/75, 7061/75, 7107/75, 7113/75, 71361/75, Ser A vol 61 (1983); *Malone v United Kingdom*, App No 8691/79, Ser A vol 82 (1985).

[7] See, for example, *R v Lord Saville of Newdigate, ex p A and others* [1999] 4 All ER 860; *R (on the application of A and others) v Lord Saville of Newdigate and others (No 2)* [2001] EWHC Civ 2048.   [8] *A and others v Secretary of State for the Home Department* [2004] UKHL 56.

of the situation, provided that such measures are not inconsistent with its other obligations under international law.

. . .

17. Nothing in this Convention may be interpreted as implying for any State, group or person any right to engage in any activity or perform any act aimed at the destruction of any of the rights and freedoms set forth herein or at their limitation to a greater extent than is provided for in the Convention.

As well as seeing human rights as restraints on official action, they can also **7.06** perform as demands for protection, as explained in Chapter 2.[9]

B. Draft Civil Contingencies Bill

(a) General position

By and large, the 2004 Act does not necessarily infringe human rights in its primary **7.07** provisions, though its regulation-making powers provide a threat of secondary impact. The Cabinet Office paper certainly downplayed any adverse human rights implications, asserting that 'The Government does not believe that there is anything in the draft Bill that conflicts with the Convention and will prevent such a statement [under s 19 of the HRA] being made.'[10] Consequently, a Ministerial declaration of compliance under s 19 of the Human Rights Act 1998 was duly issued.[11]

Part I of the Bill even received commendation overall from the Joint Committee **7.08** on Human Rights:

> In our view, the provisions of Part 1 are likely to enable public authorities to act more effectively to protect the human rights of people in their areas in an emergency. We do not consider that they give rise to a significant threat of a violation of human rights.[12]

However, the Committee was very much less sanguine about Part II:

> We conclude that the provisions of Part 2 of the Draft Bill would, if enacted, give rise to a significant risk that regulations could be made which would violate, or authorise a violation of, Convention rights, without any judicial remedy being available for a victim of the violation.[13]

(b) Clause 25

As well as general concern about Part II, one provision in the draft Bill published **7.09** in June 2003 sparked controversy above all others. By cl 25: 'For the purposes of

---

[9] See further Handmer, J and Monson, R, 'Does a Rights Based Approach Make a Difference? The Role of Public Law in Vulnerability Reduction' (2004) 22.3 *International Journal of Mass Emergencies and Disasters* 43.

[10] Cabinet Office, *The Draft Civil Contingencies Bill*, para 61.

[11] Civil Contingencies Bill 2003–04 HC No 14, Explanatory Memorandum para 69.

[12] *Scrutiny of Bills and Draft Bills; Further Progress Report* (2002–03 HC No 1005, HL No 149) para 3.4.　　　　　　　　　　　　　　　　　　　　　　　　　　[13] Ibid, para 3.35.

the Human Rights Act 1998 (c. 42) an instrument containing regulations under s 21 shall be treated as if it were an Act of Parliament.' Though cl 25 was dropped after the initial round of consultation in Autumn 2003,[14] it may still be worth exploring as elucidating the relationship between rights and emergencies.

7.10 The genesis of cl 25, which was the first ever attempt to limit the Human Rights Act by this mode, arose from the Government's wish to instil some flexibility into the legislative scheme.[15] The concern was that while primary legislation can be challenged in court, it may not be struck down nor blocked by injunction. By contrast, secondary legislation is subject to injunction and can be quashed. This fate was potentially alarming:

> In an emergency, where speed is of the essence, it is not desirable for any emergency regulations to be held up by injunctions, especially where delay may prevent effective resolution of an emergency which threatens the safety of the community. Claims that human rights are being infringed may in the end prove unfounded, but a Court might on an interim basis order that emergency regulations be suspended.[16]

Clause 25 was the rejoinder which recategorized emergency regulations as primary legislation, thereby rendering them immune from being struck down under s 3 of the HRA.[17] The worst fate that could befall them would be a declaration of incompatibility under s 4 of the HRA. The Government thus proffered cl 25 but simultaneously encouraged debate since its advocacy 'is by no means certain'.[18] By January 2004, its mind had been turned, as it bowed to the many critics of cl 25 and withdrew it from the Bill.[19]

7.11 The provision was heavily criticized by the Joint Committee on the Draft Civil Contingencies Bill,[20] by the House of Lords Select Committee on the Constitution,[21] by the Defence Committee,[22] and the Joint Committee on Human Rights.[23] As well as setting a bad precedent,[24] there were several substantive reasons for the eventual omission of cl 25.[25]

---

[14] In addition, what is now s 30(2) was added to make the position explicit: HC Debs Standing Committee F col 304 10 February 2004, Douglas Alexander.

[15] Cabinet Office, *The Draft Civil Contingencies Bill*, para 5.32.   [16] Ibid, para 5.33.

[17] Compare Pontin, B and Billings, P, 'Prerogative Powers and the Human Rights Act: Elevating the Status of Orders in Council' [2001] PL 21.

[18] Cabinet Office, *The Draft Civil Contingencies Bill*, para 5.36.

[19] Cabinet Office, *The Government's Response to the Report of the Joint Committee*, para 29; Civil Contingencies Secretariat, *The Government's Response*, 36, 37.

[20] Draft Civil Contingencies Bill (2002–03 HC 1074) paras 11, 156.

[21] Ibid, para 146   [22] Draft Civil Contingencies Bill (2002–03 HC 557) para 68.

[23] Scrutiny of Bills and Draft Bills; Further Progress Report (2002–03 HC No 1005, HL No 149) para 3.26.   [24] Joint Committee on The Draft Civil Contingencies Bill, para 148.

[25] In addition to those listed below, there was the further point that the proclamation of an emergency, as detailed in cls 18, 19 of the draft Bill of June 2003 was not subject to cl 25 and so was reviewable under the Human Rights Act: Joint Committee on The Draft Civil Contingencies Bill, para 153.

Some concerned the limited legal impact. Clause 25 did not allow for the **7.12** making of emergency regulations in breach of the European Convention— consequences would still eventually follow which would be uncomfortable for Ministers. The crisis following *A v Secretary of State for the Home Department*[26] suggests that the political impact of a 'mere' declaration of incompatibility under s 4 of the HRA is also not to be underrated.[27] Equally, nothing in cl 25 prevented the ultimate consequence of an adverse judgment under international law in Strasbourg, albeit a number of years after the event.

There was also nothing to prevent continued judicial review under the general **7.13** principles of administrative law under the Civil Procedure Rules, Part 54 either of the secondary instrument itself or of executive action taken under it.[28] Therefore, it appeared odd to prevent the impact of the HRA but not of administrative law whereunder regulations could still be invalidated for being outside the scope of the delegated legislative power, or irrational, or made without complying with procedural preconditions.[29] At the same time, the Joint Committee recognized that the possibility of *ultra vires* was more apparent than real: 'the scope of the power to make regulations is so wide as to make it virtually impossible for any regulation to fall outside the scope of the delegated legislative power, and there is little likelihood of a court holding irrational a regulation made by a Minister in response to a properly declared emergency'.[30]

Besides, the Government itself wanted emergency powers always to operate **7.14** within the confines of the HRA, a consideration eventually reflected in s 21(5)(b)(iv). It is also reflected in specific provisions in s 23 against military or industrial service (Art 4 of the European Convention), the prohibition of a strike or other industrial action (Art 11), or the establishment of a special court or tribunal to try offences (Art 6). However, this list still leaves scope for cl 25 to have curtailed the right to freedom from detention without trial (Art 5), the right to respect for private and family life and the home (Art 8), the right to freedom of expression (Art 10), freedom of peaceful assembly (Art 11), the right to the peaceful enjoyment of possessions (Art 1 of Protocol 1), and the right to take part in elections (Art 3 of Protocol 1). Indeed, interferences are expressly contemplated of the latter two by s 22(3)(b), (c), and (f).

Other arguments for ditching cl 25 concerned the scope of existing qualifications **7.15** on rights. Because the European Convention already allows for wide discretion and adaptation in an emergency, extra room for manoeuvre is not needed. Many

---

[26] [2004] UKHL 56.

[27] Compare Joint Committee on Human Rights, *Scrutiny of Bills and Draft Bills: Further Progress Report* (2002–03 HC No 1005, HL No 149) para 3.29: 'it is of little value to the victim of a violation'. [28] Subject to HRA, s 3(2)(c).

[29] Ibid, para 3.22. It was recognized that the invalidity of a whole regulation was unlikely: Civil Contingencies Secretariat, *The Government's Response*, 38. [30] Ibid.

of the Convention rights are qualified rights and so can be interfered with (though some are absolute, such as Art 3). Presumably, the express incursions envisaged by, say, s 22(3)(b), (c), and (f) can be said to be potentially compatible on this basis, though no explanations are adduced.[31] In addition to the limits in the words of the relevant articles, the European Court of Human Rights will allow a margin of appreciation to deal with grave situations, as indicated in Chapter 6. There also remains the device of derogation from the Convention under Art 15 which is one which takes 'little time' to invoke.[32] A fax to the Secretary General of the Council of Europe is sufficient and may be lodged retrospectively to the purported derogation, as described below. If the circumstances demand some temporary encroachment on human rights, it could be argued that a derogation from the European Convention would be the more Convention-compliant way of proceeding. A derogation represents an important signal both of an extraordinary incursion into rights which ought to be made explicit and also of the need to bring the derogation to a close.

7.16   Next, the Government was persuaded that the reaction of courts to emergency regulations will be respectful and that fears of injunctions at the drop of a hat were rather fanciful. Any interim injunction must pass muster under the 'balance of convenience' test, and the courts will consider the wider public interest. An injunction was gratifyingly refused in *A and others v Secretary of State for the Home Department*.[33] In the light of the eventual decision in that case,[34] it may now be an exaggeration to say that the courts are 'deferential', but most challenges have failed either under human rights standards against anti-terrorism legislation or under administrative law against regulations under the Emergency Powers Acts.[35] The deferential judicial stance is also shared by the European Court of Human Rights, wherein most challenges to the special laws in Northern Ireland have failed, as described in Chapter 6. A notable exception was *Brogan v United Kingdom*,[36] but that judgment is hardly a convincing example of the courts causing havoc for the security forces. In that case, a period of detention following arrest under special powers was found to contravene Art 5(3) of the Convention since there was no judicial hearing within the

---

[31] Joint Committee on Human Rights, *Scrutiny of Bills and Draft Bills; Further Progress Report* (2002–03 HC No 1005, HL No 149) para 3.32.

[32] Joint Committee on The Draft Civil Contingencies Bill, para 152.

[33] [2002] EWCA Civ 1502.

[34] [2004] UKHL 56. See Walker, C, 'Prisoners of "War All the Time" ' [2005] *European Human Rights Law Review* 50.

[35] See Joint Committee on The Draft Civil Contingencies Bill, para 150; Livingstone, S, 'The House of Lords and the Northern Ireland Conflict' (1994) 57 MLR 333. For the Emergency Powers Acts, see Ch 2.

[36] App Nos 11209, 11234, 11266/84, 11386/85, Ser A 145-B (1988). See Marks, S, 'Terrorism and Derogation under the ECHR' (1993) 52 *Cambridge Law Journal* 360.

seven-day period allowed to the police. The outcome was the lodging of a notice of derogation but no legislative change (or more than minor changes in operations) until the Terrorism Act 2000, over a decade later, following which the derogation was withdrawn. A new derogation notice was necessitated by the institution of detention without trial under Part IV of the Anti-Terrorism, Crime and Security Act 2001.[37] In turn, Part IV was declared by the House of Lords to be incompatible with the Human Rights Act, on grounds of discrimination and proportionality.[38] But even this thunderous clash between branches of the state did not result in dangerous international terrorists being decanted from Belmarsh prison onto the streets of London. Instead, the detentions could continue in domestic law[39] until a 'control order' system was put in place under the Prevention of Terrorism Act 2005.

Finally, one might argue that cl 25 was disproportionate, for there could be less severe proposals for dealing with any potential problem.[40] For example, there could be an automatic stay (rather than relying on the court to exercise its discretion in favour of the Government) which would disable a court from implementing any finding of invalidity until there had been an opportunity for the exhaustion of all appeal processes or for Parliament to provide for new regulations. In addition, the possibilities of maverick judgments could be reduced by requiring litigation to be brought before a higher judicial body. **7.17**

Moving from restraints to positive protections, the Joint Committee on the Draft Civil Contingencies Bill recommended that the Bill prohibit regulations which would breach any of the Convention rights from which it is not possible to derogate or any provision in the Geneva Conventions.[41] Though this was not followed to the letter, an assertion of the importance of human rights did duly appear in s 20(5)(b)(iv). Under this measure, emergency regulations under Part II can only be issued if the Minister duly declares that s/he is satisfied that the regulations are compatible with the Convention rights (within the meaning of s 1 of the HRA). **7.18**

The outcome to the cl 25 debate is enshrined in what became s 30(2) (an addition to the Bill introduced in January 2004) by which 'Emergency regulations shall be treated for the purposes of the Human Rights Act 1998 (c. 42) as subordinate legislation and not primary legislation (whether or not they amend primary legislation).' **7.19**

---

[37] See Walker, C, *A Guide to the Anti-terrorism Legislation* (Oxford University Press, Oxford, 2002) Ch 8. That derogation was lifted following the passage of the Prevention of Terrorism Act 2005.

[38] *A and others v Secretary of State for the Home Department* [2004] UKHL 56.

[39] But the continued detentions may well give rise to a claim for compensation under the Human Rights Act 1998, s 8, and applications may be brought before the Strasbourg Court.

[40] Ibid, para 155.   [41] *Draft Civil Contingencies Bill* (2002–03 HC 1074) para 12.

C. The Act—Definitions

7.20 As already indicated in Chapter 4, the definitions in ss 1 and 19 are much broader than their predecessor. Forms of legitimate political activity and protest under Art 10 or legitimate assemblies or industrial action under Art 11 will be caught by this very broad definition.

D. The Act—Part I

7.21 Relatively, few acute human rights threats arise under Part I. Indeed, the planning processes are likely to result in protection rather than threats to human rights. Nevertheless, there remains one official concern, namely that secondary legislation can be made to require Category 1 and 2 responders to provide information to each other in certain circumstances.[42] It is conceivable that such information would relate to the private life of an individual and that the sharing of information would engage Art 8 (right to private life). Of course, such an incursion may be covered by Art 8(2).

E. The Act—Part II

7.22 The broad sweep of Part II regulations is more troubling.[43] The main features of the controversies relating to rights are as follows: the protection for rights in general is inadequate; there are protections for several specified human rights, but many are unduly limited; and the relationship between derogation and the exercise of powers is obscure.

(a) Protection for rights in general

7.23 The unspecified protection for rights is reflected in the requirements of necessity (s 21(3)) and proportionality (s 23(1)(b)). A specific concession in relation to Part II in favour of rights was granted in January 2004 when it was announced that, akin to the position under s 19 of the Human Rights Act 1998:

> ...the Minister responsible for the Parliamentary debate on the regulations in each House should make a statement as to whether he considers that the regulations are compatible with the Convention rights. This reflects the undertaking that the Government has already given to volunteer a statement of compatibility in relation to instruments that are subject to the affirmative

---

[42] Cabinet Office, *The Draft Civil Contingencies Bill: Explanatory Notes etc*, para 71.
[43] Ibid, para 72.

resolution procedure. The statement will generally be made in the explanatory memorandum which is prepared for the debate or, should the urgency of the matter mean that no explanatory memorandum is available, by the Minister in charge of the debate.[44]

This concession is reflected in s 20(5)(b)(iv), by which the person making the regulations must declare satisfaction that the regulations are compatible with the Convention rights. One effect of this provision is to provide indirect protection against the abrogation of absolute human rights which cannot be affected even in emergencies and are not otherwise mentioned in the Act (as in the case of Art 4)—the right to life (Art 2), freedom from torture (Art 3) and from penalties for retrospective offences (Art 7).[45] Of course, such rights are further protected in the sense that it would be 'difficult to justify' an incursion into them on administrative law grounds.[46] In most cases, the rules of interpretation in s 2 of the HRA should ensure its predominance over the later regulatory legislation arising under Part II, subject to judicial courage. Though there are causes for optimism, it is odd and potentially inconsistent with the Ministerial statement of compatibility that no specific mention is made of these non-derogable rights[47] and yet more specific protections exist for some other non-derogable rights (such as Art 4).

Regulations may affect derogable rights such as to liberty under Arts 5 and 11 **7.24**
(through restriction on liberty and detention without trial), respect for private and family life and the home (Art 8), freedom of expression (Art 10) and freedom of assembly and association (Art 11), rights to property under Art 1 of Protocol 1 (through requisition or destruction and without compensation, which also raises an issue of a lack of a remedy contrary to Art 13), and rights to participate in elections (Art 3 of Protocol 1). Requirements to act in performance of a function (cl 21(3)(k)) could infringe Art 4 (prohibition on forced labour), though any demands are subject to s 23(3). Concerns could also arise under Art 6 (fair trial), but s 23(4)(d) does prevent alteration to criminal process. Should there be more protection for these rights?

---

[44] Cabinet Office, *The Government's Response to the Report of the Joint Committee*, para 37. See further HC Debs vol col 1400 24 May 2004, Hazel Blears. The statement might, subject to wording, serve also as the 'official proclamation' required by the International Covenant on Civil and Political Rights (New York, UNTS 171, 1966) (Cmnd 6702, London, 1977), art 4: *Brannigan and McBride v United Kingdom*, App Nos 14553/89, 14554/89, Ser A vol 258-B (1994), para 73.

[45] Note also the right not to be subject to the death penalty under Protocol No 6 to the Convention for the Protection of Human Rights and Fundamental Freedoms concerning the abolition of the death penalty (ETS 114, 1983). No derogation is permitted (Art 3), nor is any intended: HL Debs vol 665 col 1007 21 October 2004, Baroness Scotland.

[46] Joint Committee on Human Rights, *Scrutiny of Bills and Draft Bills; Further Progress Report* (2002–03 HC No 1005, HL No 149) para 3.33.

[47] Joint Committee on The Draft Civil Contingencies Bill, para 167.

7.25 Further to the mention of Art 5 in the foregoing list, the Act does not expressly forbid nor grant powers to detain without trial, an issue already discussed in Chapter 5.

7.26 The lobby in favour of rights to free expression was surprisingly weak.[48] One could argue that precedents exist for its special position[49] and that it is vital in an emergency as a protection for rights and constitutionality.[50]

7.27 More concern was raised about the loss of property without compensation which will only be excused under Art 1 of Protocol 1 in very exceptional circumstances.[51] Even the police were sympathetic to the award of compensation as of right,[52] predicting that it would lubricate public cooperation:

> ...if I were a police constable who had to requisition property or state that property would be destroyed, it would make my life a lot easier if I could say to the person who owned it, 'You will be compensated for it.'[53]

The Government indicated that a failure to provide compensation was improbable:

> ...in every civil contingency in which the 1920 legislation has been used, compensation has been paid. The Government's habit is to pay compensation. Our desire is to do that, because in our experience it lubricates the management of an emergency. It secures good will, which is necessary if things are to work.[54]

Nevertheless, the Government went on to give examples where no payment would be forthcoming.[55] One arises when the police or others requisition heavy lifting or earth-moving equipment. An extensive compensation arrangement might not be worthwhile if the requisition period is fleeting. Secondly, where a large proportion of the population is affected, such as where the water supply is cut off or the internet is suspended or business interrupted, payment on that broad scale— almost a tax— is resisted. Thirdly, it may be necessary to prevent a growing catastrophe to that person's own property by making a fire-break. Fourth, culpability can come into play, as where an owner who wilfully infected cattle and thereby caused a major

---

[48] Ibid, para 171.
[49] See Contempt of Court Act 1981, s 10; Police and Criminal Evidence Act 1984, s 13.
[50] Support is often given for the principle of free speech but less so for its primacy: *R v Secretary of State for the Home Department, ex p Brind* [1990] 1 All ER 469; *R v Secretary of State for the Home Department, ex p Simms* [2000] AC 115 at 126 per Lord Steyn; *Reynolds v Times Newspapers* [2001] 2 AC 127 at 205 per Lord Nicholls.
[51] See *Holy Monasteries v Greece* App Nos 13092/87, 13984/88, Ser A 301A, and *Loizidou v Turkey*, App No 25781/94, 2001-IV
[52] Joint Committee on The Draft Civil Contingencies Bill, para 173.
[53] Ibid, minutes of Evidence Q63 (Alan Goldsmith, ACPO).
[54] HC Debs Standing Committee F col 230 5 February 2004, Fiona Mactaggart.
[55] Ibid and Joint Committee on The Draft Civil Contingencies Bill, Appendix 9 q 37.

outbreak of an infectious disease, would not be a candidate for compensation.[56] Fifthly, the availability of insurance payments is also considered relevant.

Yet it is not certain that insurance will always be available, as there is no scheme **7.28** for state reinsurance for private property equivalent to the Pool Re scheme under the Reinsurance (Acts of Terrorism) Act 1993.[57] However, the Government refused to offer a 'blank cheque from the public purse in all circumstances'.[58] This stance is subject to the interpretation of property requisition under wartime powers as requiring compensation.[59] Compensation was not mentioned under the Emergency Powers Act 1920, though, during the state of emergency declared in 1973, regs 30 and 31 required the Government to pay compensation if they requisitioned chattels or took possession of land. Furthermore, Art 1 of Protocol 1 does not invariably require compensation, and s 23(5)(b) may be sufficient to protect such compensation as is demanded by it.[60] A further concern might arise under art 13 if the regulation excludes any possibility of redress by way of compensation.

(b) Protection for specified rights

As rehearsed in Chapter 5, by s 23(4) and (5), regulations under Part II are not **7.29** allowed to detract from a number of specified human rights. Since these have been described and analysed in Chapter 5, only an outline will be repeated here.

First, there is a ban on regulations which require that a person provide military **7.30** or industrial service (relevant to Art 4).[61]

Secondly, there is a ban on regulations which prohibit strikes or industrial **7.31** action (Art 11).

Thirdly, there is a ban on regulations which create new offences (aside from **7.32** ancillary offences under s 22(3)(i)) (Arts 6, 7, 9, 10, or 11).

Fourthly, there is a ban on regulations which alter procedure in relation to **7.33** criminal proceedings (Art 6). There is no specific mention of Art 6, but the

---

[56] HC Debs Standing Committee F col 313 10 February 2004, Douglas Alexander. The disqualification of those who are to blame for their losses is reflected in other compensation schemes such as those relating to criminal injuries: Criminal Injuries Compensation Scheme (2001) para 13.
[57] Walker, C, 'Political Violence and Commercial Risk' (2004) 56 *Current Legal Problems* 531.
[58] Joint Committee on The Draft Civil Contingencies Bill, para 175.
[59] *Cannon Brewery Co, Ltd v Central Control Board (Liquor Traffic)* [1918] 2 Ch 101; *A-G v De Keyser's Royal Hotel* [1920] AC 508; *Burmah Oil Co v Lord Advocate* [1965] AC 75.
[60] Cabinet Office, *The Government's Response to the Report of the Joint Committee*, para 33. For jurisprudence on property rights, see Coban, AR, *Protection of Property Rights within the European Convention on Human Rights* (Ashgate, Aldershot, 2004).
[61] See Ch 5 for its elaboration.

mention of criminal proceedings represents an important improvement on the draft Bill of June 2003.[62] The addition replicates s 2(2) of the Emergency Powers Act 1920 but does not rule out changes in venue and time-limits.

7.34 Fifthly, there is a ban on regulations which amend Part II of the Civil Contingencies Act. Thus, it is not possible, for example, to resurrect cl 25, should predictions about its irrelevance turn out to be misplaced.

7.35 Sixthly, there is a ban on regulations which amend the HRA itself. The debate over what is now s 23(5)(b) has been indicated in Chapter 5. The protection for rights by reference to the HRA means that other rights instruments receive no protection.[63] There is, however, a promise that 'the Government would certainly take into account the international obligations of the United Kingdom before taking action under the Bill.'[64] Otherwise, the Government believed that measures of international law are best protected in international law.[65]

(c) Derogation

7.36 The general question as to whether it is justifiable to provide for the permanence of emergency laws has already been tackled in Chapter 2. The more specific question to be answered here is what is the relationship between derogation under Art 15 and the exercise of powers under Part II of the Act? It cannot be assumed that all the circumstances within Part II are bound to meet the tests of 'public emergency' under Art 15. Then again, most invocations of Part II of the Civil Contingencies Act will not require resort to a derogation under Art 15. All depends on the focus and ferocity of the relevant regulations. Furthermore, the

---

[62] Joint Committee on The Draft Civil Contingencies Bill, para 164.
[63] For examples, see UN International Covenant on Civil and Political Rights 1966 (New York, UNTS 171, 1966) (Cmnd 6702, London, 1977), Geneva Third Convention relative to the Treatment of Prisoners of War 1949 (75 UNTS 135) (Cmnd 550, London, 1958), Geneva Fourth Convention relative to the Protection of Civilian Persons in Time of War 1949 (75 UNTS 287) (Cmnd 550, London, 1958), First Protocol Additional to the Geneva Conventions of 12 August 1949, and Relating to the Protection of Victims of International Armed Conflicts (1125 UNTS 3) (Cm 4338, London, 1999), Second Protocol Additional to the Geneva Conventions of 12 August 1949, and Relating to the Protection of Victims of Non-International Armed Conflicts (1125 UNTS 609) (Cm 4339, London, 1999); United Nations Convention on the Status of Refugees (Geneva, 28 July 1951, TS 39) (Cmd 9171, London, 1954); International Convention on the Elimination of All Forms of Racial Discrimination (New York, 7 March 1966, TS 77) (Cmnd 4108, London, 1969); Paris Minimum Standards of Human Rights Norms in a State of Emergency (1985) 79 AJIL 1072; UN Convention against Torture and Other Cruel, Inhuman or Degrading Treatment or Punishment (New York, 1984, TS 107) (Cm 1775, London, 1991); European Convention for the Prevention of Torture and Inhuman or Degrading Treatment or Punishment (Strasbourg, 1987, ETS 126) (Cm 339, London, 1988); Declaration on the Protection of All Persons from Enforced Disappearance (New York, 1992 UNGA res 47/133; 32 ILM 903).
[64] Joint Committee on The Draft Civil Contingencies Bill, Appendix 9 q 19.
[65] Cabinet Office, *The Government's Response to the Report of the Joint Committee*, para 32.

official intention is to avoid derogation: 'the Government have not sought to derogate from the provisions of the [Human Rights] Act because we do not expect any regulations made under the powers in the Bill to contravene that Act.'[66]

Conversely, if derogation is required, then the new-found rules as to geo- **7.37** graphical proportionality in Part II do not necessarily cause problems even though the wording of Art 15 of the European Convention on Human Rights envisages an emergency threatening the life of 'the nation'. After all, most derogations since 1951 within the United Kingdom have been confined to the region of Northern Ireland. Presumably the same could apply to a region of England and Wales, especially as many will be larger in population size than Northern Ireland. But it must be shown that the emergency has national repercussions—it threatens the normal life of the nation—and so it is conceivable that localized crises properly triggering Part II powers will not meet Art 15 standards. It would have been desirable to give some guidance in the Act 'to avoid error and to ensure due consideration'.[67]

A possible resolution of the lack of fit between Part II and Art 15 would have **7.38** been to provide for two distinct levels of emergency.[68] A precedent along these lines is offered by the Prevention of Terrorism Act 2005 which allows derogating control orders and non-derogating control orders, depending on the severity of the restrictions in a control order. Likewise, one could in Part II have one set of procedures under the rubric of 'national emergency' for those measures which are beyond the pale of normal times and deal with events which are said to 'threaten the life of the nation'. The lower level could be called a 'local' or 'regional' emergency.

F. Conclusions

The deliberations over cl 25 helpfully demonstrated to the Cabinet Office the **7.39** depth of feeling about rights values and the foreboding caused by the legislation in general. In that respect, there has been affirmation of the prediction by Lord Chancellor Irvine, who said in Dec 1997:

> [The Human Rights] Bill will therefore create a more explicitly moral approach to decisions and decision making; will promote both a culture where positive rights and liberties become the focus and concern of legislators, administrators and

---

[66] HC Debs vol 416 col 1179 19 January 2004, Fiona Mactaggart.
[67] Joint Committee on The Draft Civil Contingencies Bill, para 172.
[68] See Law Society, *Response to The Draft Civil Contingencies Bill* (London, 2003).

judges alike; and a culture in judicial decision making where there will be a greater concentration on substance rather than form.[69]

7.40 The removal of cl 25 was followed by many official protestations that there was no intention to come into conflict with the HRA and, indeed, no powers to do so. The sounds are certainly soothing, but one should not be lulled into a false sense of security. It should be recalled that the draft regulations under Part II have been put under wraps. If ever unpacked, they may contain some severe restrictions on rights. One then must hope that the Parliament and the courts will be forthright in their protection of rights[70] to a degree not seen in most previous bouts of emergency.

---

[69] Tom Sargant memorial lecture, December 1997, http://www.open.gov.uk/lcd/sppeches/tomsarg.htm.

[70] It is assumed that the absence of a statement equivalent to the Anti-Terrorism, Crime and Security Act 2001, s 30 would make no difference, as it is meant to restrain rather than found jurisdiction. But compare *A and others v Secretary of State for the Home Department* [2004] UKHL 56 at para 164 per Lord Rodger.

PART III

THE OPERATIONALIZATION
OF RESILIENCE

8

TOWARDS A CIVIL CONTINGENCIES FRAMEWORK

The aim of this chapter is to examine the broader institutional and organiza- **8.01** tional impact and setting of the Civil Contingencies Act. This inquiry will thereby move beyond the local and regional tiers, which lie at the heart of the legislation, into the work of central government, on which there is near silence.

A. Central Government Structures[1]

(a) Cabinet level

Cabinet committees

The Civil Contingencies Committee (CCC), chaired by the Home Secretary, is **8.02** the key Cabinet Committee. Its terms of reference and related structures are illustrated in Table 8A overleaf:[2]

---

[1] See Jeffery, K and Hennessy, P, *States of Emergency: British Governments and Strikebreaking since 1919* (Routledge & Kegan Paul, London, 1983) Ch 8; Campbell, D, *War Plan UK* (Burnett Books Ltd, London, 1982); Laurie, P, *Beneath the City Streets* (Granada, London, 1983); Hennessy, P, 'Whitehall Contingency Planning for Industrial Disputes' in Rowe, PJ and Whelan, CJ (eds), *Military Intervention in Democratic Societies* (Croom Helm, London, 1985).

[2] Source: Defence Committee, *Defence and Security in the UK* (2001–02 HC 518) para 132 (as adapted). See also http://www.cabinetoffice.gov.uk/secretariats/committees/ccc.asp.

Table 8A: Civil Contigency Cabinet Committies in 2002

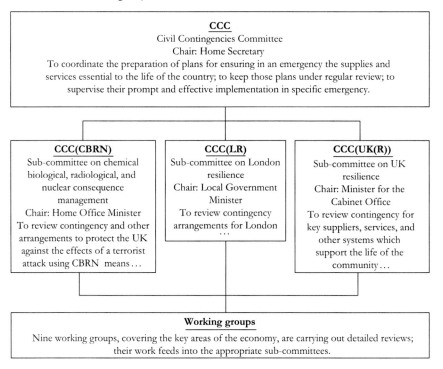

8.03 The foregoing structure was settled largely in 2001[3] but was augmented in 2002 with a range of Cabinet sub-committees based on the standing Defence and Overseas Policy Committee.[4] These committees deal with policy rather than response. Most relevant is the Sub-Committee on Protective Security and Resilience (DOP(IT)(PSR)) which is also chaired by the Home Secretary.[5] It also comprises a number of other important Ministers (including the Foreign Secretary) and places civil contingencies planning in a wider policy context, including oversight of the capabilities programme. There is additionally the Sub-Committee on International Terrorism (DOP(IT)) which keeps under review the Government's strategy for reducing the risk from international terrorism.

---

[3] Coordinating Cabinet committees, linking to regional commissioners, can be traced to the Supply and Transport Committee in 1919, the Industrial Emergencies Committee in 1945 and the Emergencies Committee in 1948: O'Brien, TH, *Civil Defence* (HMSO, London, 1955) 29; Jeffery, K and Hennessy, P, *States of Emergency: British Governments and Strikebreaking since 1919* (Routledge & Kegan Paul, London, 1983) Ch 8; Thurlow, R, *The Secret State* (Blackwell, Oxford, 1994) 288.   [4] See Ministry of Defence, *Operations in the UK*, para 215.
[5] See http://www.cabinetoffice.gov.uk/secretariats/committees/dopitpsr.asp.

Civil Contingencies Secretariat

Moving from the Ministerial to the administrative, the Civil Contingencies **8.04**
Secretariat (CCS) is the 'pivotal' organization[6] which services the relevant
Cabinet committees and other governmental bodies. It is not an executive body
in all emergencies but does provide the central focus for those crises which
demand cross-departmental and cross-agency involvement.[7] The Lead Government Department (LGD—see below) is responsible for alerting the CCS
whenever a crisis might warrant collective consideration.[8] The CCS was set up
in July 2001 within the Cabinet Office to consolidate diverse departmental
responsibilities, including those of the Home Office from where most of the
staff transferred.[9] As now constituted, the CCS reports to Ministers and
through the Security and Intelligence Co-ordinator and Permanent Secretary to
the Cabinet Office.

The work of the CCS involves: the improvement of resilience (through **8.05**
horizon scanning activity and assessment of potential challenges);[10] support to
Ministers, including through the Committees described above; advice about,
and assurance of, resilience within Government and the public sector; the
development of skills and awareness; and assistance in the management of
domestic crises.[11] These activities are expressed through the following corporate services team: Assessments and Commitments; Legislation; Capabilities and Committees; and the Emergency Planning College.[12] It operates
via six specialist desks which cover varying issue-areas: Social stability and
government continuity of government; Communications and media; Economic and financial affairs; Transport, energy and distribution; Public and
animal health and food safety; and Environment and water.[13] Horizon scans
and specific assessments will be circulated to departments, as well as to the
cross-government Domestic Horizon Scanning Committee. During a crisis,

---

[6] Cabinet Office, *Dealing with Disaster* (rev 3rd ed, Cabinet Office, London, 2003) para 1.12.
Its forerunner is the Supply and Transport Organisation, first operative in 1919: Thurlow, R, *The
Secret State* (Blackwell, Oxford, 1994) 124, 154.   [7] Ibid, para 7.17.
[8] Cabinet Office, *Dealing with Disaster* (rev 3rd ed, Cabinet Office, London, 2003) para 7.3.
[9] Defence Committee, *Defence and Security in the UK* (2001–02 HC 518) para 140. The
previous Civil Contingencies Unit was formed in 1972: Bonner, D, *Emergency Powers in
Peacetime* (Sweet & Maxwell, London, 1985) 29, 224.
[10] It oversees the departmental responses to the threat assessments produced by the Joint
Terrorism Analysis Centre (JTAC). It develops and maintains the threat assessments which feed
into the Planning Assumptions that underpin the UK Resilience Capabilities Programme and also
the risk assessments undertaken by departments: Government Reply to the House of Commons
Science and Technology Committee, *The Scientific Response to Terrorism* (Cm 6108, London, 2004)
paras 16–18.
[11] See further Government Response to the Defence Committee, *Defence and Security in the UK*
(2001–02 HC 1230) para 31.   [12] See http://www.epcollege.gov.uk/.
[13] See Civil Contingencies Secretariat, *The Lead Government Department*, para 3.2.

233

regular assessments will be produced by a Joint Analysis Cell, run by the Assessments and Commitments Team, in support of the Cabinet Office Briefing Rooms (COBR—see below).[14]

8.06 The task of horizon scanning was considered by the Strategy Unit in the Cabinet Office in its report, *Risk: Improving Government's Capability to Handle Risk and Uncertainty*.[15] It concluded that the CCS should continue to develop its role in relation to potential disruptive challenges to the running of the United Kingdom, identifying and assessing the key risks within a 12-month horizon. However, the work of the CCS should be complemented by a longer-term horizon-scanning role for the Strategy Unit. The Unit is now entitled the Prime Minister's Strategy Unit and has indeed produced in 2005 an audit of the challenges facing the United Kingdom.[16]

8.07 As well as planning and prevention, the CCS also contains a crisis coordination centre which can operate under Home Secretary chairmanship when there is a need for the actual management of the consequences of a live event.[17] In a live emergency, the CCS transacts various strategic tasks:[18] providing an assessment of needs, and securing their provision; planning for different scenarios; communicating with relevant departments; and facilitating augmentation of the LGD's resources, flows of advice and information, and public information systems.

8.08 Alongside the CCS is the post of Security and Intelligence Co-ordinator and Permanent Secretary, Cabinet Office, set up in June 2002. This official coordinates security, intelligence and consequence management themes, and deal with risks and major emergencies should they arise, as well as guiding the work of the CCS.[19]

8.09 At regional level, the position is overseen not only by the CCS but also by the Regional Coordination Unit (the Whitehall centre of operations for the Government Office Network) within the Office of the Deputy Prime Minister. More specific direction is provided by the Regional Resilience Division (now part of the Civil Resilience Directorate) which was set up in April 2003 to implement the regional 'resilience' tier by checking that current response

---

[14] Ibid, para 3.8.
[15] See http://www.strategy.gov.uk/downloads/su/RISK/REPORT/downloads/su-risk.pdf, 2002, 20, 21, 22, 23, 24: 53–55.
[16] *Strategic Audit: Progress and Challenges for the UK* (http://www.strategy.gov.uk/downloads/work_areas/strategic_audit/strategic_audit2.pdf, 2005).
[17] Government Response to the Defence Committee, *Defence and Security in the UK* (2001–02 HC 1230) para 28.
[18] Cabinet Office, *Dealing with Disasters* (rev 3rd ed, Cabinet Office, London, 2003) para 7.18.
[19] Defence Committee, *Defence and Security in the UK* (2001–02 HC 518) para 183; Government Response to the Defence Committee, *Defence and Security in the UK* (2001–02 HC 1230) para 28.

capabilities are understood, that gaps in preparedness are identified, and that plans are made to address contingencies by regional and local authorities.[20]

The CCS had a difficult birth, with initial confusion over its role[21] and pro- **8.10** blems involving the assimilation of the plans of individual departments and the need to establish relationships with them. The House of Commons Defence Committee depicted the CCS as suffering from 'the habitual "departmentalism" of Whitehall'.[22] It seems to have bedded in since 2002, attracting distinct approval for its handling of the July 2005 London bombings.

Cabinet Office Briefing Rooms (COBR)

When the crisis reaches beyond the interests or capabilities of one government **8.11** department, COBR will be the location for meetings about the involvement of central government.[23] The COBR meeting, normally chaired by the Home Secretary, will include a senior Association of Chief Police Officers (ACPO) representative, and the Police National Information Coordination Centre can be activated to give support. Where a Strategic Coordinating Group (see below) for response and recovery has been established and where a central government response is required because of scale or breadth, a Government Liaison Team (GLT) will be formed and sent to the region.[24] The GLT will be headed by a Government Liaison Officer, a senior official from the LGD or from the relevant Government Office for the Regions and will feed back information to COBR. Publicity about the central intervention will emerge via a News Coordination Centre or a specific media centre.[25]

(b) Lead Government Department

Whilst the CCS is seeking to facilitate an important unifying principle in central **8.12** government, in an emergency the concept of designated 'Lead Government Department' (LGD) (or a devolved administration for devolved matters wholly occurring in that region) remains imperative. It reflects a resilience strategy which is to spread involvement and responsibility amongst potentially all departments, a strategy which was followed even during the total war of 1939 to 1945.[26]

---

[20] See http://www.odpm.gov.uk/stellent/groups/odpm_civilres/documents/sectionhomepage/odpm_civilres_page.hcsp.
[21] Defence Committee, *Defence and Security in the UK* (2001–02 HC 518) para 149.
[22] Ibid, para 153.
[23] Cabinet Office, *Emergency Response and Recovery*, paras 12.12–12.14; Cabinet Office, *Central Government Arrangements*, para 19.
[24] Cabinet Office, *Central Government Arrangements*, para 20.
[25] Ibid, para 12.15; Cabinet Office, *Central Government Arrangements*, para 30.
[26] See O'Brien, TH, *Civil Defence* (HMSO, London, 1955) 169.

Pre-nominated LGDs for 28 different types of disaster or emergency have been delineated.[27] Selection will be based upon:

The nature of the crisis or emergency. There is usually a clear link between the nature of the events in question and the normal business of the government department. *Access to information.* Ready access to a flow of up-to-date and accurate information is essential for a Lead Government Department. The chance of achieving this is greater if officials of the department, agencies and organisations concerned know each other and have worked together . . .

The availability of facilities. Where events are large-scale or protracted, it may be necessary to use a dedicated emergency room. Those departments most likely to need to act as Lead Government Departments have such facilities.[28]

In the event of dispute, the Cabinet Office will make a recommendation to the Prime Minister.[29]

8.13 The Cabinet Office has issued guidance in the form of *The Lead Government Department and its Role—Guidance and Best Practice*.[30] The LGDs have duties of planning and preparation, including training and exercising. They are advised to maintain and equip an emergency control centre.[31] When an emergency occurs, then the range of duties include:[32] acting as the focal point for communications between central government and the local strategic coordinating group; producing a situation report for the Prime Minister, the Chairman of the CCC, and the head of the CCS; producing a handling plan which advises as to the capacity of the LGD; applying the LGD resources and coordinating the support from other departments and agencies; taking decisions and actions and providing assistance and advice; publishing information for the public and the media, and collaborating with the Government Communication Network; providing accountability to Parliament and any inquiry. But the Cabinet Office guidance emphasizes at the outset that LGD involvement is to be viewed as exceptional, where the scale or complexity of the crisis so demands,[33] and that most emergencies should be handled locally.[34]

8.14 In preparation for any intervention, there should be a risk assessment in cooperation with Departmental stakeholders and in the light of the CCS assessments and the Resilience Capability Framework (see below). There must also be a capability assessment on the current state of preparedness within the LGD.[35] The assessment will be followed by planning and preparation by Effective Planning and Crisis Management Teams which will involve Emergency Operations/Co-ordination Centres, regular training, and a public

---

[27] See http://www.ukresilience.info/handling.htm.

[28] Cabinet Office, *Dealing with Disaster* (rev 3rd ed, Cabinet Office, London, 2003) para 7.5.

[29] Cabinet Office, *Central Government Arrangements*, para 12.

[30] Civil Contingencies Secretariat, *The Lead Government Department, and its Role* (http://www.ukresilience.info/lgds.pdf, 2004). [31] Ibid, para 7.10.

[32] Ibid, para 7.14. [33] Cabinet Office, *Emergency Response and Recovery*, para 12.6.

[34] Civil Contingencies Secretariat, *The Lead Government Department*, para 1.

[35] Ibid, para 7.

information and media strategy. The final stage in the cycle is assurance on the contingency planning process and plan content, such as by exercises and audit. The mechanisms are mainly internal, but independent internal or external assurance can be obtained from the Committee on Domestic and International Terrorism (Resilience) (TIDO(R)) Exercise Working Party or external consultants, with internal audit reports being provided to the CCS.[36]

The LGD principle[37] received strong affirmation during the passage of the Civil   **8.15**
Contingencies Act:

> Our view is that resilience must be embedded within all Government organisations; putting it into a separate silo risks particular Departments not taking responsibility for their own resilience and marginalising events that are by their very nature not day-to-day but extraordinary, although they unfortunately feel more day-to-day at present. It is important to ensure that the way in which the Government go about their business includes the preparation for the extraordinary at every level.[38]

The Government offers a number of justifications for the LGD principle. In   **8.16**
short, it allows for a range of capabilities and expertise to be quickly available in any emergency, followed by knowledgeable responses which are unhampered by long chains of command or the need to convey expertise and resources into the theatre of operations, all of which may prove difficult to effect in an emergency.[39] Furthermore, there is an issue of responsibilization:

> Each of the central government departments must take responsibility for its own sectoral area and not be tempted to try to dislodge its responsibilities onto a central body.[40]

Yet, the Defence Committee was critical of the arrangement. It asked whether the   **8.17**
CCS could play a stronger role so that there is 'a one-stop shop' in the management of emergency as well as a locus for the gathering of experience.[41] In its absence, its view was that the existing structures in central government are 'unnecessarily bureaucratic, inward-looking and confused'.[42] Departmental administration is noted for cultures of caution and concern for territory—'departmentalism' in the words of the Defence Committee report.[43] The indeterminate structure beyond the local level creates real dangers because 'Emergencies . . . frequently develop into disasters because of inadequacies of command and control.'[44] This debate will be

---

[36] Ibid, para 2.3, Ch 11.
[37] HC Debs vol 389 col 1070w 23 July 2002, David Blunkett.
[38] HC Debs Standing Committee F col 181 3 February 2004, Fiona Mactaggart.
[39] See Executive Session on Domestic Preparedness. *Beyond the Beltway: Focusing on Hometown Security* (John F Kennedy School of Government, Harvard University, 2002); Cabinet Office, *Dealing with Disaster* (rev 3rd ed, Cabinet Office, London, 2003) para 1.8.
[40] See Joint Committee on the Draft Civil Contingencies Bill, Appendix 9 q 76.
[41] *Defence and Security in the UK* (2001–02 HC 518) para 173.
[42] Ibid, para 181.   [43] Ibid, para 153.
[44] Joint Committee on the Draft Civil Contingencies Bill, Appendix 7.

revisited in Chapter 9, including consideration of the model of a unified US Department of Homeland Security.

(c) Levels of engagement

8.18 Since 'Most emergencies in the United Kingdom are handled at a local level by the emergency services and by the appropriate local authority or authorities, with no direct involvement by Central Government',[45] central intervention is to be confined to three levels.[46] In case of catastrophic emergency (Level 3: one which has a high and potentially widespread impact and requires immediate central government direction and support) the response would be led from COBR under the leadership of the Prime Minister. For a serious emergency (Level 2: one which has, or threatens, a wide and prolonged impact requiring sustained central government coordination and support from many departments and agencies), the central government response would again be led from COBR but under the direction of the Home Secretary or a nominated lead Minister. During a significant emergency (Level 1: one which requires central government support primarily from an LGD or devolved administration), COBR may not be continuously involved so the focus shifts to the LGD. It will be up to COBR and/or the LGD to determine whether to convene the appropriate regional resilience tier. The CCS can provide support on impact management and recovery issues. The guiding principles in all cases are said to be preparedness, continuity, subsidiarity, clear direction, integration, communication, cooperation, and anticipation.[47]

B. Central Government Policy on Resilience

8.19 The 'Key Capabilities Programme' is the core agenda for the delivery of civil resilience within the United Kingdom. The aim is 'to ensure that a robust infrastructure of response is in place to deal rapidly, effectively, and flexibly with the consequences of civil devastation and widespread disaster inflicted as a result of conventional or non-conventional disruptive activity.'[48] The Resilience Capability Framework comprises planning assumptions about the assessed risks, capability targets, and assessments of capability readiness levels.

8.20 The Programme now consists of a total of 17 capability 'work-streams'. They are categorized into three groups: three of the work streams are structural, dealing with the central, regional, and local response capabilities; five are concerned with the maintenance of essential services (food, water, fuel,

---

[45] Civil Contingencies Secretariat, *The Lead Government Department*, para 1.
[46] Ibid, para 3; Cabinet Office, *Central Government Arrangements*, para 5.
[47] Cabinet Office, *Central Government Arrangements*, para 8.
[48] Civil Contingencies Secretariat, *The Lead Government Department*, para 4.3.

transport, health, and financial services); nine are functional (the assessment of risks and consequences, CBRN resilience, infectious human diseases; infectious animal and plant diseases; mass casualties; mass fatalities; mass evacuation; site clearance; and warning and informing the public). A department is assigned as the lead for each, working to targets agreed between the Ministers and the CCS. There is oversight upwards of the Programme through the Cabinet Office and an Official Committee on Domestic and International Terrorism (Resilience) (TIDO(R)) which reports to the Cabinet Sub-Committee Committee on International Terrorism), (Ministerial Group on Consequence Management and Resilience) (DOP(IT)(R)). There is delivery downwards of the functional work-streams through the Regional Resilience Teams in the Government Offices of the Regions or through the devolved administrations.

Alongside capabilities are the relevant strategies for dealing with specific con- **8.21** tingencies. One notable is the Counter-Terrorism Strategy (CONTEST).[49] Its elements comprise *prevention*: horizon scanning activity to identify and assess potential disruptive challenges and so develop integrated responses such as the Joint Terrorism Analysis Centre (JTAC), formed in 2003 to deal with threat intelligence assessment;[50] *pursuit*: including extra investment in security services and police and enhanced legislation; *protection*: which includes investment in border security; and *preparation for consequences*: an emphasis across government on improved business continuity arrangements plus the civil contingencies legislation. There are corresponding strategies for dealing with chemical, biological, radioactive and nuclear (CBRN) attack,[51] human pandemic disease,[52] flood and coastal erosion,[53] and flooding.[54]

[49] See http://www.homeoffice.gov.uk/terrorism//legislation/background.html.

[50] See Intelligence and Security Committee, *Annual Report 2002–03* (Cabinet Office, London, 2003) para 62; Government reply to the House of Commons Science and Technology Committee (Cm 6108, London, 2004) para 15. In 2005, the role of chair of JTAC has been combined with the post of the Security and Intelligence Co-ordinator, and Permanent Secretary to the Cabinet Office.

[51] See http://security.homeoffice.gov.uk/counter-terrorism-strategy/cbrn-resilience/; Home Office, *The Release of Chemical, Biological, Radiological or Nuclear (CBRN) Substances or Material: Guidance for Local Authorities* (http://www.ukresilience.info/cbrn/cbrn_guidance.pdf, London, 2003); Home Office, *The Decontamination of People Exposed to Chemical, Biological, Radiological or Nuclear (CBRN) Substances or Material Strategic National Guidance* (2nd ed, http://www.ukresilience.info/cbrn/peoplecbrn.pdf, London, 2004).

[52] Department of Health, *Influenza pandemic contingency plan* (http://www.dh.gov.uk/assetRoot/04/12/17/44/04121744.pdf, 2005); Health Protection Agency, *Influenza Pandemic Contingency Plan* (http://hpa.org.uk/infections/topics_az/influenza/pdfs/HPAPandemicplan.pdf, 2005).

[53] DEFRA, *DEFRA Flood Management Division, National Appraisal of Assets at Risk of Flooding and Coastal Erosion in England and Wales* (http://www.defra.gov.uk/environ/fcd/policy/NAA-R1101.pdf, 2001); *National Assessment of Defence Needs and Costs for Flood and Coastal Erosion Management* (http://www.defra.gov.uk/environ/fcd/policy/nadnac0604.pdf, 2004); *Making Space for Water* (http://www.defra.gov.uk/corporate/consult/waterspace/consultation.pdf, 2004).

[54] Environment Agency, *Strategy for Flood Risk Management 2003–2008* (http://www.ukresilience.info/frm_strategy.pdf, 2003).

8.22 One other piece of the capabilities jigsaw concerns the Critical National Infrastructure. A key player here is the National Infrastructure Security Co-ordination Centre (NISCC), established in 1999.[55] Its role is to consider the risk to the critical national infrastructure from electronic attack. Its work engages both public and private bodies, and it involves them in its work-streams on Threat Assessment, Outreach (information sharing, advice, and best practice), Response (warnings, advice, investigation, and recovery), and Research and Development.

C. Central Government and the Civil Contingencies Act

8.23 While 'local responders will almost always be first on the scene and carry the main burden',[56] central government remains a potent force in dealing with civil contingencies. Amongst its roles are: providing strategic directions; setting the legislative framework; mobilizing national assets and resources, including the military, determining national protective security and other counter-measures; determining the public information strategy, managing the international implications; overseeing local planning and assisting and advising on local response and recovery.[57]

8.24 The need for central involvement to be delineated more clearly was put forward strongly from the outset of the policy debate in the Cabinet Office consultation in 2001.[58] This opening salvo was followed by the Defence Committee in 2002, which was critical of the absence in the draft Civil Contingencies Bill of provisions relating to central government's national responsibilities.[59]

8.25 A comparison may be made with the predecessor Civil Defence Act 1948. Whilst it concentrated on the powers and duties of localities, s 1 did regulate the functions of Ministers and made it 'part of the functions of the designated Minister to take such steps as appear to him from time to time to be necessary or expedient for civil defence purposes', followed by an enumeration of more specific duties. A set of comparable duties under the 2004 Act might have comprised:[60] to assess risks; to devise plans to deal with those risks in liaison with local and regional structures; to provide a national forum for other stakeholders

---

[55] See http://www.niscc.gov.uk/niscc/index-en.html.

[56] Cabinet Office, *Central Government Arrangements*, para 4.

[57] Cabinet Office, *Emergency Response and Recovery*, para 12.5.

[58] Cabinet Office, *Emergency Planning Review: The Future of Emergency Planning in England and Wales: Results of the Consultation* (London, 2002) paras 28–32.

[59] Defence Committee, *Defence and Security in the UK* (2001–02 HC 518) para 24.

[60] This goes further than simply categorising central and devolved governments as 'responders' as suggested at HL Debs vol 666 col 819 9 November 2004, Baroness Buscombe. Compare the (failed) amendment at HL Debs vol 665 col 415 14 October 2004, Lord Lucas.

to discuss the plans and other issues of relevance; to ensure the production of regulations and guidance, with suitable mechanisms for consultation; to set performance standards; to provide structures to monitor and audit performance against those standards; to deal with enforcement; to set standards of training, education, and qualifications and to provide a lead for education through the Emergency Planning College; to ensure government business continuity management; to ensure that specialist equipment and materials (such as vaccines) are made available as well as a suitable level of finance for appointments, training, and exercises; to give suitable advice and warning to the public; to plan for the maintenance of essential services provided at a national level; to report to Parliament.[61] It is argued later in this chapter that a key element would be to set up a central agency by which most of these demands could be delivered. Such transparency would have boosted the importance of the civil contingency agenda and the transparency and accountability of its processes.

The Government defended the status quo by reference to a number of factors **8.26** which are said to ensure effective central involvement in civil contingencies work,[62] including responsibility through the LGD principle, coordination and delivery through the CCS, delivery through the Capabilities Programme, training via the Emergency Planning College, exercising, and audit.[63] The Consultation Paper accompanying the Civil Contingencies Bill concluded that

> ... a key focus for resources at central government level should be to audit and provide practical support to government departments in helping them develop sound arrangements in relation to their Lead Government Department responsibilities; that additional resources should be devoted at a regional level through Office of the Deputy Prime Minister (ODPM) to ensuring a sound response capability; and that, in the short-run, the priority at local responder level was to introduce the new framework, which would then support a possible assessment of the need for greater capacity.[64]

Yet, at the same time, the Government recognized that 'It no longer seemed appropriate to legislate principally for local responders: legislation needed to accommodate every level and every conceivable and inconceivable circumstance.'[65] In fact, none of the options in its paper integrated the legal regulation of the roles of the central authorities.[66] As a result, the Joint Committee on the Draft Civil Contingencies Bill[67] reiterated the wide range of representations for

---

[61] As suggested ibid, col 816, Lord Lucas.
[62] Joint Committee on the Draft Civil Contingencies Bill, Appendix 9 q 31. For the key role of the CCS, see further Government Response to the Defence Committee, *Defence and Security in the UK* (2001–02 HC 1230) para 33.
[63] Joint Committee on the Draft Civil Contingencies Bill, Appendix 9 q 40.
[64] Civil Contingencies Secretariat, *Full Regulatory Impact*, para 9.
[65] Cabinet Office, *The Draft Civil Contingencies Bill*, para 6.
[66] See Cabinet Office, *The Draft Civil Contingencies Bill: Explanatory Notes etc*, paras 26–34.
[67] 2002–03 HC 1074, HL 184, para 94.

the inclusion of central and regional government as Category 1 responders so vital to the achievement of 'the comprehensive national framework that the Government hopes to attain through this Bill'.[68] Some respondents (including ACPO) also suggested that a National Resilience Forum should be established. Others called for the creation of a National Emergency Management Agency.[69]

8.27 Further attempts at reasoned response were offered. Douglas Alexander, then Minister of State at the Cabinet Office, arguing that 'It is difficult to see how a sensible, meaningful duty could be imposed on central Government by way of statute.'[70] In addition, the Civil Contingencies Secretariat contended that:

> No reference to the role of central government is needed within the Bill as no statutory provision is required for the Government to engage in the full range of civil protection duties. Its efforts are agreed by Ministers who are accountable for them to Parliament and are therefore not appropriate for the face of a legal instrument. This is standard practice.[71]

8.28 In detail, none of these arguments stands up to scrutiny. The Ministerial quote ignores the fact that to confer a statutory duty on a Minister is commonplace. The duty in s 1 of the Police Reform Act 2002 might be a telling example—the Secretary of State is obliged to prepare a national policing plan—for 'policing plan', read 'nation civil protection plan'?[72] By s 172 of the Energy Act 2004, the Secretary of State must annually publish a report on the availability of electricity and gas and lay that report before Parliament. Next, the Secretariat's quote confuses a power to act with a duty to act. In this respect, central government is in no different position to local government. The point of Part I is to ensure the powers are used consistently and appropriately—and that they are used. Exactly the same desires apply to central government Ministers. Finally, to say that because Ministers are members of Parliament and so can be held to account without statute ignores that a statute can be a vehicle for greater clarity and greater accountability, such as through reporting requirements.

8.29 Another response was that 'there is a particular legal difficulty with [imposing a legal duty], in that Government Offices of the Regions do not have a separate legal personality and therefore to place a distinct legal burden upon them would be, to say the least, difficult'.[73] An easy retort is that imposing a statutory duty upon Secretaries of State to take responsibility for them would overcome any difficulty.

---

[68] Joint Committee on the Draft Civil Contingencies Bill, para 101.
[69] Civil Contingencies Secretariat, *The Government's Response*, 50.
[70] (2002–03 HC 1074, HL 184) para 98.
[71] Civil Contingencies Secretariat, *The Government's Response*, 50.
[72] See also Environment Act 1995 s 80—publication of the National Air Quality Strategy.
[73] Joint Committee on the Draft Civil Contingencies Bill, para 100.

The failure to address by law the roles and organization of central government **8.30** in civil contingencies is very unsatisfactory from a number of perspectives.[74] For example, since the structure is not defined within the terms of the Act, it does not benefit from the certainty of how the equivalent of Category 2 responders will be involved. It might sound far-fetched that major utilities or the power suppliers would not wish to assist in an emergency. However, the experience of the fuel crisis in 2000, when oil companies acted rather coyly about the sympathetic actions of their employees in response to protests about the level of taxes and duties, should give pause for thought.

It is suggested that the reluctance to address the 'bottom-heavy' design of the civil **8.31** protection regime under the Act is motivated by three factors. First, Ministers wish to retain flexibility for themselves to an extent not allowed unto others. But with flexibility comes unaccountability. Secondly, it reflects the lack of agreement at central level as to which department should be ultimately responsible. This battle has been fought for decades, with the Home Office, Prime Minister's Office, and Cabinet Office all involved. Thirdly, it reflects a wish to limit the commitment of central government and place it on the locality. A flavour of the resulting uncertainty (which the 2004 Act should ameliorate) can be garnered from an example given by the Defence Committee in 2002:

> Following 11 September, we were told, a request was put in to the CCS for assistance with providing facilities to provide accommodation for a large number of people stuck at Heathrow airport because their flights were cancelled. The CCS response was 'a categoric flat no, not our responsibility, nothing to do with us' . . . [75]

The Joint Committee demanded that the role and responsibilities of gov- **8.32** ernment departments, the National Assembly for Wales, and regional government be outlined on the face of the Bill and that they be required by statute to undertake planning responsibilities. The Committee equally argued for a unified system of enforcement and inspection under Part I of the Act through a national Civil Contingencies Agency (CCA).[76] It would include officials, as well as seconded experts. At national level, it could provide a focus for ongoing dialogue and liaison between central government departments, the police, the military, the National Health Service, the national representatives of local Responders, and other national representative organizations of importance.[77] It could have representatives based in the

---

[74] See further Joint Committee on the Draft Civil Contingencies Bill, Appendix 4.

[75] Defence Committee, *Defence and Security in the UK* (2001–02 HC 518) para 149.

[76] Ibid, paras 257, 260.

[77] To give a flavour, the following might be considered: Association of Electricity Producers, Association of Train Operating Companies, Electricity Association, Environmental Services Association; Food and Drink Federation, National Voluntary Aid Society Emergency Committee, Nuclear Industry Association, Society of British Gas Industry, UK Petroleum Industry

Government's regional offices to advise and support designated Regional Nominated Coordinators. It could be answerable to a Civil Contingencies Commissioner appointed by, and reporting to, a Cabinet Minister and Parliament. The Agency could incorporate an inspectorate,[78] and it would set national guidelines and standards for Responders. A fuller outline of its functions would include: the measuring of capacity, setting training and operational standards and ensuring consistent compliance; to offer expert advice to localities; to collate technical, scientific, and other information and to consider strategies for dealing with wide area emergencies; to 'horizon scan'; to advise on LGD and emergency service involvement.[79]

8.33 In response to the Joint Committee, the Government stood firm:

> For a dedicated inspectorate to be effective, the costs of setting it up and main-taining it are likely to be disproportionate to the benefits. Resources in the area of civil protection are limited and stakeholders have also indicated that if a choice were to be made between more funding for the basic duty and the establishment of an inspectorate, they would prefer the former. A dedicated inspectorate is deemed to be unnecessary because audit can be carried out by the existing inspectorate bodies for each of the responder organisations, and regular meetings can be arranged across the inspectorates at national and regional levels to ensure that there is a consistency in their approach.[80]

The preference remained for reliance upon established audit and regulatory bodies such as (for local authorities) the Best Value and Comprehensive Performance Assessment regime operated by ODPM and the Audit Commission.[81]

8.34 In conclusion, one may question the Government's commitment to engineer the statutory implementation of the civil contingency agenda did not extend to the structures of central government (with the exception of two Category 1 responders), as a result of 'The draft Bill's provisions have negligible expenditure provisions for Government Departments and negligible impact on public sector manpower.'[82]

Association Limited, United Kingdom Airlines Emergency Planning Group United Kingdom Flight Safety Committee, United Kingdom Telecommunications Association, Water UK.

[78] Based on its experience of Crime and Disorder Reduction Partnerships, ACPO also favoured a specialist inspectorate: Joint Committee on the Draft Civil Contingencies Bill, paras 248, 250.

[79] Joint Committee on the Draft Civil Contingencies Bill, para 258. See further Appendix 5 para 1.12.

[80] Cabinet Office, *The Government's Response to the Report of the Joint Committee*, para 48. See also Civil Contingencies Secretariat, *The Government's Response*, 25, 26.

[81] Cabinet Office, *The Draft Civil Contingencies Bill: Explanatory Notes etc*, para 70; Cabinet Office, *Emergency Preparedness: Guidance on Part 1*, paras 13.4–12.

[82] *Draft Civil Contingencies Bill: Explanatory Notes etc*, para 58.

D. The Role of the Military

(a) Military assets

The military are an alluring and important resource in an emergency as part of a **8.35**
coordinated response,[83] albeit as 'a last resort'.[84] Regular forces may possess
dedicated assets specifically related to emergencies, such as bomb disposal
teams,[85] search and rescue facilities,[86] and mountain rescue teams.[87] Non-
dedicated assets might comprise non-specialist manpower or bridging and
pumping equipment.

The desire to augment this structure was signalled in the Ministry of Defence **8.36**
White Paper of 2002.[88] It proposed an enhanced ability for Army Regional
headquarters to plan, liaise, and operate continuously and to the establishment
of a Reserve Reaction Force of, on average, some 500 Volunteer Reserves for
each region (a total of 6,000) who would be available to deploy in a crisis. It also
proposed formalizing the role of 2 Signals Brigade, and its predominately
Territorial Army support units, in supporting those personnel deploying. Any
mobilization can be backed by orders under the Reserve Forces Act 1996, s 52
(which allows a call-out where 'it appears to Her [Majesty] that national danger
is imminent or that a great emergency has arisen'[89] or 'in the event of an actual
or apprehended attack on the United Kingdom'), s 54 (where 'warlike operations are in preparation or progress'), or s 56 ('on operations outside the United
Kingdom for the protection of life or property' or 'on operations anywhere in
the world for the alleviation of distress or the preservation of life or property in
time of disaster or apprehended disaster'). Likely tasks include: 'reconnaissance,
assistance with mass casualties, site search and clearance, transport and communications, the operation of water and feeding points, control and co-ordination functions, access control, the control of movement of large numbers of
the public, guarding or other tasks at the request of the civil police'.[90]

The Defence Committee has expressed concern that the 'Reserves rather than **8.37**
the regulars may now be the first port of call' in an emergency especially as they

---

[83] Ministry of Defence, *Operations in the UK*, para 107.   [84] Ibid, para 213.
[85] See, further, Ministry of Defence, *Operations in the UK*, Annex 2D. See further http://
www.royal-navy.mod.uk/static/pages/8075.html; http://www.reserve-forces-london.org.uk/units/
1090/; http://www.army.mod.uk/royalengineers/org/33regt/index.htm.
[86] See Ministry of Defence, *Operations in the UK*, Annex 2E; http://www.kinloss-raf.co.uk/
arcc.html; http://www.royal-navy.mod.uk/static/pages/2009.html; http://www.dasa.mod.uk/nat-
stats/sar/annual/intro.html. The Maritime and Coastguard Agency also plays an important role:
Search and Rescue Framework for the United Kingdom of Great Britain and Northern Ireland
(Southampton, 2002).   [87] Http://www.raf.mod.uk/news/sarstats.html.
[88] *The Strategic Defence Review: A New Chapter* (Cm 5566, London, 2002) Ch 5.
[89] The call-out order must be reported to Parliament: s 52(7).
[90] Defence Committee, *Defence and Security in the UK* (2001–02 HC 518) para 232.

will have limited training and their availability is not instant.[91] Nevertheless, in the later paper, *Delivering Security in a Changing World*,[92] the Ministry of Defence reported that the 14 Civil Contingencies Reaction Forces were being established by the end of 2003, though the numbers have since been depleted by operations in Iraq.[93]

8.38 Aside from the limitations in the military assets available for emergency work, their use in what is termed 'United Kingdom Operations' (UK Ops) raises issues concerning the laws and other processes governing their involvement. None of these matters is covered by the 2004 Act. Consequently, there is an absence of statutory regulation of joint planning for extraordinary scenarios for which the central state is inevitably involved, such as terrorist or hostile attack.[94]

(b) The legal and practice rules relating to military involvement

8.39 The obscure common law basis for the deployment of Crown forces has been described and criticized in Chapter 2, ending with the suggestion of a new Defence Act to inject some clarity and control into the situation.[95] More straightforward as a basis for military intervention is s 2 of the Emergency Powers Act 1964[96] (plus the Emergency Powers (Amendment) Act (Northern Ireland) 1964).[97] As described in Chapter 2, s 2 perpetuates a power first granted by Defence (Armed Forces) Regulations 1939 (reg 6) by providing that 'The Defence (Armed Forces) Regulations 1939 in the form set out in Part C of Schedule 2 to the Emergency Laws (Repeal) Act 1959 . . . shall become permanent.'[98] By reg 6:

> The Admiralty, the Army Council or the Air Council may by order authorise officers and men of Her Majesty's naval, military or air forces under their respective control to be temporarily employed in agricultural work or such other work as may be approved in accordance with instructions issued by the Admiralty,

---

[91] Ibid, para 226.   [92] (Cm 6043, London, 2003) para 3.9.

[93] HC Debs vol 421 col 1362 24 May 2004, Hazel Blears. See further Ministry of Defence, *Operations in the UK*, Annex 2F.

[94] See Cabinet Office, *Emergency Planning Review: The Future of Emergency Planning in England and Wales: Results of the Consultation* (London, 2002) para 6.

[95] Arguments in favour of reforming s 2 as a miscellaneous amendment to the 2004 Act were rejected: Cabinet Office, *The Draft Civil Contingencies Bill*, para 37. But note the addition of s 22(3)(l) and (m).

[96] See further Whelan, C, 'Military Intervention on Industrial Disputes' (1979) 18 *Industrial Relations Journal* 222; Peak, S, *Troops in Strikes* (Cobden Trust, London, 1984) Ch 2; Rowe, PJ and Whelan, CJ (eds), *Military Intervention in Democratic Societies* (Croom Helm, London, 1985); Bonner, D, *Emergency Powers in Peacetime* (Sweet & Maxwell, London, 1985) 229; Morris, GS, *Strikes in Essential Services* (Mansell, London, 1986) Ch 4; Thurlow, R, *The Secret State* (Blackwell, Oxford, 1994) 339.

[97] See Creighton, WB, 'Emergency Legislation and Industrial Disputes in Northern Ireland' in Wood, JC, *Encyclopaedia of Northern Ireland Labour Law and Practice* (Labour Relations Agency, Belfast, 1983).   [98] SR&O No 1304, as amended in 1942.

the Army Council or the Air Council (QR J853), as the case may be, as being urgent work of national importance, and there upon it shall be the duty of every person subject to the Naval Discipline Act, military law or air-force law to obey any command given by his superior officer in relation to such employment, and every such command shall be deemed to be a lawful command within the meaning of the Naval Discipline Act, the Army Act, 1955 or the Air Force Act 1955 as the case may be.

Section 2 serves two legal purposes. One is to ensure that there is a power to use **8.40** the armed forces in circumstances which may fall outside of the circumstances of disturbance of the peace at the heart of the common law. Therefore, third parties affected by the military action cannot complain that its impact is per se unlawful. The other is to ensure not just a power for commanders but also a duty imposed on soldiers by way of military discipline to do the bidding of the commanders. Section 2 is not accompanied by any special measures as to the military use of force, so the vagaries of s 3 of the Criminal Law Act 1967 must be endured.[99] Equally, there are no powers of requisition—hence the use of Army ambulances and 'Green Goddess' fire engines.

Whilst the laws remain rather obscure, the practice rules about military inter- **8.41** ventions in emergencies have become more manifest under the Ministry of Defence's 2005 publication of *Operations in the UK*.[100] In administrative terms, the involvement of the military will mainly fall under the generic heading of Military Assistance to the Civil Authorities (MACA).[101] Within that concept, there are three categories of relevance.[102]

One involves Military Aid to the Civil Power (MACP) where it is necessary **8.42** to assist the civil power (nowadays embodied by the police rather than magistrates)[103] in the maintenance or restoration of law and order or public

---

[99] See *Report of a Tribunal Appointed to Inquire into the Events on Sunday 30th January, 1972 etc* (1971–72 HC 220); *Reference under s 48A Criminal Appeal (Northern Ireland) Act 1968 (No 1 of 1975)* [1976] 2 All ER 937; Babington, A, *Military Intervention in Britain* (Routledge, London, 1990) Ch 13.

[100] *Operations in the UK: The Defence Contribution to Resilience* (http://www.ukresilience.info/ contingencies/defencecontrib.pdf, London, 2005).

[101] Other forms of UK Ops are Military Operations (such as support for overseas operations and the security of defence key points) and Military Home Defence (defence of the UK from external attack): Ministry of Defence, *Operations in the UK*, paras 113, 114.

[102] See *Manual of Military Law* (9th ed, Ministry of Defence, London, 1969) Section V; *Queen's Regulations for the Army* (rev ed, Ministry of Defence, London, 1996) Ch 11; *Queen's Regulations for the Navy* (Ministry of Defence, London, 1997) Ch 48; *Manual of Air Force Law* (6th ed, Ministry of Defence, London, 1994) Ch IX.

[103] The change was set in the Queen's Regulations of 1895 following failures by the magistrates in the Featherstone riots of 1893: *Report on the Featherstone Disturbances* (C 7234, London, 1893); Peak, S, *Troops in Strikes* (Cobden Trust, London, 1984) Ch 1; Babington, A, *Military Intervention in Britain* (Routledge, London, 1990) Ch 10. It is assumed that *R v Pinney* (1832) 172 ER 962 does not lay down a legal precedent for the power to be conferred on magistrates but rather refers to the local civil authority in more generic terms.

safety.[104] Soldiers have been employed in Northern Ireland since 1969 on this basis. More commonplace exercises of MACP involve the disposal of explosive ordnance, which involved, for example, 957 calls in 2001 and 431 in 2002; other requests average between 30 and 40 per year, including operations against terrorism.[105] There is also a specific statutory duty to help with Customs operations (such as against drugs smugglers) under the Customs and Excise Management Act 1979, s 11.

8.43 Military Aid in the Civil Community (MACC) is the provision of unarmed support in emergencies under three categories.[106] Category A involves assistance to the emergency services in times of emergency such as environmental disasters, major accidents, or major search and rescue operations. Common examples include severe weather conditions or major accidents. Category B involves short-term routine assistance on special projects of significant value to the civil community. Category C involves the deployment of volunteers to social services (or similar) organizations for specific periods. Dozens of instances of MACC have occurred since 1983,[107] including flood relief, help at air crashes (including searches for bodies and wreckage after the Lockerbie crash in 1989), searches for missing people, and tackling heath and forest fires. At the other end of the scale is the Quick Reaction Alert (Interceptor) Force which protects against airborne threat, such as a hijacked plane being used as a weapon by terrorists.[108] A special case of MACC relates to maritime disasters. The Secretary of State's Representative for purposes of maritime salvage and intervention[109] may call upon the Navy for aid.

8.44 Military Aid to Other Government Departments (MAGD) comprises urgent work of national importance or in maintaining supplies and services essential to the life, health, and safety of the community.[110] Section 2 of the Emergency Powers Act 1964 will be the legal basis for the operation. Between 1984 and 2003, there were 15 instances of MAGD.[111] For example, during the March 1984 to March 1985 National Union of Mineworkers' strike, logistic assistance was provided to the police. Soldiers have also been deployed during ambulance and fire service disputes (in 1977, 1999, 2000, and 2002) and have also been called upon to set up temporary prisons at times of overcrowding and industrial dispute. During the September to November 2000 fuel crisis, the military made

---

[104] See Ministry of Defence, *Operations in the UK*, Ch 4.

[105] HC Debs vol 400 col 437w 25 February 2003, Adam Ingram.

[106] Ministry of Defence, *Operations in the UK*, Ch 5.

[107] Joint Committee on the Draft Civil Contingencies Bill, Appendix 9 Annex C.

[108] Ministry of Defence, *The Strategic Defence Review: A New Chapter* (Cm 5566, London, 2002) para 5.86; http://www.rafmarham.co.uk/relations/stories/f3-at-marham.htm. Any decision to shoot down an aircraft requires Ministerial approval: House of Commons Defence Committee, *Defence and Security in the UK* (2001–02 HC 518) para 45.

[109] See Ch 4.   [110] See Ministry of Defence, *Operations in the UK*, Ch 3.

[111] Joint Committee on the Draft Civil Contingencies Bill, Appendix 9 Annex B.

available tankers and refuelling vehicles and trained to operate civilian vehicles. There was very extensive military involvement in the Foot and Mouth outbreak from March to October 2001.

Mechanics of military involvement

Government guidance underscores that military involvement must be exceptional: **8.45**

> . . . assistance is provided on an availability basis and the armed forces cannot make a commitment that guarantees assistance to meet specific emergencies. It is therefore essential that Category 1 and 2 responders do not base plans and organise exercises on the assumption of military assistance.[112]

This edict is reinforced in the guidance document, *Emergency Response and Recovery* which states, 'The Armed Forces maintain no standing forces for MACA tasks. There are, by definition, no permanent or standing MACA responses.'[113] The military are also advised that 'Neither the production of contingency plans nor the participation in planning exercises guarantees the provision of MACC support.'[114] The demands regarding defence, including foreign interventions, must take priority.

Deployments of the military must be duly authorized. The Prerogative powers **8.46** in relation to the military are exercised on behalf of the Crown by the Defence Council. The Defence Council consists of military leaders and Government ministers and officials. It is based within the Ministry of Defence and is normally chaired by the Secretary of State for Defence.[115] A Board for each Service, (the Admiralty, Army, and Air Force) reports to the Defence Council. Contingency planning and preparation for UK Ops takes place also in the J5 Standing Joint Command at Wilton[116] and in Joint Services Coordination Groups, which meet at least annually and will invite civilian representatives.[117] It is not revealed what central plans have been devised, but there is an expectation to coordinate with the civilian Integrated Emergency Management approach (described later in this chapter).[118]

The starting point for intervention is likely to be a request from a local **8.47** responder, which should be made to the Joint Regional Liaison Officer at the nearest Regional Brigade Headquarters of the Army.[119]

---

[112] Cabinet Office, *Emergency Preparedness: Guidance on Part 1*, para 15.5.
[113] Cabinet Office, *Emergency Response and Recovery*, para 3.44.
[114] Ministry of Defence, *Operations in the UK*, para 505.
[115] See http://www.armedforces.co.uk/mod/listings/l0004.html.
[116] Ministry of Defence, *Operations in the UK*, para 601.
[117] Ibid, para 238. ACPO also funds a Police Military Liaison Officer: para 243.
[118] Ibid, Ch 7.
[119] Cabinet Office, *Emergency Response and Recovery*, para 3.46; Ministry of Defence, *Operations in the UK*, paras 232, 236.

8.48 In cases of MACP, the chain of authorization involves the military commander contacting the Ministry of Defence (Counter Terrorism and UK Operations) which, if in agreement, will seek formal approval from the Minister of State for the Armed Forces.[120] The source of the request is also vital. The prime agency to deal with public order is the police, and so deployments will only be authorized if there is clear support from the police and the agreement of the Home Office. With the advent of the 2004 Act, it has been suggested that, as well as the police, Regional and emergency coordinators would be able to request military assistance under MACA arrangements.[121] The same arrangement was put in place for Regional Commissioners during the Second World War.[122] Where an emergency is so pressing that there is no time for these procedures, unit commanders are authorized to provide support (generally to the police or other emergency services) on their own authority. These instances will be confined to the case of a 'grave and sudden' emergency . . . which . . . demand . . . immediate intervention to protect life and property.'[123] It must be shown that lives are in danger, that the military are using resources under their direct command, and that they are following a clear request from the police or other emergency services for that support.[124] They must subsequently seek full authorization through their command chain, up to MOD Ministers, and this becomes obligatory for the continuance of support when lives are no longer in danger. In deciding whether to grant a request, Ministers will consider the nature and feasibility of the request and whether it can be said to be a last resort, policy and cost issues, and the impact on the Armed Forces. Another exception is that individual approval for each investigation into explosive ordnance is not required since there is a standing Ministerial approval, though there is reporting requirement to the MOD.

8.49 MAGD deployments will usually be governed by s 2 of the Emergency Powers Act 1964.[125] The starting point will be a request from a LGD or the Home Office on its behalf. The procedures within the MOD demand a specific Defence Council Order to be signed by two members of the Defence Council (including a Minister), on the same day, prior to any deployment.[126] The Defence Council will then issue instructions to the military.[127] For more major emergencies (such as may now fall within Part II of the 2004 Act), the Ministry

---

[120] *Queen's Regulations for the Army* (rev ed, Ministry of Defence, London, 1996) J11.002.
[121] Joint Committee on the Draft Civil Contingencies Bill, Appendix 9 q 65.
[122] O'Brien, TH, *Civil Defence* (HMSO, London, 1955) 638–40.
[123] *Queen's Regulations for the Army* (rev ed, Ministry of Defence, London, 1996) J11.002. See further Nicolson, D, 'The Citizen's Duty to Assist the Police' [1992] Crim LR 611; Ministry of Defence, *Operations in the UK*, para 414.
[124] Joint Committee on the Draft Civil Contingencies Bill, Appendix 9 q 53.
[125] See Ministry of Defence, *Operations in the UK*, para 306.
[126] Joint Committee on the Draft Civil Contingencies Bill, Appendix 9 q 53.
[127] *Queen's Regulations for the Army* (rev ed, Ministry of Defence, London, 1996) J11.004.

of Defence will be responsible for planning and control.[128] Emergency work can again be undertaken exceptionally, and there is standing authority under the Defence Council Order of 17 January 1983.[129]

MACC deployments may be under s 2 but are often delivered (so far as they fall **8.50** under Category A) without reference to the Ministry of Defence but with notification to headquarters.[130]

Once a deployment has been given ministerial approval, the local military **8.51** commander will respond on the ground. The organizational boundaries of the Army are based around regional brigade areas (East, East Midlands, London, North East, North West, South East, South West, West Midlands, Yorkshire and the Humber, Northern Ireland, Scotland Highland, Scotland Lowland, and Wales).[131] On deployment, the military will as a matter of practice act under the supervision of the police, but, as indicated earlier, the legal position is not clear and it is certainly not the legal position that 'The primacy of the police is recognised at all times'.[132]

MACA is normally charged by the Ministry of Defence at full cost to the **8.52** requesting authority since it strays beyond exclusively defence purposes.[133] Standing operations are partially funded by departments (for instance, the Home Office contributes to the cost of explosive ordnance disposal support to the police).[134] Military support in Northern Ireland is a special case for these purposes. But the costs can be waived by the Ministry of Defence in consideration of whether spare capacity was used, whether there was detriment to core defence tasks, and whether there was a training value.[135] Charges for MACC assistance were simplified in 2001. Assistance can be given free where there is a danger to life (not necessarily an 'immediate' danger), but the full cost regime resumes once the danger has passed.

These rules and arrangements are wholly unrelated to the Civil Contingencies Act. **8.53** Hence, there is no need to wait for the invocation of Part II before MACA can be offered, nor, conversely would the advent of a state of emergency under Part II always necessitate military involvement, though its existence 'would also influence the Secretary of State for Defence when judging the "reasonableness" of a request from the police or a central government Department to provide support.'[136]

---

[128] Ibid, J11.006.   [129] Ibid, J11.008.
[130] Ministry of Defence, *Operations in the UK*, para 504.
[131] See http://www.mod.uk/issues/sdr/new_chapter/reserves/map.htm.
[132] HC Debs vol 400 col 437w 25 February 2003, Adam Ingram.
[133] See further Ministry of Defence, *Operations in the UK*, para 246.
[134] Joint Committee on the Draft Civil Contingencies Bill, Appendix 9 q 52.
[135] See the examples ibid at Appendix 9 Annex D.
[136] Joint Committee on the Draft Civil Contingencies Bill, Appendix 9 q 53.

E. Practical Guides for Local Planners

8.54 It is not the purpose of this book to provide a 'how to do it' manual for civil contingency operatives.[137] However, it will outline the broad themes which arise from operationalizing civil resilience to show whether or not the legislative scheme is practicable.

(a) *Dealing with Disaster*

8.55 The key Government guide in this area has since 1992 been the booklet, *Dealing with Disasters*.[138] With just 70 substantive pages, the approach is to provide an aide-memoire more than a blueprint. It emphasized an integrated approach, which would be based around assessment, prevention, preparation, response, and recovery management—collectively labelled as an 'Integrated Emergency Management' (IEM) approach.[139]

8.56 In dealing with live major emergencies, at least those of sudden-impact, it was the police who were to be the lead agency.[140] The practice grew of ensuring a structure which covered Operational, Tactical and Strategic aspects, and these are commonly labelled 'Bronze', 'Silver', and 'Gold' (see Table 8B).

Table 8B: Framework of command/management[141]

Strategic	Establish strategic objectives and overall management framework. Ensure long-term resourcing/expertise
Tactical	Determine priorities in obtaining and allocating resources; plan and coordinate overall response
Operational	The 'Doers': manage front-line operations

---

[137] See, for example, Lakha, R and Moore, T, *Tolley's Handbook of Disaster & Emergency Management: Principles & Practice* (2nd ed, LEXIS-NEXIS, Croydon, 2004).

[138] See now revised 3rd ed, Cabinet Office, London, 2003). For the equivalent in Scotland, see Scottish Executive, Dealing with Disasters Together (http://www.scotland.gov.uk/library5/government/dealdisasters.pdf, Edinburgh, 2003). Other important guides included Home/Office, Standards for Civil Protection in England and Wales (http://www.ukresilience.info/contingencies/pubs/sfcpew.pdf, 1999). [139] Ibid, para 2.3.

[140] Ibid, para 2.12.

[141] Source: Cabinet Office, *Dealing with Disaster* (rev 3rd ed, Cabinet Office, London, 2003) para 3.6.

Though not originally officially approved for all purposes,[142] the labels have **8.57**
proven to be enduring as have the associated structures.

(b) 2004 Act guidance

The *Dealing with Disasters* booklet has been replaced by more voluminous guid- **8.58**
ance under the 2004 Act, though it is recognized that the predecessor's overall
structure and much of its detail remain pertinent.[143] The two volumes are organized around the six phases of IEM. *Emergency Preparedness*[144] outlines the first
four stages of Prevention Anticipation, Assessment, Prevention and Preparation.
Emergency Response and Recovery[145] details the latter two. The overall feel is meant
to be 'purposively flexible'.[146] Within that milieu, the former is statutory guidance
(within ss 3 and 6) and the latter non-statutory. The effect of this distinction is
modest. There are several legal requirements to have regard to statutory guidance
which should place it higher in the hierarchy of bureaucratic thought processes
than a relevant consideration in administrative law (or, in time, a legitimate
expectation of compliance)[147] as applies to the latter guidance.[148] Certainly, the
Government means business with *Emergency Preparedness*: 'If the Minister considers that a Category 1 or 2 responder has failed to comply with its obligations
under the Act, he/she may take proceedings against that responder in the High
Court.'[149] As for *Emergency Response and Recovery*, its lesser status reflects that its
contents go beyond the terms of the Act (for example, into the realms of central
and regional government).[150]

The *Scottish Guidance on Preparing for Emergencies* is also being developed.[151] **8.59**
The chief equivalent guidance for Northern Ireland emerged as a statement
from the Central Emergency Planning Unit (CEPU), issued in 2004 as *A Guide
to Emergency Planning Arrangements in Northern Ireland.*[152]

---

[142] Ibid, para 3.5.
[143] Cabinet Office, *Civil Contingencies Act 2004: Consultation on the Draft Regulations and Guidance* (Cm 6401, London, 2004) para 30.
[144] Cabinet Office, *Emergency Preparedness: Guidance on Part 1 of the Civil Contingencies Act 2004, its associated Regulations and non-statutory arrangements* (http://www.ukresilience.info/ccact/emergprepfinal.pdf, 2005). Its draft title was 'Preparing for Emergencies'.
[145] Cabinet Office, *Emergency Response and Recovery* (http://www.ukresilience.info/ccact/emergresponse.pdf, 2005). Its draft title was 'Responding to Emergencies'.
[146] Cabinet Office, *Civil Contingencies Act 2004: Consultation on the Draft Regulations and Guidance* (Cm 6401, London, 2004) para 21.   [147] See Ch 6.
[148] See *South Lakeland District Council v Secretary Of State for the Environment* [1992] 2 AC 141; *B v Calderdale Metropolitan Council* [2004] EWCA Civ 134.
[149] Cabinet Office, *Emergency Preparedness: Guidance on Part 1*, para 16.11.
[150] Ibid, para 1.46.
[151] Scc http://www.scotland.gov.uk/consultations/justice/crgcca-03.asp.
[152] See http://cepu.nics.gov.uk/pubs/emerplanarrange.pdf. Other relevant documents include *A Guide to Plan Preparation* (on emergency plans), *A Guide to Evacuations in Northern Ireland, Northern Ireland Standards in Civil Protection, A Guide to Risk Assessment* and *Business Continuity Management in the Public Sector*. See http://cepu.nics.gov.uk/pubs.htm.

(c) **The local tier**

Contextual overview

8.60 The focus here will be on how the guidance seeks to contextualize the provisions of the Act and thereby outlines a coherent and consistent operational framework for civil protection.

8.61 In the introduction to *Emergency Preparedness*, comparatively little attention is paid to developing the concept of 'emergency' beyond a basic and brief restatement of the definitions contained in the Act. In one sense, this concision might be considered startling, given that the guidance states, 'all the duties specified in Part 1 of the Act are contingent on the definition of emergency.'[153] However, it is this very notion of 'contingency' that might explain why the guidance only cursorily examines the definition. This is because the scale and scope of the obligations and demands arising from the new duties cannot be predicted with certainty because they are determined by unforeseen future events and circumstances.

8.62 The true breadth of scope is revealed when the discussion moves on to contextualize the emergency provisions of the Act in terms of IEM. Here it is stated that 'The wide concept of IEM within and across Category 1 responders is geared to the idea of building greater overall resilience in the face of a broad range of disruptive challenges.'[154] Six activities are regarded as fundamental to an integrated approach: anticipation; assessment; prevention; preparation; response; and recovery management.[155] The Act focuses on just two of these activities—assessment and preparation—which are consequently 'covered extensively in this volume of guidance on preparing for emergencies.'[156] Prevention is said to be not within the Act, 'to any great extent because it is largely a matter for other legislation',[157] while response and recovery management are discussed in *Emergency Response and Recovery* 'because they are not covered directly in the Act.'[158] Although anticipation is linked with 'horizon scanning' in both volumes of guidance, little further is said of this function despite it being considered 'crucial in both the pre-emergency and post-emergency phases.'[159] Instead, in *Emergency Preparedness* 'anticipation' is very generally defined as meaning that responders 'should aim to be aware' of new hazards and adjust their plans accordingly[160] while, in *Emergency Response and Recovery*, it is considered primarily in relation to responders gaining the initiative 'amidst the pressure and urgency of events.'[161]

---

[153] Ibid, para 1.13.   [154] Ibid, para 1.48.   [155] Ibid, para 1.42.
[156] Ibid, para 1.43.
[157] Ibid, para 1.45. Although it is noted that plans can be put into effect prior to the occurrence of an emergency in the attempt to mitigate its impact.
[158] Ibid, para 1.46.   [159] Cabinet Office, *Emergency Response and Recovery*, para 2.26.
[160] Cabinet Office, *Emergency Preparedness: Guidance on Part 1*, para 1.44.
[161] Cabinet Office, *Emergency Response and Recovery*, para 2.26.

The definition of the civil protection functions contained in the legislation **8.63**
arises from a context of building resilience in the face of a broad range of
challenges using an IEM framework. But, key here, is that it is *Emergency
Preparedness* which explicitly draws them together and describes how each
relates to the others in an overarching framework of civil protection. According
to the guidance, Category 1 responders must engage in the seven principal
duties, whose relationships are represented in Table 8C.[162]

Table 8C: Relationships between civil protection duties[163]

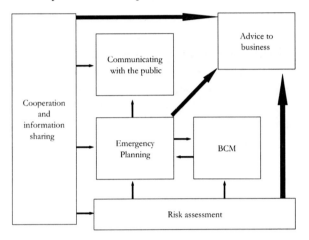

The guidance then devotes a discrete chapter to each of the specific duties and **8.64**
seeks to develop best practice within each.

Cooperation

Chapter 2 provides a more detailed description of the structure and role of the **8.65**
Local Resilience Forum (LRF) as the 'principal mechanism for multi-agency
cooperation' between responders.[164] The guidance considers a range of issues,
including possible sub-groups[165] and how the 'lead responder principle' can be
effectively applied.[166] Significantly, the Guidance also emphasizes that the LRF
should be regarded as, 'a process by which the organisations on which the duty
falls co-operate with each other.'[167] This focus on the cooperative function is
further reinforced by the observation that the LRF, although based essentially

---

[162] Cabinet Office, *Emergency Preparedness: Guidance on Part 1*, paras 1.30–1.32.
[163] Source: ibid, p 6. Figure 1.1.   [164] Ibid, para 2.4.
[165] The following are suggested at ibid, paras 2.2, 2.60: general working; capabilities; areas;
sectoral; specialist; humanitarian assistance including the operation of family assistance centres
(on which see ACPO and Cabinet Office, *Humanitarian Assistance in Emergencies—Guidance on
Establishing Family Assistance Centres* (http://www.ukresilience.info/contingencies/facacpogui-
dance.pdf, 2005).   [166] Ibid, paras 2.32–38.
[167] Ibid, para 2.4.

on local police force areas, is intended to sit alongside other elements of multi-agency planning at other levels and that this relationship is not a 'hierarchy' in that the LRFs are not subordinate to regional groupings. Instead it is envisaged that the LRF will act as a two-way conduit passing issues up to regional groups and receiving information which is 'cascaded downwards'.[168]

8.66 Given their recent establishment, it is premature to rush to judgement regarding the effectiveness of the LRF as a coordinating mechanism for local civil protection. Nonetheless, some general (and perhaps rather obvious) observations can be offered. Not least of these is the fact that the LRF occupies a critical position since it operates as the nexus for coordinating local arrangements for civil protection and, as the Government has consistently made clear, the 'local response capability is the building block of our ability to deal with emergencies.'[169] However, as the guidance has also made clear, the LRF depends upon the cooperation of its members in order to function and, in this regard, there are some structural features of the LRF, such as the leadership role, that might impact on how the wider culture of cooperation actually develops. In this regard, the leadership component of LRF function might be viewed by some responders as more 'naturally' falling under a police service remit than any other. After all, the LRFs are organized on local police force areas, coupled with the fact that the police service has traditionally taken the lead in coordinated multi-emergency service responses in the past. If this leadership pattern replicates itself under the 2004 Act, then it is possible that the LRF process will be influenced by the relative dominance of one service culture in comparison to the others, though there may be some counter-balance provided by the fact that the secretariat will probably be located within a local authority emergency planning unit.

Information sharing

8.67 Chapter 3 of *Emergency Preparedness* stresses that information sharing is 'crucial'[170] since it underpins all other forms of cooperation. But delivery is likely to be more complicated by commercial or security sensitivities or misperceptions between Responders. As a result: 'These . . . competing factors point towards a framework in which the initial presumption is that information should be shared, but that some information should be controlled if its release would be counterproductive or damaging in some other way.'[171]

8.68 The guidance offers the following general precepts as relevant to decisions about information sharing.[172] All Category 1 and 2 responders should work on the presumption that information requested should be disclosed. Where the Category

---

[168] Ibid, para 2.45.   [169] Cabinet Office, *The Draft Civil Contingencies Bill*, 16, para 1.
[170] Cabinet Office, *Emergency Preparedness: Guidance on Part 1*, para 3.1.
[171] Ibid, para 3.6.   [172] Ibid, para 3.18.

1 or 2 responder knows that the information has originated from the intelligence services and that disclosure to the public would threaten national security, then the information must not be disclosed. In considering national security implications, the test is whether disclosure to the public would threaten national security, not whether disclosure to the requesting Category 1 or 2 responder would threaten national security. Where a request relates to information, part of which is sensitive and part of which is not, the sensitive information should be redacted. Where possible, open sources of information should be used in the first instance and, beyond this, requests should generally be limited to areas relating to local configurations, facilities, and contact details for staff.[173] Requests should also be channelled through as small a number of routes as possible and the guidance suggests using the LRF/RRF or sector agencies as coordinating bodies in this regard.[174] There is also encouragement for responders to abandon crude distinctions between 'public' and 'private' information in favour of a more 'nuanced' picture, 'with a spectrum which runs from limited-access information (even within organisations) through to information intended to be absorbed and understood by the public.'[175] These varying degrees of access and control on the free flow of information make it possible to identify different 'information loops' in the local response with each loop drawing in different ranges of organizations and individuals, as well as having particular information characteristics pertaining to it and, consequently, different information handling procedures.[176]

By emphasizing limitation in the scope of requests and in suggesting ways of rationalizing channels of communication so as to avoid unnecessary duplication, the guidance seeks to address a key concern expressed by utilities and telecommunications companies during the consultation exercises on the draft Civil Contingencies Bill.[177] **8.69**

It appears that the LRFs will be faced by some serious 'structural' asymmetries in the flow of information particularly in relation to intelligence derived from the security services. In its 2002 report, *Defence and Security in the UK*, the House of Commons Defence Select Committee identified a number of problems regarding the dissemination of intelligence to a wider audience. The Committee observed that: 'Outside government the Security Service's assessments are made available to the police through Special Branches . . . and to others in a number of ways including what is known as the "vellum" System.'[178] **8.70**

---

[173] Ibid, para 3.39.   [174] Ibid, para 3.41.   [175] Ibid, para 3.35.

[176] Ibid, paras 3.35–3.37.

[177] See, for example, Joint Committee on the Draft Civil Contingencies Bill, *Memorandum from Western Power Distribution*, EV 141, and *Memorandum from CE Electric UK*, EV 197, Mr. Doug Turner, Head of Network Continuity and Emergency Planning, BT, Q326, EV155, and United Utilities, CC EV 73.

[178] House of Commons Defence Committee, *Defence and Security in the UK* (2001–02 HC 518–1) para 85.

While the Security Service–Special Branch nexus was regarded as being well established and effective, the Committee identified a series of serious intelligence gaps in relation to other spheres of activity. In particular, the Committee observed that although private sector economic targets were now just as likely to be the targets of terrorist attack as military or government ones:

> The memorandum from the Home Office stated that 'Vellum' assessments were available to companies in a wide range of sectors, but it also explained that those assessments were set at a very low level of classification—not higher than 'restricted'. Assessments at that level can only be couched in very general terms.[179]

The Defence Committee also noted that local authorities did not receive sensitive information from Government and that local authority Emergency Planning Officers do not have security clearance.[180] Some emergency planning officers, especially in London, are now receiving security clearance, but this privilege does not apply to all or to local authority chiefs.[181] The Government refused to impose greater information transmission:

> Decisions about vetting of individuals are made on the basis of individual need. Local responders receive security clearance only where that is necessary, and to different levels as appropriate. This Government has no plans at present to change this approach.[182]

8.71 Clearly, *Emergency Preparedness* does seek to build on the provisions of the Act by endeavouring to facilitate a 'culture' of cooperation and a systematization of information sharing between responders. However, it is not certain whether the asymmetries identified above will be resolved. And asymmetries in access to information could have potentially serious implications for the efficacy of the LRF-based local civil protection function.

Risk assessment

8.72 Chapter 4 of *Emergency Preparedness* examines the duty on all Category 1 responders to carry out risk assessment. The purposes of this duty comprise: to ensure that Category 1 responders have an accurate and shared understanding of the risks that they face so that planning has a sound foundation and is proportionate; to provide a rational basis for the prioritization of objectives and work programmes and the allocation of resources; to enable Category 1 responders to assess the adequacy of their plans and capabilities; to facilitate joined-up local

---

[179] Ibid, para 86.

[180] Ibid, para 89. Yet Chief Fire Officers and other nominated principal Fire Officers of Shire County Fire Brigades have been security cleared: para 90.

[181] House of Commons Defence Committee, *Draft Civil Contingencies Bill* (2002–03 HC 557) Memorandum of the Emergency Planning Society.

[182] Joint Committee on the Draft Civil Contingencies Bill, Appendix 9 q 59. There is a reminder of existing guidelines on information secrecy in Cabinet Office, *Emergency Preparedness: Guidance on Part 1*, para 3.39. With RRTs in mind, CEPOs are receiving Clearance.

planning, based on consistent planning assumptions; to enable Category 1 responders to provide an accessible overview of the emergency planning and business continuity planning context; and to inform and reflect regional and national risk assessments.[183] It also is meant to deliver a consistent risk-based approach across the local tier.

As part of the LRF, Category 1 responders must also cooperate in the maintenance **8.73** of a Community Risk Register (CRR). The guidance encourages collaboration to produce a single, collective risk assessment. The responders are also encouraged to solicit the advice of third parties on particular risks, the potential benefits being 'increased stakeholder engagement, a deeper understanding of the risk assessment, and enhanced credibility when communicating and explaining the assessment.'[184]

The guidance then offers an extensive survey of how the risk assessment process **8.74** should be defined and developed. The guidance adopts a six-stage conceptualization of the risk assessment process (Contextualization, Hazard/Threat Identification, Risk Analysis, Risk Evaluation, Risk Treatment, Monitoring and Review).[185] It is emphasized that risk assessments should drive a standard emergency planning process that will support the creation of both Emergency Plans and Business Continuity Management Plans in the LRF and that it is anticipated that LRFs will set up a Risk Assessment Working Group to undertake this work.[186]

The risk assessment framework that is outlined in the guidance is overtly quan- **8.75** titative in approach and ambitious in scope, as it seeks to draw in many different stakeholders and diffuse issues. A key test for this aspect of the duty is whether the technical rigour outlined will be sustained across a complex organizational environment or will cultural, perceptual, and communications factors hamper the assessment process. For example, much emphasis is placed in the guidance on facilitating a 'culture' of cooperation within the LRF. While such arrangements impart organizational flexibility, they also allow for varying degrees of input into the process. On one level, this could prove advantageous for local bodies such as those county and shire district councils who, in evidence to the Joint Parliamentary Committee, expressed the view that they lack the resources to take full responsibility for local authority civil protection work.[187] However, it also introduces the possibility that 'collaboration' will really mean permanent inequality in burden-sharing which, in turn, will undermine the collective credentials of the assessment process.

Finally, no matter how technically rigorous, Quantified Risk Assessment (QRA) **8.76** processes can only ever provide a quantified estimate of risk, they cannot give an indication of whether such risks are 'acceptable'. As the Health and Safety Executive

---

[183] Cabinet Office, *Emergency Preparedness: Guidance on Part 1*, para 4.1.
[184] Ibid, para 4.13.   [185] Ibid, paras 4.29–4.30.   [186] Ibid, paras 4.32–4.33.
[187] See Joint Committee on the Draft Civil Contingencies Bill, paras 103, 223.

notes in relation to its study of QRA and input into decision-making: 'No uniform "upper envelope" of acceptability of societal risk is deducible' because the factors upon which such judgements are made 'neither could, nor we think should, be ranked or weighted'.[188] Since risk decisions remain matters of judgement, they are subject to a vast array of perceptual, cultural, and social factors whose influence and operation lies outside the remit of even the most advanced QRA technique. Standards in risk assessment are considered further below.

Emergency planning

8.77 Chapter 5 of *Emergency Preparedness* examines the process of Emergency Planning which 'is at the heart of the civil protection duty on Category 1 responders.'[189] The larger part of this chapter is devoted to outlining what is meant by the 'cycle of emergency planning' (Table 8D):[190]

Table 8D: The cycle of emergency planning[191]

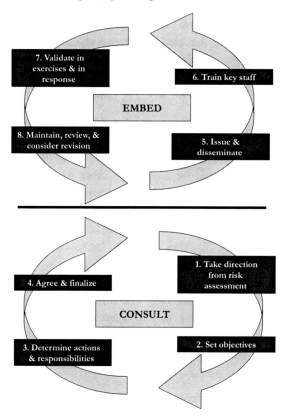

---

[188] Health and Safety Executive, *Quantified Risk Assessment: Its Input to Decision Making* (HMSO, London, 1993) 18.

[189] Cabinet Office, *Emergency Preparedness: Guidance on Part 1*, 47.

[190] Ibid, para 5.45.   [191] Source: ibid, 54.

The guidance examines each of the eight stages of the planning cycle. As with the earlier chapter on risk assessment, the intention is to provide clear step-by-step information for responders on how to negotiate the entire planning cycle. However, although the intent remains the same, there is a discernible difference in emphasis between the two chapters regarding the Local–Regional–National level relationship. For example, Chapter 4 notes that 'considerable benefits' will accrue through having a UK-wide standardized risk assessment approach via the *Local Risk Assessment Guidance*, so that 'top-down and bottom-up risk assessment processes should become increasingly integrated.'[192] However, in Chapter 5 of *Emergency Preparedness*, the Local–Regional–National level relationship is characterized in very different and more diffuse terms. For example, when highlighting, 'Multi-level plans and the role of the lead government Department' it is noted that: 'Ownership of their part of these plans is the responsibility of each of the relevant Category 1 Responders, but co-ordination or leadership in the development and execution of these plans is *likely* to be taken by a national or regional organisation, *perhaps* relying on one of the local partners . . . to co-ordinate.' The conditional language here (subject to the *Local Risk Assessment Guidance*) allows potential disjunctive with the UK Capabilities Programme:

> The programme relies on the framework of the Act to provide a basic structure for civil protection and resilience at the local level. However, the expectations of the UK Capabilities Programme are a matter of government policy and not a requirement of the Act.[193]

Business Continuity Management (BCM)

Chapter 6 examines the Category 1 responder duty to maintain BCM plans. **8.78** The Guidance reiterates that the Act only requires responders to construct BCM arrangements 'so far as is reasonably practicable', which the guidance suggests involves three elements: Criticality (Category 1 responders should focus on ensuring that they can deliver critical functions); Service Levels (the Act does not require Category 1 responders to continue to deliver their functions at ordinary levels during an emergency; some critical functions may need to be scaled up, while others (which are non-critical) may need to be scaled down or suspended); Balance of Investments (responders need not commit unlimited resources to BCM but must decide the level of protection sought in the light of resources and appetite for risk).[194]

---

[192] Ibid, 43. See Box 4.3 of Ch 4.
[193] Ibid, 61, 64. See Boxes 5.7 and 5.9 of Ch 5. (Emphasis added.)
[194] Ibid, para 6.12.

8.79 Other components of the BCM duty as required by the Act and the associated regulations are then described, though the guidance is here expressed to be less directive than in other areas as responders 'may use other models to deliver statutory requirements in this area where there are compelling reasons for doing so.'[195] The guidance adopts as its template, and seeks to build upon, the Business Continuity Institute's, 'five-stage process, which has become widely accepted and has been incorporated into a British Standards Institute Publicly Available Specification—PAS 56.'[196]

Communicating with the public

8.80 In Chapter 7 of *Emergency Preparedness*, it is observed that, 'Category 1 responders' duties to communicate with the public under the Act are based on the belief that a well-informed public is better able to respond to an emergency and to minimise the impact of the emergency on the community.'[197] The guidance then focuses on how the Government believes the duties can be best carried out.

8.81 In relation to public communications, the guidance is primarily concerned with setting clear parameters for any communications strategy. This includes clearly defining the key objectives for responder plans, evaluating how much the public needs to know and how (as well as how much) information should be published. Other advice includes describing processes for developing effective warning regimes, providing a structured definition of 'the public' and audience 'types' along with identifying various warning methods that might be of use.[198]

8.82 The endeavour to achieve clarity of purpose may result in a degree of over-simplification, particularly in the guidance's homogenous and simplistic definition of target audiences. Nor does it venture into how psychological, social, and cultural factors in risk perception might affect the communications process,[199] factors that could have a direct bearing on the level of public trust in the institutions of risk management and the messages they seek to put into the public domain.

Advice and assistance to business and voluntary organizations

8.83 In Chapter 8, the guidance returns to the issue of BCM, this time to examine the duty placed on local authorities to provide general advice on BCM to local

---

[195] Ibid, para 6.38.
[196] Cabinet Office, *Emergency Preparedness: Guidance on Part 1*, para 6.43. For details of PAS 56, see British Standards Institution, *Publicly Available Specification 56: Guide to Business Continuity Management* (http://www.bsi-global.com/Fire/Business/pas56.xalter, 2003).
[197] At 93, Chapter Summary. See also ibid, para 7.4.   [198] Ibid, 103–6.
[199] See Ch 1.

commercial and voluntary organizations. The guidance points out that not all local organizations would want or profit from such advice and therefore reiterates the need for prioritization and definition of target groups.[200] It is unclear how far generalist local authorities are skilled for such activity, but, as a start, they can of course explain their own plans and mechanisms.

As is also the case with the earlier BCM chapter, the role of the LRF appears **8.84** undervalued as a potential mechanism for Category 1 and 2 cooperation in this sphere. Such cooperation could easily include fostering information flows from Business Continuity professionals in the private sector to their public sector counterparts. Neither of the BCM chapters in *Emergency Preparedness* nor the BCM duty in the Act and Regulations, appear to acknowledge this possibility.

(d) Devolution and London

Chapters 9 (London), 10 (Scotland), 11 (Wales), and 12 (Northern Ireland) of **8.85** *Emergency Preparedness* are primarily geared towards a more descriptive survey of variations in institutional structures and relationships. The details have already been related in Chapter 4.

(e) Monitoring and enforcement

Chapter 13 of the Guidance provides more information on the oversight of **8.86** delivery under Part I. As already discussed, the Government has decided not to establish a formal inspectorate but 'will rely on current good practice in performance management and on established audit and regulatory bodies across the Category 1 and 2 organisations to assess performance.'[201]

There are estimated to be around eight relevant audit bodies, and they include **8.87** the Audit Commission which will utilize the Comprehensive Performance Assessment for local authorities to ask questions around local authority planning for 'Safer and Stronger Communities'.[202] As for police forces, through its 'Capability Review process . . . [Her Majesty's Inspectorates of Constabulary] will undertake force-level assessments testing capability in this area against an established set of standards.'[203] Other inspectorates are to undertake collaborative work. For example, the Audit Commission and the Fire Service Inspectorate have a memorandum of understanding, and the main bodies responsible for inspecting, regulating, and auditing healthcare signed a concordat in June 2004.[204]

---

[200] Ibid, paras 8.31, 8.42.   [201] Ibid, para 13.4.
[202] Ibid, para 13.10. A three-year cycle applies.   [203] Ibid, paras 13.11–13.12.
[204] Ibid, para 13.13. The Audit Commission will act alone when the FSI ceases in 2007.

8.88 The guidance emphasizes the important role of self-assessment in the monitoring process and highlights the self-assessment tools that have been developed by the Audit Commission.[205] The guidance document itself provides a further set of self-assessment templates that cover each of the seven civil protection activities.

8.89 In summary, perhaps with a view to gaining acceptance, the auditing of the civil protection function is to be conducted with a light regulatory touch.

(f) Role of the voluntary sector

8.90 Chapter 14 of *Emergency Preparedness* suggests four models of engagement with the voluntary services.[206] Model 1: engagement through the LRF; Model 2: establishing a voluntary sector sub-group of the LRF; Model 3: bilateral links on the basis of functions; and Model 4: bilateral links on the basis of capabilities. It is also observed that the voluntary sector should be represented at a regional level on the RRFs and that a voluntary sector-CCS working group already exists to discuss collaborative issues at the national level.[207]

8.91 Support is accorded to studies by the British Red Cross,[208] and their message that volunteers have an important part role to play in civil protection arrangements. In addition, a concordat was signed by the Local Government Association, the Association of Chief Police Officers, British Association for Immediate Care, British Red Cross, the Chief Fire Officers' Association, CRUSE Bereavement Care, the Salvation Army, Samaritans, St John Ambulance, Victim Support, and the WRVS on 7 July 2005 at the Local Government Association's Annual Conference to recognize that the voluntary sector must be involved in emergency planning.[209]

8.92 The 2004 Act engages some key elements of the private sector as Category 2 responders, but could the private sector undertake more of the responsibility for risk, for example through insurance mechanisms? According to the Civil Contingencies Secretariat:

> In the UK, insurance has achieved a high level of penetration in the domestic and business markets (87% of all home-owning householders have buildings cover, 75% of all households have contents cover). All Risks cover is common (standard on Household policies, and accounting for an estimated 80% of business property policies). In such circumstances insurance costs might be expected to represent around 25–33% of total economic costs arising from a catastrophic event, with

---

[205] See http://www.audit-commission.gov.uk/emergencyplanning/index.asp.
[206] Ibid, paras 14.11–14.15. [207] Ibid, paras 14.28–14.29.
[208] Wood-Heath, M and Annis, M, *The Role of Non-Governmental Organisations' Volunteers in Civil Protection in European Member States & European Economic Area Countries* (British Red Cross, London, 2002). See also Wood-Heath, M and Annis, M, *Working Together to Support Individuals in an Emergency or Disaster* (British Red Cross, London, 2004).
[209] See http://www.egovmonitor.com/node/1918.

much of the remainder falling as costs to Government, costs arising from disruption to transport systems and costs retained by individuals due to lack of insurance, under-insurance, or because insurance is not available. It should be noted that insurance costs relate to property and business interruption claims only, and exclude any personal injury, life insurance, health, liability or motor claims, for example, or any costs compensated by public schemes such as the Criminal Injuries Compensation Scheme.[210]

Following an agreement in 1961 (revised in 2002), household insurance against flooding is available throughout the United Kingdom, though premiums are such that, as revealed in the quotation, not all households actually purchase insurance.[211] As regards terrorism, the state has intervened to ensure coverage for aircraft, commercial property and, in Northern Ireland, all property.[212] Maybe more could be shifted to the private sector if insurance were compulsory in more situations, for example, for householders in flood-plains.

Another way of shifting liability, explored by the Defence Committee, would be the institution of security regulations—the equivalent of fire or health and safety regulations for private property owners.[213] **8.93**

There are also opportunities for the involvement of the private security industry.[214] The overt partnership of public and private is now a commonplace in policing circles[215] and has extended even to target-hardening against terrorism.[216] But these modes of engagement are not through the responder system of Part I of the 2004 Act. Arrangements operate primarily through the police, mainly on a voluntary basis but including community safety accreditation schemes under Part IV of the Police Reform Act 2002.[217] **8.94**

Another direct way of shifting the burden from public to private would be the creation of an emergency volunteer reserve force.[218] In bygone years, volunteers were prized in quantitative terms—for sheer manpower—as during the **8.95**

---

[210] Civil Contingencies Secretariat, *Full Regulatory Impact Assessment*, para 20.

[211] Huber, M, 'Insurability and Regulatory Reform' (2004) 29 *Geneva Papers on Risk and Insurance* 169; Priest, SJ, Clark, MJ, and Treby, EJ, 'Flood Insurance' (2005) 37 *Area* 295.

[212] See Walker, C, 'Liability for Acts of Terrorism: United Kingdom Perspective' in *European Centre for Tort And Insurance Law Liability for Acts of Terrorism* (Springer, Vienna, 2004); Walker, C, 'Political Violence and Commercial Risk' (2004) 56 *Current Legal Problems* 531.

[213] Defence Committee, *Draft Civil Contingencies Bill* (2002–03 HC 557) paras 48, 57.

[214] See Defence Committee, *Defence and Security in the UK* (2001–02 HC 518) para 212; Joint Committee on the Draft Civil Contingencies Bill, para 135.

[215] See Jones, T and Newburn, T, *Private Security and Public Policing* (Clarendon Press, Oxford, 1998).

[216] See Walker, C, 'Political Violence and Commercial Risk' (2004) 56 *Current Legal Problems* 531.

[217] Cabinet Office, *The Government's Response to the Report of the Joint Committee*, Annex A para 14.

[218] HC Debs Standing Committee F col 88 29 January 2004 and HC Debs vol 421 col 1323 24 May 2004, Patrick Mercer.

General Strike of 1926 when Volunteer Services Committees were set up to recruit the general public (around 100,000 did enlist), and their activities were organized by the government-approved Organisation for the Maintenance of Supplies.[219] Special constables were also increased from 98,000 to 240,000.[220] Yet the precedent of the Civil Defence Act 1948 (described in Chapter 2) is not encouraging, and the current modes of engagement with volunteers, perhaps owing to the complexities of modern industry and services, emphasize their special expertise and skills rather than numbers. The Government argues that there is likely to be more public commitment to more voluntary bodies, such as the Red Cross,[221] and that 'Setting up a new force would be burdensome and bureaucratic and could undermine existing voluntary organisations.'[222]

8.96 No doubt, private sector involvement is vital, but the Civil Contingencies Act does too little to engender it. Its mode of governance is primarily hierarchical rather than heterarchical. The sponsoring of a volunteer force would not change this stance, but the opening of forums and information flows could do so. BCM assistance is a start, but it hardly treats the private sectors as equals worthy of 'smart' regulation,[223] even less does it speak to the general public.

(g) Organizations not covered by the Act

8.97 Although the 2004 Act aims to bring into a single framework those organizations involved in civil protection, it is acknowledged that it has not captured every relevant organization. According to Chapter 15 of *Emergency Preparedness*, those left outside include: the armed forces; retail companies; insurance companies; transport companies; media and communications networks; offshore industries; security firms; internal drainage boards; and general practitioners and chemists.[224] It is suggested that all should be 'encouraged' to participate.[225] The other principal ghost in the machine is central government, the relevant powers of which are outlined in Chapter 16 of *Emergency Preparedness*.[226]

---

[219] O'Brien, TH, *Civil Defence* (HMSO, London, 1955) 29.

[220] HC Debs vol 196 col 597wa 1 June 1926, Sir W Joynson-Hicks; Morgan, J, *Conflict and Order: The Police and Labour Disputes in England and Wales, 1900–1939* (Clarendon, Oxford, 1987) 120.

[221] HC Debs Standing Committee F col 98 29 January 2004, Douglas Alexander.

[222] HC Debs vol 421 col 1361 24 May 2004, Hazel Blears. Compare the USA Citizen Corps (http://www.citizencorps.gov/about.shtm), launched in January 2002 to makes use of volunteers to assist in disaster relief and including the Community Emergency Response Team (CERT) program. The scheme is coordinated by the Department of Homeland Security (see Ch 9). See Simpson, DM and Strang, W, 'Volunteerism, Disasters and Homeland Security' (2004) *Journal of Homeland Security and Emergency Management* Vol 1: No 4, Article 404.

[223] See Gunningham, N and Grabosky, P, *Smarter Regulation* (Clarendon Press, Oxford, 1998).

[224] Ibid, para 15.4.   [225] Ibid, para 15.3.

[226] Ibid, paras 16.3–16.5, 16.7–16.10, 16.11–16.12.

The role of central government has been described earlier in this chapter. In Chapters 17 and 18 of *Emergency Preparedness*, some outlines are given of the regional bodies which emanate not from the 2004 Act but from the corridors of the Government Offices in the English Regions, as related in Chapter 4.

(h) Response and recovery

Response and recovery are the subjects of the second guidance volume, *Emergency Response and Recovery*. 'Response' encompasses the actions taken to deal with the immediate effects of an emergency; 'recovery' is more long term.[227] The management and coordination of local operations involve the familiar Gold, Silver, and Bronze tiers. Gold command for response, often headed by the police, takes the form of a Strategic Coordinating Group (SCG) when there is an emergency with a substantial impact or lasting over a prolonged period; they form a kind of operational arm of LRFs and will link to RCCCs.[228] The police might also form a Joint Intelligence Group (for incidents involving criminality such as terrorism), while the health experts might form a Joint Health Advisory Cell.[229]

8.98

Other issues include the care of those affected. A local authority crisis/emergency centre may be formed, along with a Survivor Reception Centre, a Family and Friends Reception Centre (both of which in the immediate aftermath of an incident reunite relatives with survivors and provide temporary shelter), and a Family Assistance Centre (for emergency information, advice, and support).[230] Consideration is given to the roles of Police Family Liaison Officers and police casualty bureaux[231] and media liaison points with a designated media liaison officer.[232] The guidance is silent about welfare payments[233] despite the likely

8.99

---

[227] See further Toigo, J, *Disaster Recovery Planning: Preparing for the Unthinkable* (Prentice Hall, London, 2000).

[228] The SCG should include a Government Liaison Officer: Cabinet Office, *Emergency Response and Recovery*, paras 4.20, 7.7. See further Ministry of Defence, *Operations in the UK*, para 9.18; Cabinet Office, *Central Government Arrangements*, para 15; Association of Chief Police Officers, *Local Emergency Centres: Good Practice Guide* (ACPO, London, 2004). For a comparison, see FEMA (*Responding to Incidents of National Consequence: Recommendations for America's Fire and Emergency Services Based on the Events of September 11, 2001, and Other Similar Incidents* (Washington, 2002) 36.   [229] Ministry of Defence, *Operations in the UK*, paras 926, 929.

[230] See ACPO and Cabinet Office, *Humanitarian Assistance in Emergencies—Guidance on Establishing Family Assistance Centres* (http://www.ukresilience.info/contingencies/facacpoguidance.pdf, 2005); Newburn, T, *Disaster and After* (Jessica Kingsley, London, 1993). For disasters abroad, the Foreign & Commonwealth Office Emergency Response Team will provide a central contact and information point (Cabinet Office, *Emergency Response and Recovery*, para 7.16).

[231] See also Wood-Heath, M and Annis, M, *Working Together to Support Individuals in an Emergency or Disaster* (British Red Cross, London, 2004).

[232] See Shearer, A, *Survivors and the Media* (John Libbey, London, 1991) Harrison, S, *Disaster and the Media* (Macmillan, Basingstoke, 1999).

[233] See Social Security (Payments on account, Overpayments and Recovery) Regulations 1988 SI No 664, art 2.

een disaster impact and socio-economic status.[234] No indication
behaviour patterns, which are often 'prosocial' in disasters.[235]
out dealing with fatalities and with the sensitivities of faith
left to other Home Office/Cabinet Office guides.[236] Some
areness of human issues involved in a disaster response.[237]
importance of counselling is overblown.[238]

cated, liaison with the public is an important part of emergency
se. In major incidents, there is a National Casualty Bureau which acts as
the main contact between the police casualty bureaux and the public.[239] The
police will appoint a Senior Identification Manager (SIM) to manage the
identification of victims.[240] The task of identifying the dead is undertaken by an
Identification Commission, which comprises the supervising pathologist, a
police Senior Identification Manager and a Family Liaison Co-ordinator, plus
forensic specialists.[241] Reports by Lord Justice Clarke on the Marchioness
sinking of 1989 [242] and Lord Cullen on the Dunblane Shootings in 1996[243]
highlight difficulties with practices then in place.

---

[234] See Hewitt, K, 'Excluded Perspectives in the Social Construction of Disaster' in
Quarantelli, EL (ed), *What is a Disaster?* (Routledge, London, 1998); Enarson, E and Morrow,
BH (eds), *The Gendered Terrain of Disaster* (Praeger, Westport, 1998).

[235] Quarantelli, EL, 'Epilogue' in Quarantelli, EL (ed), *What is a Disaster?* (Routledge,
London, 1998) 241.

[236] Home Office and Cabinet Office, *Guidance on Dealing with Fatalities in Emergencies*
(http://www.ukresilience.info/contingencies/pubs/fatalities.pdf, London, 2004); Home Office
and Cabinet Office, *The Needs of Faith Communities in Major Emergencies* (http://www.ukresi-
lience.info/contingencies/pubs/faith-communities.pdf, 2005). Under the Public Health Act
1936, s 198, local authorities can be directed to provide a public mortuary. Consideration has also
been given to the avoidance of overlap between coroners, inquests and other inquiries: Home
Office, *Report of the Disasters and Inquest Working Group* (HMSO, London, 1997).

[237] Emergency Planning Society Professional Issue Groups—Human Aspects, http://
www.emergplansoc.org.uk/index.php?tab=groups&area=11&group=73.

[238] Furedi, F, *Therapy Culture* (Routledge, London, 2003).

[239] HC Debs vol 436 col 2572w 12 September 2005. The computerization of information
about survivors, casualties, and missing persons was recommended by Lord Cullen, *The Ladbroke
Grove Rail Inquiry* (HSE Books, Norwich, 2001) Pt I, para 4.119.

[240] Cabinet Office, *Dealing with Disaster* (rev 3rd ed, Cabinet Office, London, 2003) para 4.32.

[241] Ibid, para 4.38. See further Health and Safety Executive, *Disasters Working Party, Disasters:
Planning for a Caring Response* (TSO, Norwich, 1991); Association of Train Operating Companies, *Joint Industry Provision of Customer Care following a Major Passenger Rail Accident* (http://
www.rgsonline.co.uk/docushare/dsweb/Get/Rail-10401/acop011.pdf, 2002).

[242] The severing of hands from victims for identification purposes was the subject of particular
distress: Lord Justice Clarke, *Public Inquiry into the Identification of Victims following Major
Transport Accidents* (Cm 5012, HMSO, London, 2001). See further Lord Justice Clarke, *Thames
Safety Inquiry: Final Report* (Cm 4558, London, 2000); Davis, H and Scraton P, 'Institutionalised
Conflict and the Subordination of "Loss" in the Immediate Aftermath of UK Mass Fatality
Disasters' (1999) 7.2 *Journal of Contingencies and Crisis Management* 86.

[243] Lord Cullen, *Public Inquiry into the Shootings at Dunblane Primary School on 13 March
1996* (Cm 3386, HMSO, London, 1996) para 3.31–3.35 and Government Response (Cm 3392,
HMSO, London, 1996).

In recovery phases, Gold Command may pass to the local authority.[244] There **8.101**
should also be a Recovery Working Group.[245] Eight operational principles
pertain: continuity with existing functions; preparedness; subsidiarity; clear
direction; integration; cooperation; good communication; and anticipation.[246]
These considerations are not confined to responders. For example, insurance
staff can be 'a key enabler in the recovery process'.[247] Otherwise, as already
mentioned, the role of insurance is left unexplored, even though it can be
decisive in achieving early recovery.

It is also disappointing that SCGs are so poorly illuminated. There was no **8.102**
debate in Parliament, nor any mention in the legislation. Their institutional
composition, roles, and relations to other bodies (including LRFs) should be
more transparent.

F. Standards of Delivery

Under this heading, it is intended to explore the targets being set for achieve- **8.103**
ments under the Act. It is not intended to examine here the cognitive processes
by which risk is perceived and weighted.[248] The short answer to this inquiry is
that the 2004 legislation and guidance is wholly silent as to the expected
standards of delivery. One must therefore consider approaches elsewhere.

(a) Risk assessment

Under the Health and Safety Act 1974, the common standard is that the risk **8.104**
must be reduced 'As Low As Reasonably Practicable' (ALARP): that a risk
reduction measure should be implemented unless there is a 'gross disproportion'
between the cost of a control measure and the benefits of the risk reduction that
will be achieved.[249] ALARP has become a widespread rule of risk management. It
is a flexible standard, and, as noted by the BSE Inquiry, 'It involves an exercise in
proportionality that often calls for nice judgement.'[250] It also involves normative

---

[244] Cabinet Office, *Central Government Arrangements*, para 15.
[245] Ibid, paras 4.20–4.24. See also Office of the Deputy Prime Minister, *Guidance on Development of a Site Clearance Capability in England and Wales* (http://www.odpm.gov.uk/embedded_object.asp?id=1123763, London, 2005) para 3.10. The Group will work under the strategic direction of the SCG.
[246] Cabinet Office, *Emergency Response and Recovery*, Ch 2.
[247] Ibid, para 3.57.
[248] See Royal Society, Risk: *Analysis, Perception and Management* (London, 1992); Douglas, M, *Risk and Blame: Essays in Cultural Theory* (Routledge, London, 1992).
[249] See http://www.hse.gov.uk/risk/theory/alarp.htm; *Edwards v National Coal Board* [1949] 1 KB 704.
[250] BSE Inquiry, *Report* (1999–00 HC 887) vol 1 Ch 2, para 164. The application is considered in vol 15 Ch 5. See also Government Response to the report on the BSE inquiry (Cm 5263, London, 2001).

choices, and, especially in the public sector, cost–benefit analysis may have to be subordinate to an equity approach in which all individuals are entitled to a base level of protection rather than being sacrificed for the maximal position. ALARP is not the only legal framework to inform the management of risks. Other principles, which all seek to achieve rational proportionality, appearing in legislation, include ALARA ('as low as reasonably achievable') and BPM ('best practicable means'), both terms used in connection with radiation, as well as BPEO ('Best Practicable Environmental Option'), and BATNEEC ('Best Available Techniques Not Entailing Excessive Cost'), used in environmental contexts.[251] NOAEL (No Observable Adverse Effects Level) has been used when licensing medicines or assessing risks in food from additives or residues. The Interdepartmental Liaison Group on Risk Assessment (ILGRA) considered policy development and practical applications of risk assessment in Government Departments. Its report[252] in 1996 reached no finding as to the best standard, and, clearly, context and objectives are not constants.

8.105 If a risk is evident, then preventive measures can be applied with one of the foregoing standards in mind. But where a hazard or threat is not well understood and the risk unproven, as will often prevail in contingency planning, attention focuses on the 'precautionary principle'.[253] It is associated with heightened (and perhaps reverse) burdens of proof of the risky activity's costs and benefits. In its Third Report in 2002,[254] the ILGRA noted the emphasis in the *BSE Report* on the importance of applying and enforcing precautionary measures.[255] The principle is elaborated in a special paper, *The Precautionary Principle: Policy and Application*, in 2002.[256] It outlines policy guidelines which encourage an official decision where there is good reason to believe that harmful effects may occur to human, animal or plant health or to the environment, notwithstanding scientific uncertainty about the nature and extent of the risk:

> The purpose of the Precautionary Principle is to create an impetus to take a decision notwithstanding scientific uncertainty about the nature and extent of the

---

[251] See DEFRA, *The Air Quality Strategy for England, Scotland, Wales and Northern Ireland* (Cm 4548, London, 2000).

[252] *Use of Risk Assessment within Government Departments*, http://www.hse.gov.uk/aboutus/meetings/ilgra/minrpt1.htm, 1996.

[253] See *This Common Inheritance* (Cm 1200, London, 1990) para 1.15; *R v Secretary of State for Trade and Industry, ex p Duddridge* [1995] 3 CMLR 231, [1996] 2 CMLR 361; Harding, R and Fisher, E (eds), *Perspectives on the Precautionary Principle* (Federation Press, Annandale, 1999); Alder, J and Wilkinson, D, *Environmental Law and Ethics* (MacMillan, Basingstoke, 1999) 149; Fisher, E, 'Is the Precautionary Principle justiciable?' [2001] 13 *Journal of Environmental Law* 315.

[254] See http://www.hse.gov.uk/aboutus/meetings/ilgra/minrpt3.pdf.

[255] BSE Inquiry, *Report* (1999–00 HC 887).

[256] ILGRA (2002), http://www.hse.gov.uk/dst/ilgra/pppa.htm. In the *Interim Response to the BSE Inquiry* (DEFRA, London, 2001) para 6.14, it is stated that 'The Government is committed to taking a precautionary approach where appropriate.'

risk, i.e. to avoid 'paralysis by analysis' by removing excuses for inaction on the grounds of scientific uncertainty.[257]

The ILGRA's agenda was taken over by the Risk Support Team in HM **8.106**
Treasury in 2001. Its work drew upon the report from the Strategy Unit in the Cabinet Office, *Risk—Improving Government's Capability to Handle Risk and Uncertainty*.[258] It recommended that consideration be given to the extension of systematic approaches to strategic policy making, such as the ALARP principle,[259] adapting them as necessary to recognize the less quantifiable nature of the data involved. The Risk Support Team oversaw the implementation in a two-year programme to improve risk management across government, including through departmentally based Risk Improvement Managers. A variety of guides on risk management,[260] sponsored by HM Treasury, have now been developed, including *Managing Risks to the Public*,[261] *Communicating Risk*,[262] and *The Orange Book: Management of Risk—Principles and Concepts*.[263] The Strategy Unit continues its work under the title of the Prime Minister's Strategy Unit.[264]

The 2004 Act and guidance is silent on these complex but highly relevant issues. If **8.107**
responders are to engage in planning, then surely the standards by which risks should be judged as actionable and the standards to be attained by resilience efforts should be as rational as possible. Further work is required on this topic.[265]

---

[257] Ibid, para 11. Compare the (New Zealand) Civil Defence Emergency Management Act 2002 s 7: 'All persons exercising functions in relation to the development and implementation of civil defence emergency management plans under this Act may be cautious in managing risks even if there is scientific and technical uncertainty about those risks.' The ILGRA more effectively discouraging chronic inactivity which is a present danger (see Furedi, F, *Culture of Fear: Risk-taking and the Morality of Low Expectation* (rev ed, Continuum, London, 2002) 107).

[258] See http://www.strategy.gov.uk/downloads/su/RISK/REPORT/downloads/su-risk.pdf, 2002.

[259] Paragraph 4.2.41.

[260] See http://www.hm-treasury.gov.uk/documents/public_spending_and_services/risk/pss_riskportal.cfm. It is natural that these are often viewed as supplementary to the principal Treasury guidance on cost/benefit analysis, *Appraisal and Evaluation in Central Government* (the 'Green Book': http://www.hm-treasury.gov.uk/economic_data_and_tools/greenbook/data_greenbook_index.cfm). For Northern Ireland, see CEPU, *A Guide to Emergency Planning in Northern Ireland* (http://cepu.nics.gov.uk/pubs/RISK%20ASSESSMENT%20DOC%20VER%203.doc, 2005).

[261] See http://www.hm-treasury.gov.uk/media/3E3/EE/managingrisks_appraisal220705.pdf.

[262] See http://www.ukresilience.info/risk/communicatingrisk.pdf. In the *Interim Response to the BSE Inquiry* (DEFRA, London, 2001) para 6.5, it is stated that 'The Government's general approach to risk is to take action where appropriate, in proportion to the risk and to make available to the public sufficient information about a risk, in a form that is easily understood, so that individuals can make their own choices.'

[263] See http://www.hm-treasury.gov.uk./media/FE6/60/FE66035B-BCDC-D4B3–11057A7707D2521F.pdf, 2004. See also Audit Commission, *Worth the Risk* (London, 2001).

[264] See http://www.strategy.gov.uk/.

[265] A guide is offered the by European Commission, *Communication on the Precautionary Principle* (Brussels, COM(2000)1).

(b) Training and exercises

8.108 What about standards in drill and practice?[266] Exercises may be 'table-top'
role-pay exercises or live.[267] The Exercise Working Party has been set up to
address the requirements for cross-Departmental coordination of planning and
exercising.[268] Major exercises have included Exercise Trump Card in 2000,
which simulated a release of chemicals in London.[269] This was one of at least
three full-scale live terrorist attack simulations and approximately 12 to 15
table-top exercises that take place each year. Other recent live exercises have
included Operation Osiris, at the Bank underground station in 2003, which
revealed a number of shortcomings, such as the inability of those wearing
protective suits to communicate under difficult conditions.[270] There was also
an anti-terrorist chemical attack exercise at the Birmingham NEC in 2004,
highlighting the need to improve speed of response and cordon controls.[271]
Exercise Hornbeam was a DEFRA Foot and Mouth Disease exercise in
2004.[272] Exercise Triton 04 was staged by the Environment Agency to test
responses to coastal flooding, with the timely release of information to the
public a key lesson.[273] Exercise East Wind, sponsored by the Health Protection Agency (Emergency Response Division), raised the problem of large
numbers of people self-referring to hospitals following a dirty radiological
bomb in Cambridge.[274] Its Exercise Magpie in Newcastle in 2004 tested
responses to chemical or biological attack and emphasized the need for speedy
public health messages.[275] The Atlantic Blue Exercise was conducted in the
financial sector in 2005.[276]

8.109 In the absence of formal standards, it is difficult to be objective about what
has been achieved by these efforts either in terms of how much of the apparatus
has been tested or whether procedures and routines have been rigorously
scrutinized.

---

[266] For practical advice, see Home Office, *Exercise Planners Guide* (http://www.ukresi-
lience.info/contingencies/business/exercise_planners_guide.htm, London, 1998). For military
perspectives, see Ministry of Defence, *Operations in the UK*, Ch 10.

[267] HC Debs Standing Committee F col 76 27 January 2004, Douglas Alexander.

[268] Civil Contingencies Secretariat, *The Lead Government Department*, para 5.19.

[269] House of Commons Defence Committee, *Defence and Security in the UK* (2001–02 HC
518) q 1224, 7 May 2002, David Veness.

[270] See http://www.ukresilience.info/contingencies/exercise.htm; HC Debs vol 405 col 120w
15 December 2003, Alistair Darling For exercises in 2004, see http://www.homeoffice.gov.uk/
docs4/220205tp4.pdf.   [271] *Birmingham Post*, 19 July 2004 at 11.

[272] See http://www.defra.gov.uk/footandmouth/hornbeam/finalrpt-hornbeam.pdf.

[273] See http://www.environment-agency.gov.uk/commondata/acrobat/geho0305birqe-
p_1106396.pdf.   [274] See http://hpa.org.uk/hpa/erd/pdf/east_wind.pdf.

[275] Exercise Magpie, *Executive Summary* (http://www.hpa.org.uk/hpa/erd/pdf/magpie.pdf).

[276] *Financial Sector Business Continuity Annual Report 2005* (http://www.fsc.gov.uk/upload/
public/Files/5/64791_1.pdfannualrep.pdf).

G. Finance

(a) Planning costs

Since the Act aims towards greater consistency in resilience planning and **8.110** response, a Regulatory Impact Assessment[277] duly determined that no inequality or unfairness would result from their implementation nor any anti-competitive impact.[278] However, there will be compliance costs, and the Government was particularly concerned about those falling upon businesses in Category 2. These costs were estimated as being in the region of 'at most one day a month of the time of a relatively senior manager . . .', based upon requests for information, possible attendance at meetings, and 48 hours of exercises per year.[279] Existing costs may be offset, especially in the utilities sector where emergency planning existed prior to the 2004 Act.

A Cabinet Office paper of June 2003 initially worked out the impact of three options **8.111** relevant to Part I of the Act.[280] The figures were reworked by the final risk impact assessment in January 2004 after some criticism from Category 2 utilities (Table 8E):[281]

Table 8E: Costs per LRF area for Category 2 responders [282]

Activity	Cost per annum (£)		
	Option 1	Option 2	Option 3
Obtaining LA BCM advice	n/a	100–2,000	100–2,000
Meeting all duties as a Category 1 responder	n/a	n/a	6,580–13,150
Supply of information, cooperation, training and exercising as a Category 2 responder	1,740	2,700	3,290
Total	1,740	2,800–4,050	3,390–15,150

Explanation of options:

- Option 1: Continuation of current regime without new statutory duties—not a serious contender.
- Option 2: Category 1 duties confined to a limited range of organizations and Category 2 range also limited. This vision was that of the draft Bill.

---

[277] See Regulatory Impact Unit and Cabinet Office, *Better Policy Making: A Guide to Regulatory Impact Assessment* (London, 2003). The result is akin to cost–benefit analysis 'with a thumb on the cost-side': Posner, RA, *Catastrophe: Risk and Response* (Oxford University Press, New York, 2004) 148.

[278] Cabinet Office *The Draft Civil Contingencies Bill: Explanatory Notes etc*, para 24, 67.

[279] Ibid, para 48. [280] Ibid, paras 27–33.

[281] See Civil Contingencies Secretariat, *Full Regulatory Impact*, paras 45–83; Joint Committee on the Draft Civil Contingencies Bill, para 216. [282] Source: ibid, para 78.

• Option 3: This option would include a larger range of organizations within Category 1 (such as the utility companies) and would widen Category 2 to include large voluntary organizations.

Aside from any pure policy gains, the economics seemed to point in favour of Option 2.

8.112   A final iteration of the calculations was provided in January 2004.[283] It was there revealed that potential Category 2 responders, while entertaining concerns about the additional costs, did not object to their designation.[284] Clearly, the inclusion of some private sector bodies (especially utilities and transport) rather than others could be seen as unequal, but it reflects the principle, 'the greater the risk, the tighter the regulation' and so is proportionate to their roles in society.[285] In any event, the Office of Fair Trading was unable to identify any markets likely to be significantly affected.[286] Accordingly, the Government resisted any scheme to pay any compliance costs incurred by Category 2 responders.[287]

8.113   The Regulatory Impact Assessments had focused on private sector involvement under Category 2 and gave no detailed estimates of the burdens of local authorities, the primary responders. As regards the pre-2004 Act funding in England and Wales,[288] the Civil Defence Grant[289] has been terminated, and the Government undertook to find the resources to cover the costs of the new statutory duties out of the general Revenue Support Grant.[290] This new financial disposition is contended to secure three policy gains. First, it allows a wider range of activities to be funded since the Civil Defence Grant was confined to the agenda of the previous Acts. Secondly, it reflects a general move away from ring-fenced funding which reflects the policy of the government and of the Local Government Association in favour of greater autonomy for local government. Thirdly, it betokens the mainstreaming of the resilience agenda.[291]

---

[283] Civil Contingencies Secretariat, *Full Regulatory Impact*.   [284] Ibid, para 11.
[285] Ibid, para 33.   [286] Ibid, para 85.
[287] See the amendment proposed at HL Debs vol 665 col 440 14 October 2004, Baroness Buscombe.
[288] The Civil Defence Grant was distributed as a flat rate payment for 65% of the share (in 2003–04, each authority received £53,000, with an extra £12,000 per shire district within each country council) plus 35% being based on population size. There was no calibration of local risk. The grant ended in Scotland in 2001 and was replaced by Civil Protection Grant Aided Expenditure under the Civil Defence (Scotland) Regulations 2001 SI No 139 (see Scottish Parliament, Justice 2 Committee col 28 28 February 2001). In Wales, local authorities are funded through the National Assembly, which distributes the block grant system. In Northern Ireland, funding of civil protection activities has always been part of the normal public expenditure process: Northern Ireland Office, *Civil Protection in Northern Ireland: The Implications of the Civil Contingencies Bill* (Belfast, 2003).
[289] Civil Defence Act 1948, s 3, as replaced by the Civil Defence (Grant) Act 2002 (for the reasons, see HC Delos, vol 305 col 1013, 28 November 2001, Christopher Leslie).
[290] Cabinet Office, *The Draft Civil Contingencies Bill: Explanatory Notes etc*, para 64.
[291] Cabinet Office, *The Draft Civil Contingencies Bill*, para 3.33.

While there was to be symbolic change to the source of financing, there was no **8.114** promise that the amount was to change, since it was viewed as 'sufficient to support the basic responsibilities for local authorities that flow from the Bill.'[292] The final Risk Impact Assessment likewise depicted the changes as cost neutral for public sector organizations.[293] But it is indubitable that extra costs will be incurred, since there are new duties, including: warning the public; promoting business continuity management; taking action to prevent and respond to emergencies; training; and participation in the new local resilience forums. The definition of emergency has also expanded, and another enlargement concerns the involvement of shire district councils.[294] As a result, many questioned the Government's commitment to improving the delivery of resilience and wondered whether the exercise is more concentrated on a lowest common denominator rather than excellence.[295] According to the Local Government Association, 'Based on the LGA's surveys, what can be stated with some certainty is that local authorities in England and Wales spend at present £36m annually, whilst receiving only £19 m from central government in Civil Contingencies Grant.'[296]

One Government response was that the Civil Defence Grant was only £14.4m **8.115** in each of 1998–99, 1999–2000, and 2000–01, so the increase to £19m was itself generous.[297] This reply was disingenuous, for the grant had previously been reduced (from a heady £24.5m in 1991) as part of a 'peace dividend' at the end of the Cold War.

Eventually, the Government relented. The 2004 Spending Review confirmed **8.116** the doubling of the central government contribution to the cost of local authority civil protection activities to £40m per year, running from 2005 through to 2008.[298] Spending levels in the early years of the new Act will form part of the evidence base for the SR 2006 exercise and so on. In Scotland, the Scottish Executive announced in 2005 funding for a coordinator in each SCG area, with additional funding rising overall from £1.7m in 2004–05 to £2.7m in 2007–08.[299] Secondly, as well as funding directly related to civil defence or civil

---

[292] Cabinet Office, *The Draft Civil Contingencies Bill*, para 3.35.

[293] Civil Contingencies Secretariat, *Full Regulatory Impact*, para 80.

[294] Joint Committee on the Draft Civil Contingencies Bill, para 223.

[295] Defence Committee, *Draft Civil Contingencies Bill* (2002–03 HC 557) para 58; Civil Contingencies Secretariat, *The Government's Response*, 22.

[296] Civil Contingencies Secretariat, *The Government's Response*, 23. The work of district councils is estimated by the LGA at £4m out of this total. See further Local Government Association, *Emergency Planning* (Research Briefing 4.03, London, 2003).

[297] HC Deb 12 December 2002 c 405W. See also Cabinet Office, *The Government's Response to the Report of the Joint Committee*, para 43.

[298] HM Treasury, *2004 Spending Review* (Cm 6237, London, 2004) Ch 6.

[299] Letter from Scottish Executive Justice Department, 30 June 2005, http://www.scotland. gov.uk/Resource/Doc/55971/0015631.pdf. A forum is being established for the coordinators.

contingencies, a wide array of other expenditure has been afforded to planning against terrorism and other sources of chemical, biological, radiological, and nuclear (CBRN) attack or contamination.

8.117 As well as the arguments about quantum, the ending of a specific grant mechanism created its own concerns. The fear is that the element in the Revenue Support Grant will be absorbed by more urgent or politically prominent local authority services.[300] The Joint Committee considered that temporary ring-fencing for at least two years would be a wise compromise.[301] The Government conceded some monitoring of the position, so that if the predictions of funding leakage come to pass, adjustments can be made:

> The overall amount provided by Government to local authorities for civil contingencies will be identifiable, as a Cabinet Office transfer to ODPM. The amount will be set out in the material provided to local government at the time of the announcement of the Government's proposals for funding of local authority expenditure.[302]

(b) Action costs

8.118 Moving to the financial implications of actions in response to emergencies, considerable expenditure could be incurred, but no ready estimates can be made since everything depends on the nature and scale of the emergency and the response and recovery strategies adopted.[303]

8.119 The most relevant source of extra funding for emergency action is the 'Bellwin Scheme'.[304] Local authorities (including police and fire authorities and Transport for London) can obtain discretionary financial assistance towards the revenue costs (manpower, services, and materials) and costs of short-term repairs incurred in dealing with an emergency where those costs were not normally insurable. It must be shown that the incident was exceptional by local standards and that it has inflicted an undue financial burden. There is a claims threshold of 0.2 per cent of the authority's annual budget and 15 per cent of marginal costs beyond this level, and any payments are at a rate of 85 per cent of spending above the threshold. Local authorities must accordingly bed some contingency reserve within their normal budgets.

8.120 In England, the scheme is administered by the Office of the Deputy Prime Minister (in Wales, by the Welsh Assembly Government and in Scotland by the

---

[300] Joint Committee on the Draft Civil Contingencies Bill, para 230; Civil Contingencies Secretariat, *The Government's Response*, 21, 22; Local Government and Public Services Committee, *Report on the Civil Contingencies Bill* (National Assembly for Wales, Cardiff, 2004) para 11.
[301] Ibid, para 233.   [302] Ibid, Appendix 9 q 43.
[303] Cabinet Office, *The Draft Civil Contingencies Bill: Explanatory Notes etc*, para 7.
[304] See *Guidance Notes and Thresholds*, http://www.local.odpm.gov.uk/finance/bellwin.htm; Cabinet Office, *Emergency Response and Recovery*, Annex 1B.

Scottish Executive) under s 155 of the Local Government and Housing Act 1989 (as amended):

(1) In any case where
    (a) an emergency or disaster occurs involving destruction of or danger to life or property, and
    (b) as a result, one or more local authorities incur expenditure on, or in connection with, the taking of immediate action (whether by the carrying out of works or otherwise) to safeguard life or property, or to prevent suffering or severe inconvenience, in their area or among its inhabitants, the Secretary of State may establish a scheme under this section for the giving of financial assistance to those authorities in respect of that expenditure.

. . .

(2) Financial assistance given pursuant to a scheme under this section shall take the form of grants paid by the Secretary of State with the consent of the Treasury and, subject to that, the terms and conditions of a scheme shall be such as the Secretary of State considers appropriate to the circumstances of the particular emergency or disaster concerned.
(3) Without prejudice to the generality of subsection (2) above, a scheme under this section may
    (a) make the payment of grants conditional upon the making of claims of a description specified in the scheme;
    (b) make provision with respect to the expenditure qualifying for grant and the rates and amounts of grants;
    (c) make provision in certain specified circumstances for the repayment of any grant, in whole or in part; and
    (d) make different provision for different local authorities or descriptions of authority and for different areas. . . . '

The system was applied to Northern Ireland in 1992.[305]

8.121 Between 1987–88 and 2000–01, 23 events or incidents warranted payments arising from the Bellwin Scheme to authorities in England and Wales.[306] By far the great majority were weather-related—floods and storms—and included the 'Great Storm' in southern England on 16–17 October 1987 which also produced by far the greatest payment (up to £20m). In Scotland, all bar one payments have been for flooding and storm damage.[307]

8.122 The Bellwin Scheme reimburses local authorities for immediate action (with payment made within 15 days) to safeguard life and property or to prevent suffering or severe inconvenience in their area following an emergency or disaster.[308] It should be emphasized that the Bellwin scheme is wider than the circumstances legitimately within Part II of the 2004 Act, and few if any payments in the last 15 years would have related to disasters within Part II.[309]

---

[305] See Local Government (Miscellaneous Provisions) (Northern Ireland) Order 1992 (SI No 810, NI6) Art 26.

[306] Joint Committee on the Draft Civil Contingencies Bill, Appendix 9 Annex A.

[307] Scottish Executive, *Review of the Bellwin Scheme* (http://www.scotland.gov.uk/Resource/Doc/69582/0017087.pdf, 2005) para 2.2.

[308] Joint Committee on the Draft Civil Contingencies Bill, Appendix 9 q 47, 48.

[309] Ibid, Appendix 9 q 13.

Even grants following the London and Manchester terrorist bombings in 1996 came from main spending programmes. The fact that there is disjunction between the scheme and the 2004 Act may be one of the reasons why it is left untouched.

8.123 More specific funding is available in some sectors. Government departments have various grants and funding schemes, such as the Department for Transport Supplementary Transport Grant (s 87 of the Local Government Finance Act 1988) and the Fuel Security Code 1990[310] (under s 6 of the Electricity Act 1989) allows for compensation for abnormal operational requirements in the event of an emergency.

8.124 There have been many criticisms of the Bellwin scheme over the years. It is seen as overly bureaucratic and dilatory in making payments; it is said to be miserly when account is taken of the thresholds; the thresholds do not take account of relevant factors, such as susceptibility to loss and damage because of topography or otherwise; it is too limited—it only covers uninsurable, revenue costs and, in particular, does not cover (since they are not 'immediate' action) prevention measures or capital costs.[311] The Bellwin Scheme was reviewed in England and Wales and largely confirmed in 2001, though the Joint DTLR/LGA Bellwin Review Group did recommend a statutory fund.[312] Furthermore, the local government White Paper, *Strong Local Leadership—Quality Public Services* promised that 'we will reform the [Bellwin] scheme along the lines recommended by the joint DTLR/local government review group, by setting its funding on a sounder footing'.[313] The Government made no moves to address the scheme during the legislative process leading to the Civil Contingencies Act, but reviewed the details again in 2004.[314] The Office of the Deputy Prime Minister report concluded that it remained 'fit for purpose' but considered the possibility of an 'advance funding stream', more unified guidance on the various sources of funding available to meet the costs of emergencies, and added guidance to local authorities on the incorporation of emergency planning in their risk management strategies.

---

[310] See http://www.dti.gov.uk/energy/domestic_markets/security_of_supply/fuel_code.pdf p 32. The Code is under review.

[311] Scottish Executive, *Review of the Bellwin Scheme* (http://www.scotland.gov.uk/Resource/Doc/69582/0017087.pdf, 2005) para 3.1. The absence of capital funding may mean that assets are restored to inadequate levels of resilience.

[312] Department for Transport, *Local Government and the Regions, Bellwin Scheme of emergency financial assistance to local authorities* (http://www.local.dtlr.gov.uk/finance/bellwin/bellrprt.doc, 2001). The group agreed that the current guidance should be rewritten to bring out more clearly the underlying principles of the scheme and to reflect the improved administrative arrangements (such as interim payments). [313] (Cm 5326, London, 2001) para 6.16.

[314] Office of the Deputy Prime Minister, *The Financial Management of Local Disasters* (http://www.odpm.gov.uk/embedded_object.asp?id=1137068, London, 2004). The scheme is currently under review in Scotland.

Finally, the Government has promised that if existing mechanisms for funding **8.125** prove insufficient, then regulations, presumably either under Part I or II, could be used to provide for alternatives.[315] This offer notwithstanding, it remains to be seen whether current funding will allow the achievement of a civil resilience culture.

H. International Assistance

The European Union Solidarity Fund[316] was set up in 2002 to aid in two cases. **8.126** One is major natural disaster, provided the estimated cost of the direct damage is over EUR 3 billion (2002 prices) or 0.6 per cent of the gross domestic product of the state. Secondly, exceptionally, payment can be made in the event of an extraordinary regional disaster that affects the majority of the population and has serious and lasting effects on its economic stability and living conditions. Special attention is given to outlying or isolated regions.

Aside from financial assistance, the amount of European Union activity is modest, **8.127** as civil protection is primarily a matter for Member States under the subsidiarity principle.[317] Initiatives at Community level are considered by the Management Committee for Civil Protection and the Permanent Network of National Correspondents. Relevant activities comprise a Community Action Project to stimulate research and projects,[318] major projects (involving exchanges of experts, common principles and guidelines on prevention and crisis management, medical cooperation in disasters, and a global system of public information), other specific projects (for example, water pollution, flood warnings, and ferry accidents), training exercises, and finally mutual assistance and joint working.[319] Mutual assistance is facilitated by the Monitoring and Information Centre (MIC),[320]

---

[315] Joint Committee on the Draft Civil Contingencies Bill, Appendix 9 q 39.

[316] See http://europa.eu.int/comm/regional_policy/funds/solidar/solid_en.htm. See Council Regulation (EC) 2012/2002, OJ L 311 11 November 2002.

[317] See Resolution of the Council and the Representatives of the Governments of the Member States, Meeting within the Council of 25 June 1987 on the introduction of Community Cooperation on Civil Protection (87/C 176/01); Vade-mecum of Civil Protection in the European Union (http://europa.eu.int/comm/environment/civil/pdfdocs/vademec.pdf, 1999); European Commission, EU focus on civil protection (http://europa.eu.int/comm/environment/civil/pdfdocs/focus_en.pdf, Luxembourg, 2002). Subsidiarity in this sphere is recognized further in Arts 58(1)b and 297 of the Treaty establishing the European Community.

[318] See Council Decision of 9 December 1999 establishing a Community action programme in the field of civil protection (1999/847/EC); 2005/12/EC: Council Decision of 20 December 2004 amending Decision 1999/847/EC as regards the extension of the Community action programme in the field of civil protection.

[319] This includes 1-1-2 as the single European emergency call number for the European Union: Council Decision 91/396/EEC of 29 July 1991; http://www.sos112.info/.

[320] See Council Decision of 23 October 2001 establishing a Community mechanism to facilitate reinforced cooperation in civil protection assistance interventions (2001/792/EC,

based in the Civil Protection Unit of the Directorate General for Environment of the European Commission in Brussels, which can receive requests for assistance and forward them to national contact points in participating States who decide their response. Further, if requested by the affected country, the MIC can mobilise small teams of experts from the Commission.[321] The MIC also acts as an information centre.

8.128 In the field of marine pollution, Decision 2850/2000 of the European Parliament and the Council sets up a Community framework for cooperation. It was augmented by Regulation (EC) No 1406/2002 of the European Parliament and of the Council of 27 June 2002 establishing the European Maritime Safety Agency which offers technical and scientific assistance to the European Commission and Member States.

8.129 The Council of Europe's EUR-Open Partial Agreement (EUR-OPA) of 1987 promotes cooperation between Member States on prevention, protection, and organization of relief in the event of major natural or technological disasters, including by the development of the European Alarm System for earthquakes and specialist research centres.[322]

8.130 NATO has operated a Civil Defence Committee since 1951 (it became the Civil Protection Committee (CPC) in 1995). It further created in 1998 the Euro-Atlantic Disaster Response Coordination Centre (EADRCC)[323] for information sharing, liaison, and coordinating disaster relief efforts amongst the Euro-Atlantic Partnership Council (EAPC) nations in cases of natural or technological disasters or terrorist attack. The CPC informs the Secretary General and the EAPC as well as the Senior Civil Emergency Planning Committee (SCEPC) about disasters and requests for international assistance and promotes participation in the Euro-Atlantic Disaster Response Unit (which is revived as necessary by the donation of national personnel and assets). The CPC is the focal point for pre-disaster planning and post-disaster analysis.

8.131 Within the United Nations, the Convention on the Transboundary Effects of Industrial Accidents 1992[324] derives from the Economic Commission for

Euratom); 2004/277/EC: Commission Decision of 29 December 2003 laying down rules for the implementation of Council Decision 2001/792/EC, Euratom establishing a Community mechanism to facilitate reinforced cooperation in civil protection assistance interventions.

[321] See the more ambitious Proposal for a Council Regulation establishing a Rapid Response and Preparedness Instrument for major emergencies COM(2005) 113 final, dated 6 April 2005.

[322] Committee of Ministers Resolution 87(2). See http://www.coe.int/T/DG4/MajorHazards/Default_en.asp.

[323] See http://www.nato.int/eadrcc/home.htm. See further EADRCC, *NATO's Role in Disaster Assistance* (http://www.nato.int/eadrcc/mcda-e.pdf, Brussels, 2001).

[324] See http://www.unece.org/env/teia/welcome.htm . See also Protocol on Civil Liability for Damage and Compensation for Damage Caused by Transboundary Effects of Industrial Accidents on Transboundary Waters, adopted in Kiev on 21 May 2003.

Europe (ECE) of the United Nations and addresses industrial accidents capable of causing transboundary impacts by fostering international cooperation. The UN/ECE Industrial Accident Notification System can be availed upon to provide early warning of an industrial accident.[325] More generally, the UN Office for the Coordination of Humanitarian Affairs (UN-OCHA) can give humanitarian aid in response to complex emergencies and natural disasters,[326] overseen by the Under-Secretary-General for Humanitarian Affairs and Emergency Relief Coordinator, who heads the Office for the Coordination of Humanitarian Affairs. The Inter-Agency Standing Committee deals with needs assessments, consolidated appeals, field coordination arrangements, and the development of humanitarian policies. Responses on the ground can be provided by the UN Disaster Assessment and Coordination Teams (disaster management professionals who are nominated and funded by member governments and other international bodies and can carry out rapid assessments)[327] and the International Search and Rescue Advisory Group (which directs its attention to information exchange and defined standards). The International Strategy for Disaster Reduction,[328] also headed by the UN Under-Secretary General for Humanitarian Affairs, promotes awareness of disaster reduction as an integral component of sustainable development.

I. Conclusions

Not surprisingly, since it is about putting law into practice, the evidence in this **8.132** chapter echoes the findings in previous chapters concerning the commendable effort of localities and Cabinet Office but also the shortcomings of the legislation, such as the gaps in it (at regional and central tiers). It has also highlighted other limitations and uncertainties—that the response phases are less advanced than the resilience stages and that there is a further need to build capability in terms of planning, testing, equipment, and appointments. There is also debate, albeit quelled after 2004, about levels of funding for implementation and a reluctance to confront standard-setting. More fundamentally, only recently has resilience become an important sector of government. The sector is at the start of a process of growth, but that growth must be accompanied, in order to encourage high standards, by shared professionalism, with thought being given to training standards, ethics, independent engagement with policy-makers, and

---

[325] ECE/CP.TEIA/2, annex II, decision 2000/1.
[326] UNGA Res.46/182, http://ochaonline.un.org.
[327] See http://ochaonline.un.org/webpage.asp?MenuID=2893&Page=552.
[328] See http://www.unisdr.org/.

even institutional formation and recognition, all the normal fare of professions. If risk has grown and is endemic, these shortcomings need to be addressed, at the least to give assurance and at best to seek to reduce risk or its consequences. One of the problems is that some of these objectives can most appropriately be tackled by central coordination where statutory impact has been at its most faltering.

9

COMPARISONS AND FINAL CONCLUSIONS

Having explored the legal and organizational reforms heralded by the Act, this **9.01** chapter will compare the UK legislation with structures and approaches developed in two comparable jurisdictions. The purpose is not to give a full survey of alternatives but to highlight possible developments and shortcomings in the domestic legislation. Some final conclusions will then be offered.

A. International Comparisons

Though the Cabinet Office dismissed the value of comparative research in **9.02** preparation of the Bill,[1] there is a profusion of pertinent overseas experience. The two jurisdictions below are selected as amongst the most comparable (in the case of New Zealand) and instructive (in the case of the United States).[2]

(a) New Zealand

A study of emergency powers by the New Zealand Law Commission was **9.03** requested following the repeal in 1987 of the Public Safety Conservation Act 1932 (the equivalent of the British Emergency Powers Act 1920).[3] It was not clear

---

[1] See Cabinet Office, *The Draft Civil Contingencies Bill*, para 1.8; Annex C, 41.
[2] Those omitted include Australia (see http://www.ema.gov.au/) and Canada (see Canadian Emergencies Act 1988, Emergency Preparedness Act 1988, and http://www.ocipep.gc.ca/home/index_e.asp).
[3] See Deane, ED, 'Public Safety Conservation Act' (1985) *New Zealand Law Journal* 266; Walker, C and Reid, K, 'Military Aid in Civil Emergencies: Lessons from New Zealand' (1998) 27 *Anglo-American Law Review* 133.

whether the statute provided the only basis for domestic military intervention, replacing the prerogative (and common law duties[4]) entirely, or whether the prerogative powers remained a viable alternative.[5] The Commission's *First Report on Emergencies* of 1990[6] and *Final Report on Emergencies* in 1991[7] centrally recommended that when emergency powers are required they should be conferred in 'sectoral legislation'—legislation deliberated upon and designed in advance of the emergency and tailored strictly to the needs of each particular kind of emergency.[8]

9.04 The *First Report on Emergencies* was largely implemented by the Defence Act 1990 which had already been drafted as the replacement for the Defence Act 1971.[9] A new, more tightly regulated Civil Defence Act was proposed by the Commission to replace an existing Act of 1983.[10] A proposed Biosecurity Act to deal with 'Public Welfare Emergencies' was an idea emanating from the Ministry of Agriculture and Fisheries and was duly assented to in 1993.[11] Likewise in considering pollution and hazardous substances, the Law Commission built on existing legislation.[12]

9.05 In greater detail, the Defence Act 1990 deals with the sources of the power for military intervention, albeit in limited detail. What one commentator dubs the 'mission statement' of the armed forces[13] is contained in s 5 of the Defence Act 1990. The parts relevant to civil emergencies are '(e) The provision of assistance to the civil power either in New Zealand or elsewhere in time of emergency; (f) The provision of any public service.' The procedures for calling upon military aid are then set out in s 9. One would have hoped that this codified set of powers and procedures would have been sufficient, but the *First Report* of the New Zealand Law Commission, here perhaps lacking the courage of its convictions, recommended the retention of underlying residual powers in the prerogative or common law since it feared that reliance solely on statutory powers could create the possibility of challenge in court.[14] Accordingly, by s 6(2) of the Defence Act

---

[4] For the New Zealand position under s 53 of the Police Act 1958, see Robertson, B, 'The Defence Act 1990 and Military Assistance to the Civil Power' (1991) 14 *New Zealand Universities Law Review* 254 at 268–9.

[5] Robertson, B, 'The Defence Act 1990 and Military Assistance to the Civil Power' (1991) 14 *New Zealand Universities Law Review* 254 at 256–8.

[6] New Zealand Law Commission, *Report No 12: First Report on Emergencies* (Wellington, New Zealand, 1990).

[7] Ibid; New Zealand Law Commission, *Report No 22: Final Report on Emergencies* (Wellington, New Zealand, 1991).   [8] *Final Report on Emergencies*, x, para 4.12.

[9] Ibid, 1–2.   [10] Ibid, para 9.80.   [11] Ibid, para 8.5.   [12] Ibid, para 9.80.

[13] Robertson, B, 'The Defence Act 1990 and Military Assistance to the Civil Power', (1991) 14 *New Zealand Universities Law Review* 254 at 255.

[14] *First Report on Emergencies*, para 93. Compare *Final Report on Emergencies*, para 4.55. But, by reference to s 9(3), the 1990 Act may be exclusive at least in the area of military aid in civil emergencies: Robertson, B, 'The Defence Act 1990 and Military Assistance to the Civil Power' (1991) 14 *New Zealand Universities Law Review* 254 at 259.

1990, it is stated that 'Nothing in . . . section 5 of this Act shall affect any power vested in the Governor-General apart from this Act.'

Next, following up the proposal to replace the Civil Defence Act of 1983, the **9.06** Civil Defence Emergency Management Act 2002 provides for regional, local, and national planning and management structures and policies across a broad range of potential emergencies without distinction. The Civil Defence Emergency Management Act 2002, s 8, provides for the appointment of a Director of Civil Defence Emergency Management[15] who advises the Ministry of Civil Defence and Emergency Management (which was set up in 1999), develops and monitors the National CDEM Plan, technical standards and guidelines, and monitors the performance of local Civil Defence Emergency Management Groups (CDEMGs).

The Minister of Civil Defence and Emergency Management must complete the **9.07** Civil Defence Emergency Management Strategy[16] (s 31) which must be presented to the House of Representatives. By s 39, the Government must also by Order in Council make and present to Parliament a national civil defence emergency management plan,[17] which deals with the hazards and risks to be managed at the national level, the civil defence emergency management necessary at the national level to manage those hazards and risks, and the objectives of the plan and its relationship to the national civil defence emergency management strategy. Finally, all government departments must prepare plans to continue functioning during and after an emergency (s 58).

Akin to the local resilience forums in the United Kingdom, every regional **9.08** council and every territorial authority within that region must unite to establish a CDEMG (s 12). They are required under s 17(1) to assess relevant hazards, engage in planning, make preparations for resilience, carry out response and recovery activities, and raise public awareness. A more specific duty, under s 48, is to prepare and approve a civil defence emergency management group plan; this plan is submitted to the Minister for comment and public notice must also be given. Next, the CDEMG must appoint at least one person as a person authorized to declare a state of local emergency for its area (s 25). They must also appoint a suitably qualified and experienced person to be the Group Controller for its area, and that person will take charge in a declared emergency (ss 26 and 28).[18] Section 29 provides for the appointment of a 'Recovery

---

[15] For details, see http://www.mcdem.govt.nz/memwebsite.nsf.

[16] See *National Civil Defence Emergency Management Strategy 2003–2006: Resilient New Zealand* (2002): Goal 1: To increase awareness, understanding, and participation in CDEM; Goal 2: To reduce the risks from hazards; Goal 3: To enhance New Zealand's capability to manage emergencies; Goal 4: To enhance New Zealand's capability to recover from disasters.

[17] See *National Civil Defence Plan* (2002).

[18] There may also be more limited 'Local Controllers' under s 27.

Co-ordinator'; they are appointed for 28 days at a time for undertaking recovery activities.

9.09 Local authorities must also plan and provide for civil defence emergency management within its district and ensure their own resilience (s 64). Likewise, by s 63, emergency services must participate in the development of strategies and plans and must provide members for CDEMGs.

9.10 Next, by s 60, and reminiscent of the concept of Category 2 responders in the United Kingdom, every 'lifeline utility' must ensure its own resilience, must make available to the Director in writing, on request, its contingency plan, must participate in the development of the national civil defence emergency management strategy and civil defence emergency management plans, and must provide, free of charge, any technical advice to any CDEMG or the Director.

9.11 Moving from resilience planning to live responses (the equivalent to Part II of the 2004 Act), the Minister may declare that a state of national emergency exists over the whole of New Zealand or any areas or districts if at any time it appears to the Minister that (a) an emergency has occurred or may occur; and (b) the emergency is, or is likely to be, of such extent, magnitude, or severity that the civil defence emergency management necessary or desirable in respect of it is, or is likely to be, beyond the resources of the CDEMGs whose areas may be affected by the emergency (s 66). The House of Representatives shall be informed as soon as practicable and Parliament must then meet. The person appointed under s 25 (or the Minister) may declare a state of local emergency in the whole of the CDEMG area or in one or more districts or wards within the area (ss 68 and 69). States of emergency last for just seven days (s 70) but may be renewed. The CDEMGs are invested with wide executive powers during the currency of a state of emergency, with explicit powers, inter alia, to evacuate areas (s 86), to enter premises, to close roads and to requisition property[19] (ss 85 to 94).

9.12 With 121 sections, the Act is far more comprehensive and transparent than the Civil Contingencies Act. It will be noted that it addresses the role of central government as much as local agencies. There is much to commend this model.

(b) United States

9.13 The striking change in contingency planning of recent origin in the United States has been institutional rather than instrumental,[20] namely, the establishment of a

---

[19] There is compensation in all cases: s 107.

[20] The National Emergencies Act 1976 (50 USC s 1601) allows the President to declare an emergency (s 1621) and to specify powers under it by Executive Order (s 1631). The Act's controls are shockingly lax: Lobel, J, 'Emergency Power and the Decline of Liberalism' (1989) 98 *Yale Law Journal* 1385; Cole, D, 'The Priority of Morality: The Emergency Constitution's Blind Spot' (2004) 113 *Yale Law Journal* 1753; Block, F, 'Civil Liberties during National Emergencies'

Department of Homeland Security (DHS),[21] at first as an administrative unit within the White House and then as a fully fledged Federal office under the Homeland Security Act of 2002.[22] Its aim is to develop and coordinate the implementation of the US National Strategy For Homeland Security,[23] which includes an emphasis on intelligence, contingency planning and preparedness. The Act divides the DHS into four units: information and infrastructure protection; border and transportation security; science and technology; and emergency response (s 1). The tasks arising involve the building of assessments, surveillance and protective guarding. The DHS has subsumed within it 22 agencies and around 180,000 officials, including the Animal and Plant Health Inspection Service, and the Federal Emergency Management Agency (FEMA).

Should there be a DHS in the United Kingdom? On the face of it, the diffusion of **9.14** potential responsibility via the LGD system seems to contrast with the consolidated DHS model. But the contrast is not straightforward. The Cabinet Office can be said to be a meta-LGD in emergencies in that it is the keeper of the systems which others must follow and can also set in motion forms of communication and coordination between departments. It can also oversee the developing response and seek to arrange further consideration if an LGD is being overwhelmed. Conversely, even in the United States, there are plenty of other Federal departments and agencies with roles in emergencies alongside the DHS, including the Department of Justice, as well as state and local players. Indeed, the Federal/state divide actually produces a more fragmented system than in the United Kingdom. The Homeland Security Act establishes an Office for State and Local Government Coordination within the DHS to help negotiate these boundaries.

The DHS is controversial in its own polity and so may not be a shining example **9.15** to others. Despite the Office just mentioned, it is not clear that it has effectively engaged with local responders, and the case of New Orleans (below) suggests deep problems. Next, it is simply too large and unwieldy, taking in disparate functions which do not really gel. This forced marriage may cause damage to the smaller agencies—the case of FEMA, which sought to distance itself from security matters in the 1990s and has a relatively small share of the DHS budget,[24] may be a case in point, though the budget of FEMA remains

---

(2005) 29 *New York University School of Law Review of Law and Social Change* 459. Even wider powers have been claimed, as Commander in Chief, to intercept communications, detain suspects, and set up military trials. See also International Emergency Economic Powers Act of 1977, 50 USC 1701.

[21] See Cmar, T, 'Office of Homeland Security' (2003) 39 *Harvard Journal on Legislation* 455; Thessin, J, 'Department of Homeland Security' (2003) 40 *Harvard Journal on Legislation* 513.

[22] Pub L No 107–296, 116 Stat 2135

[23] See http://www.dhs.gov/interweb/assetlibrary/nat_strat_hls.pdf, 2002.

[24] The budget for FEMA is around $5bn (13%) out of a budget of $38.5bn: DHS, *Budget in Brief Fiscal Year 2006* (http://www.dhs.gov/interweb/assetlibrary/Budget_BIB-FY2006.pdf, 2005) 15.

impressive in absolute terms. Next, the holders of intelligence, the FBI and the CIA might not trust such an outsized entity with the sensitive intelligence needed to found anticipatory action.

9.16 As for lessons for the United Kingdom, the House of Commons Select Committee on Defence[25] felt that a more modest body along the lines of a National Counter-Terrorism Service merited further attention. The House of Commons Select Committee on Science and Technology favoured an even more limited (but grander sounding), research-based Centre for Home Defence.[26] However, it would seem that the administrative changes made since 9/11 in the United Kingdom are viewed as conducing against the considerable disadvantages that could flow from setting up a UK DHS—such as the seismic disruption to existing structural formations, possible loss of attention to areas such as contingency planning for chemical spillages compared to the headline-grabbing terrorism issues, and the dislocation of existing expert formations. Consequently, the idea is not supported by the Government[27] which has preferred to shuffle key backroom administrative formations, rather than make bold political appointments. Its handiwork is evidenced by the establishment of the Civil Contingencies Secretariat and the Permanent Secretary-level Security and Intelligence Co-ordinator and in the Cabinet Office. It has also preferred sectoral and focused agency formation, such as the Health Protection Agency and the Joint Terrorism Analysis Centre (JTAC). As already noted, these arrangements court two problems.

9.17 First, they may leave central government uncoordinated. There is no over-arching agency or ministry to take charge when the crunch comes, especially where crises cut across ministerial boundaries or slowly grow beneath the radar of LGD attention. Secondly, the contingency and emergency field is faceless. There is no visible Minister in the Cabinet Office, and the closest to a Minister for Homeland Security has been a succession of junior Home Office Ministers, such as Beverley Hughes and Hazel Blears, who have handled a wide range of issues besides contingency planning and emergency response. The Conservative Party has shown sympathy for the idea of a Minister of Homeland Security,[28] as did the House of Commons Defence Committee.[29] But the Government has opposed these proposals, arguing that the appointment of a Security and

---

[25] 2001–02 HC 518 para 81.

[26] See House of Commons Science and Technology Committee, *The Scientific Response to Terrorism* (2003–04 HC 415) para 48; Government reply (Cm 6108, London, 2004) para 4. In the same session, the Home Affairs Committee gathered evidence on 'Homeland Security' (2003–04 HC 417) but did not report.

[27] See Defence Committee, *Defence and Security in the UK* (2001–02 HC 518) para 179; Government Reply to the House of Commons Science and Technology Committee, The Scientific Response to Terrorism (Cm 6108, London, 2004) para 44.

[28] Patrick Mercer was appointed in June 2003 as (Conservative Party) Shadow Minister for Homeland Security.   [29] (2001–02 HC 518) para 182.

Intelligence Co-ordinator and Permanent Secretary in the Cabinet Office is vital[30] and that the Home Secretary (not the Cabinet Office minister) has all-embracing responsibility.[31]

Nevertheless, the Home Office could find itself being reformulated because of a **9.18** coincidence of factors. Once the Department for Constitutional Affairs can be more mainstreamed as a consequence of the changes in the Constitutional Reform Act 2005, it might gather force as a Ministry of Justice, taking further functions from the Home Office (community and race, immigration and nationality). Furthermore, it has been suggested that the security responsibilities of the Home Office Minister should go to a Cabinet Office Minister.[32] The Home Office would remain as the Ministry for Community Security and Safety. But these changes will always encounter the difficulty that the Home Secretary is always a very powerful operator in government.

If the behemoth of the DHS is considered to be indigestible within a UK **9.19** administrative setting, what about one of its most relevant components, the Federal Emergency Management Agency?[33] FEMA has the task of planning for, recovering from, and mitigating against disasters. FEMA predates the Department of Homeland Security. It was first established by Presidential Executive Order in 1979, when it absorbed several other relevant functions and agencies. Its powers derive from the Robert T Stafford Disaster Relief and Emergency Assistance Act ('the Stafford Act') of 1974.[34] As in the United Kingdom, civil defence tended to be downgraded after the end of the Cold War, but defence against terrorism attack became a predominant concern after 9/11 with a massive influx of funding. It now consists of around 2,500 full-time employees and more than 5,000 disaster reservists.

In the planning phase, FEMA acts as the lead federal agency for the Federal **9.20** Executive Branch in seeking to ensure there is a Continuity of Operations (COOP) capability across Federal departments and agencies.[35] FEMA's Office of National Security Coordination drives and manages the COOP program.[36]

---

[30] HC 1203, para 35–6.   [31] Ibid, para 34.

[32] *The Guardian* 24 October 2005 at 4.

[33] See http://www.fema.gov/; Tucker, DG and Bragg, AO, 'Florida's Law of Storms' (2001) 30 *Stenson Law Review* 837; Anderson, CV (ed), *The Federal Emergency Management Agency (Fema)* (Nova Science Publishers, New York, 2003).

[34] 42 USC s 5121, as amended by the Disaster Mitigation Act of 2000, Pub L No 106–390, 114 Stat 1552 (2000). The details of organization and work are set out in 44 CFR See further Bentz, JA, 'The National Response Plan' and 'Government and Voluntary Agencies' in Ghosh, TK, Prelas, MA, Viswanath, DS, Loyalka, SK, (eds), *Science and Technology of Terrorism and Counterterrorism* (Marcel Dekker Inc, New York, 2002).

[35] Presidential Decision Directive, Enduring Constitutional Government and Continuity of Government Operations (PDD-NSC-67, 1998); Federal Preparedness Circular (FPC-65).

[36] See http://www.fema.gov/onsc/.

9.21 In times of crisis, assistance under the Stafford Act is triggered either by 'major disaster' (especially environmental disasters) or 'emergency' (such as terrorism).[37] As for 'major disaster', this arises when the State Governor requests that the President declare a 'major disaster' in that State.[38] The request from the Governor must confirm that the emergency is of such a 'severity and magnitude' that it exceeds the capabilities of the State to meet the emergency with its own resources.[39] The Federal intervention allows the President to commandeer the personnel and resources of any Federal agency to support the State and locality as well as offering specified categories of relief, such as food and supplies, works and services such as clearance, rescue, temporary facilities, and technical advice, and military assistance (for up to ten days).[40] The President is also conferred with powers to coordinate all disaster relief assistance to the State.[41] If assistance is given, then the President shall designate a Federal Coordinating Officer (FCO) from FEMA and shall also require the Governor to designate a State Coordinating Officer (SCO).[42] In this way, there are the equivalent of two Regional Coordinators and not one as under the UK model.

9.22 As well as the immediate disaster response, the amending Disaster Mitigation Act of 2000 allows for Federal intervention to assist with longer term consequences and recovery in the context of 'major disaster'. The President may make contributions to a State or local government for the repair, restoration, reconstruction, or replacement of public facilities.[43] The Federal Government may also render financial assistance and in-kind services to individuals and households (in some cases in consultation with the Governor). Relief on offer might include temporary housing, repairs to dwellings or replacement of dwellings, unemployment assistance, medical and funeral expenses, food distribution and transportation costs.[44] The Stafford Act also allows Federal grants to states and to owners of property at risk to help finance protection measures,

---

[37] 42 USC s 5122:' "Emergency" means any occasion or instance for which, in the determination of the President, Federal assistance is needed to supplement State and local efforts and capabilities to save lives and to protect property and public health and safety, or to lessen or avert the threat of a catastrophe in any part of the United States. "Major disaster" means any natural catastrophe (including any hurricane, tornado, storm, high water, winddriven water, tidal wave, tsunami, earthquake, volcanic eruption, landslide, mudslide, snowstorm, or drought), or, regardless of cause, any fire, flood, or explosion, in any part of the United States, which in the determination of the President causes damage of sufficient severity and magnitude to warrant major disaster assistance under this chapter to supplement the efforts and available resources of States, local governments, and disaster relief organizations in alleviating the damage, loss, hardship, or suffering caused thereby.'

[38] 42 USC s 5170. The process of request for major disaster relief is set out in 44 CFR Pt 35. The process of request for military assistance is set out at 44 CFR Pt 206.34.

[39] See 44 CFR s 206.48 (1999).

[40] 42 USC s 5170a and b. Fire management assistance is dealt with under 44 CFR Pt 204; other disasters fall within Pt 206.   [41] 42 USC s 5170a(2).

[42] 42 USC s 5143. Their duties are set out in 44 CFR Pt 206.42.

[43] 42 USC s 5172.   [44] 42 USC ss 5174, 5177–86.

such as better flood defence works.[45] There is also Federal funding for local mitigation planning.[46] The Disaster Mitigation Act also allows more sweeping Federal intervention by the provision of technical and financial assistance to States and local governments to implement hazard mitigation measures.[47]

An 'emergency' is likewise declared when the situation is of such severity and **9.23** magnitude that effective response is beyond the capabilities of the State and the affected local governments and that Federal assistance is necessary.[48] The emergency Federal assistance is very similar to that in major disasters, though without reference to military assistance.[49] There is also provision for emergency preparedness—such as the procurement and stockpiling of necessary materials and supplies, the provision of suitable warning systems, the construction or preparation of shelters, shelter areas, and control centres.[50] But the longer term recovery support mechanisms are not available in this context.

The DHS has developed the (Federal) *National Response Plan* (NRP)[51] as a **9.24** comprehensive approach to managing domestic hazards and incidents. It details best practices and procedures and explains how Federal departments and agencies will work together and how the Federal Government will intervene in state or local crises. The philosophy is that incident response is to be handled at the lowest possible organizational and jurisdictional level.[52]

The NRP incorporates 15 'Emergency Support Functions' (ESF), each with an **9.25** ESF coordinator and comprising draft regulations, guidelines, and protocols. These relate to the support, resources, and services provided by the Federal Government in an emergency and the relevant organizational structures. Each Federal agency can be the supplier of resources and capabilities, which will be channelled through the Federal Coordinating Officer (FCO) who is appointed by the President on the recommendation of the Secretary of Homeland Security. The ESFs, which can deployed in multiples together, as the need arises, comprise: ESF 1: Transportation (Lead agency: Department of Transportation); ESF 2: Communications (DHS);[53] ESF 3: Public Works and

---

[45] 42 USC s 5170c. See further the Hazard Mitigation Grant Program (44 CFR Pt 206.430.)

[46] 42 USC s 5165. see further 44 CFR Pts 201, 300.

[47] 42 USC s 5133. There was a sunset date of the end of 2003 but it has been renewed annually since that date.   [48] 42 USC s 5191. see further 44 CFR Pt 206.35.

[49] 42 USC s 5192.

[50] 42 USC s 5195. There is an Emergency Food and Shelter National Board Program which was created in 1983: http://www.efsp.unitedway.org/.

[51] See http://www.dhs.gov/dhspublic/interapp/editorial/editorial_0566.xml, 2004. Compare the Occupational Safety and Health Administration, Hazardous Waste and Emergency Response Standard (29 CFR 1910.120). See Lippy, B, 'Protecting the Health and Safety of Resilience and Recovery Workers' in Levy, BS and Sidel, VW, *Terrorism and Public Health* (Oxford University Press, New York, 2003).   [52] Ibid, 15.

[53] Note also the *National Strategy to Secure Cyberspace* (http://www.whitehouse.gov/pcipb/cyberspace_strategy.pdf, 2003).

Engineering (US Army Corps of Engineers and Department of Defense); ESF 4: Fire Fighting (US Forest Service and Department of Agriculture); ESF 5: Emergency Management (DHS and FEMA); ESF 6: Mass Care, Housing and Human Services (DHS, FEMA, and the American Red Cross); ESF 7: Resource Support (General Services Administration); ESF 8: Public Health and Medical Services (Department of Health and Human Services); ESF 9: Urban Search and Rescue (DHS and FEMA); ESF 10: Oil and Hazardous Materials (Environmental Protection Agency); ESF 11: Agriculture and Natural Resources (Department of Agriculture); ESF 12: Energy (Department of Energy); ESF 13: Public Safety and Security (DHS and Department of Justice); ESF 14: Long-term Community Recovery and Mitigation (DHS and FEMA); ESF 15: External Affairs (DHS and FEMA). There are also Incident Annexes which describe the policies, plans, responsibilities, and coordination processes for situations requiring specialized application of the NRP. These cover: Biological Incident, Catastrophic Incident, Cyber Incident, Food and Agriculture Incident, Nuclear/Radiological Incident, Oil and Hazardous Materials Incident Annex, and Terrorism Incident.

9.26   As for the processes of implementation, the Homeland Security Operations Center (HSOC) provides the national level multi-agency hub for domestic situational monitoring and operational coordination. The HSOC also includes DHS components, such as the National Infrastructure Coordinating Center, which coordinates communications during an incident. The Interagency Incident Management Group consists of senior Federal officials, who provide strategic advice to the Secretary of Homeland Security during an actual or potential 'Incident of National Significance'. The National Response Coordination Center, another component of the HSOC, is a multi-agency center that provides overall Federal response coordination. The FEMA Operations Center provides situation monitoring, communications, and notification services. There are also Regional Response Coordination Centers (RRCC) which coordinate regional response efforts and implement local federal program support until a Joint Field Office (JFO) is established. The RRCC will be staffed by FEMA and representatives from the relevant agencies. Once a response is under way, an Emergency Response Team (ERT) will be established in the affected area. The ERT is the interagency group that provides administrative, logistical, and operational support to regional response activities. It includes staff from FEMA and other agencies who support the Federal Coordinating Officer (FCO) in carrying out inter-agency activities. The ERT will be regionally based, but a National Emergency Response Team (ERT-N) can be deployed for catastrophic disasters or WMD incidents. It will be preceded by an Emergency Response Team, Advance Element (ERT-A), which will form the nucleus of the full ERT and can be deployed on the advice of the FEMA

Regional Director until the FCO is in place. The Federal Incident Response Support Team (FIRST) is the forward component of the ERT-A and will quickly assess the situation and oversee on-scene initial Federal assistance. The Emergency Support Team (EST) is the inter-agency group that coordinates national-level Federal disaster response from the FEMA National Interagency Emergency Operations Center (NIEOC) in Washington. The EST assists the Catastrophic Disaster Response Group (all relevant agencies and departments) and provides resource coordination support to the FCO and regional response operations. The FCO, working out of the JFO, manages and coordinates Federal resource support activities related to Stafford Act disasters and emergencies (in non-Stafford Act situations, a Federal Resource Coordinator (FRC) coordinates support through interagency agreements). The FCO must work with any State emergency structures. A Principal Federal Official (PFO) may be designated by the Secretary of Homeland Security during a potential or actual Incident of National Significance. The PFO does not direct or replace the incident command structure established at the incident, nor does the PFO have directive authority, but is to assist with coordination and communication and to feed information back to the HSOC. In multiple State disasters, the RRCCs remain operational to assist the EST with the prioritization and deployment of resources, until FCOs and JFOs are established within each affected State. A JFO is a temporary facility established locally to provide a central point for the ERT and FCO to coordinate resources in support of state, local, and tribal authorities. Overall Federal support to the incident command structure on-scene is coordinated through the JFO, which is organized into four sections: Operations Section, Planning Section, Logistics Section, and Finance/Administration Section. There are numerous specialist Federal support teams which can be called upon, such as the Nuclear Incident Response Team, National Medical Response Teams, Scientific and Technical Advisory and Response Teams, and Urban Search and Rescue task forces.

One cannot but be impressed by the effort which has gone into the homeland **9.27** security sector since 9/11.[54] On paper it appears very through and comprehensive. Yet, the Defence Committee expressed itself as opposed to a UK version of FEMA.[55] The Joint Committee on the Draft Bill accepted this rejection but reiterated its call for a more modest Civil Contingencies Agency, as discussed in Chapter 8.[56]

---

[54] The foregoing by no means gives a comprehensive picture. For example, the Intelligence Reform and Terrorism Prevention Act 2004 (Pub L 108–458, 118 Stat 3638) encourages resilience in the private communications sector, interoperability of communications for public safety providers, and research into an emergency alert system. [55] Paragraph 181.
[56] Joint Committee on the Draft Civil Contingencies Bill, para 256.

9.28 It may seem perverse to consider the worth of FEMA as a model for the United Kingdom in the light of its apparently abject response to the disaster of Hurricane Katrina in the US Gulf States on 29 August 2005. The post-mortem on that terrible catastrophe is awaited. The event tragically highlighted weaknesses in the US system for civil contingencies.

9.29 These include an undue concentration on terrorism emergency, with much spending on the Transport for Security Administration.[57] By contrast, spending on flood relief, such as the strengthening of levées in New Orleans, does not seem to have been accorded the same priority.[58]

9.30 Secondly, there are allegations of undue political influence in key appointments, such as that of the then head of FEMA, Michael D Brown, who resigned under heavy criticism about his performance.[59] The same trait affects FEMA's programmes which emphasize politically influenced grants rather than risk avoidance and mitigation.[60]

9.31 Thirdly, and perhaps most significant of all, the US system does have an identifiable agency at the centre, of the kind that, as has been argued in this book is signally missing in the United Kingdom. Furthermore, the US Constitution consciously fragments power between Federal, State, and local tiers of government and thereby hobbles the delivery of governmental services. Federal intervention is limited since it has no inherent police power (subject to wartime powers)[61] and is circumscribed in the use of the military as active participants in civilian law enforcement[62] by the Posse Comitatus Act 1878.[63] Therefore,

---

[57] See Aviation and Transportation Security Act 2001 (PL 107–071).

[58] See (New Orleans) *Times-Picayune* 28 May 2005 at 1.

[59] *The Times* 13 September 2005 at 32.

[60] See Platt, RH (ed), *Disaster and Democracy* (Island Press, Washington DC, 1999); Mileti, DS (ed), *Disasters by Design* (Joseph Henry Press, Washington DC, 1999).

[61] *Hamilton v Kentucky Distilleries & Warehouse Co* 251 US 264 (1919) at 156.

[62] But passive support (planning and logistics) has been allowed: *United States v Red Feather*, 392 F Supp 916 (DC SD 1975) ; *State v Nelson*, 298 NC 573, 260 SE 2d 629, *cert den*; 446 US 929, 100 S Ct 1867, 64 L Ed 2d 282 (1980). There are exceptions for anti-drugs and border control operations (10 USC ss 371–381) and for civil disturbance (10 USC ss 331–334—the Insurrection Act) where the state has requested assistance or is unable to protect civil rights and property.

[63] See 10 USC s 375 ('The Secretary of Defense shall prescribe such regulations as may be necessary to ensure that any activity (including the provision of any equipment or facility or the assignment or detail of any personnel) under this chapter does not include or permit direct participation by a member of the Army, Navy, Air Force, or Marine Corps in a search, seizure, arrest, or other similar activity unless participation in such activity by such member is otherwise authorized by law') and 18 USC s 1385 ('Whoever, except in cases and under circumstances expressly authorized by the Constitution or Act of Congress, willfully uses any part of the Army or the Air Force as a posse comitatus or otherwise to execute the laws shall be fined under this title or imprisoned not more than two years, or both'). See Rice, PJ, 'New Laws and Insights Encircle the Posse Comitatus Act' (1984) 104 *Military Law Review* 109; Abel, CA, 'Not Fit for Sea Duty: the Posse Comitatus Act, the United States Navy, and Federal Law Enforcement at Sea' (1990) 31 *William & Mary Law Review* 445; Hammond, MC, 'The Posse Comitatus Act' (1997) 75 *Washington University Law Quarterly* 953; Banks, WC, 'Troops Defending the Homeland' (2002) 14 *Terrorism & Political Violence* 1.

Federal intervention is usually dependent upon request from the State and locality which are expected to be the primary responders and to be the mechanism for triggering a central intervention. Taking New Orleans as an example, the Governor (in Louisiana, Kathleen Blanco) is empowered under the Louisiana Homeland Security and Emergency Assistance and Disaster Act[64] to declare a disaster or emergency for up to 30 days at a time, subject to being overruled by the State legislature. The declaration activates the State's emergency response and recovery programme under the command of the Director of the Louisiana State Office of Homeland Security and Emergency Preparedness.[65] A wide variety of powers are conferred upon the Governor, including the power to 'direct and compel'[66] the evacuation of all or part of the population from any stricken or threatened area within the state if deemed necessary for the preservation of life or other disaster mitigation, response, or recovery. The declaration also allows the use of the Louisiana State Police and the Louisiana National Guard[67] for purposes including law enforcement. A state of emergency was declared on 29 August 2005, on public health grounds. The Governor can direct the local authority (in this case, New Orleans Mayor Ray Nagin) to use the National Guard and New Orleans police to use force in order to overcome the public health emergency, such as by forcibly removing residents. The 'parish president' (in this case, the Mayor) is also allowed to declare a local emergency[68] and is given powers to take action necessary for mitigation, response, or recovery measures, including the evacuation and the cordoning of areas. An evacuation was ordered by the Mayor on 6 September 2005.

9.32 This muddle is an inevitable feature of a Federal system with an overweening distrust of government, but it could be more easily overcome within the United Kingdom if the Civil Contingencies Act had been allowed to address governmental hierarchies. As it stands, the United Kingdom has placed trust in local providers with no sure-footed pathway to regional or central involvement.

B. Final Conclusions

9.33 The avowed aim of the Act is to improve resilience 'by laying the foundations for enhanced resilience to incidents at the local level while enhancing regional capacity and ensuring central government has the powers necessary to deal with

---

[64] La RS 29:721 (1993, as amended in 2003).
[65] La RS 29:724. See http://www.ohsep.louisiana.gov/.   [66] La RS 29:724(D)(5).
[67] The National Guard is roughly the equivalent of the UK Territorial Army. The National Guard is not covered by the Posse Comitatus Act when acting in this state status; see 10 USC s 311; 32 USC s 101. Note that there are in addition 25 State Defence Forces (but not including Louisiana) (see, for example, the New York Guard at http://www.dmna.state.ny.us/nyg/nyg.html).   [68] La RS 29:727.

the most extreme of disruptive challenges.'[69] One might also add the need to embed resilience as a normal part of planning and operations rather than a stand-alone issue,[70] an ambitious aim which seeks an eventual change of official culture. That aim is far off, and it will not be assisted by the limited attempts in the legislation to reach out to third parties by way of governance techniques. Nevertheless, from a tangled morass of previous laws and practices, the Civil Contingencies Act 2004 is equipped to perform an excellent role in convincing several sectors of public life to take risk seriously. It is clear that previous legislation was inadequate, and this deficiency was all the more stark given the apparently increasing traumas which face society—from tornados to terrorism. Given the starting points, the legislation has advanced far, but like all major legislation, it represents a compromise. The compromise is not so much in this case the highly publicized clashes with civil liberties advocates relating to Part II. Though it is true that important concessions were made, the Act remains a massive threat to the niceties of civil society. Rather, the main compromises were with the interests and proclivities of bureaucrats and the private sector. Thus, Part I was designed as a wispy skeleton to avoid treading on too many toes all at once. Those with existing responsibilities or interests for contingency planning—the shire and regional grandees, whether in local authorities, police, or government regional offices—are to remain in place. Central government operators—the Cabinet Office, the Civil Contingencies Secretariat, the military, and political ministerial masters—can maintain their ethereal detachment. Elements of private enterprise, which now controls much more of 'essential services' than in 1920, are invited the table, but, out of respect for market freedom and profitability, their duties are given a light touch, even for the Category 2 responders for whom the Act merely repeats regulatory requirements already in place in their sector. The public is ignored but will be told in a crisis that its role is to obey.

9.34 In view of this hesitant entry of the legislated response, the prediction is that, in time, the legal demands will grow. The practices of Local and Regional Resilience Forums will cohere and settle to a point where legislative form can be enhanced. Those sectors largely let off the hook will get used to the idea of the statutory duties, while it is inevitable that one or other crisis will emerge which, if not solved as if by magic overnight, will conduce to the argument that events could be handled better if only there were more legal powers and duties. As a result, the legislation will eventually expand, and two sectors in particular can expect to feel the impact. One is the private sector. With the impact of neo-liberal economics, the need for major food supply, information and communications technology, transport and other major corporations to be resilient is as

---

[69] Cabinet Office, *The Draft Civil Contingencies Bill: Explanatory Notes etc*, para 1.
[70] Cabinet Office, *The Draft Civil Contingencies Bill*, Ministerial Foreword by Douglas Alexander, at 3.

important as the need for the public sector to be resilient. It is also the chilling truth that the new terrorist threats are as likely to be directed towards the private as the public sector. Given the growth of financial audit requirements in the same direction, it will eventually represent no extra regulatory burden to require the formal devising and lodging of plans. The other sector to come under the legislative spotlight will be central government which should not remain the ghost in the machine. The need for oversight and accountability is evident and must be made concrete by the establishment of an identifiable agency.

The legislative agenda must also expand by turning its attention to response and recovery. The response Strategic Coordinating Groups should be more firmly anchored to LRFs and the statutory apparatus, and statutory duties and powers should apply to these aspects of emergency just as they apply to planning and preparation. **9.35**

Finally, more should be directly offered to the public. They are the silent, indirect beneficiaries of much of the planning work in Part I, but there is little direct engagement. Arising out of the disaster caused by Hurricane Katrina, at least one commentator derived the lesson that the only safe form of contingency planning is self-reliance in a way which does not depend one whit or jot upon the state: **9.36**

> Every American kid should be required to watch video of the poor in New Orleans and how they suffered because they couldn't get out of town.... For centuries charlatans have been telling Americans that government will provide, and you deserve to be provided for. Bull! Depend on yourself—get educated, get smart, and get personal resources. That is the lesson of Katrina.[71]

This bleak view is not shared by the authors of the Civil Contingencies Act or of this book. Some risks are so widespread or so serious as to be beyond the capabilities of businesses, insurers, or individuals and yet should be responded to in the interests of social solidarity which has a care for restoration of society no matter what the cost. But the statute could have been more explicit and more outward reaching, though one must guard against the opposite dangers, that civil contingency effort will divert from other pressing social policies, that the expense will be ruinous and that the inconvenience and intrusion of security will serve as constant reminders of our insecurity.[72] As discussed in Chapter 1, the threat of flooding through climate change, pandemic disease, and suicidal terrorism are contemporary preoccupations which offer a paradigmatic insight into the endemic nature of risk and growth of governance in late modern society. The Civil Contingency Act 2004 is testament to those trends. The matter will not rest there, nor should it rest if the state is to fulfil its duty to protect its citizens.

---

[71] Bill O'Reilly, *The O'Reilly Factor*, Fox TV Network, http://www.billoreilly.com/show?action = viewTVShow&showID = 443/1#, 7 September 2005.

[72] See Crawford, A, (ed), *Crime, Insecurity, Safety and the New Governance* (Willan, Cullompton, 2002).

APPENDIX

The Civil Contingencies
Act 2004

Chapter 36
Long Title: An Act to make provision about civil contingencies.

PART I
LOCAL ARRANGEMENTS FOR CIVIL PROTECTION

1. Meaning of 'emergency'

(1) In this Part 'emergency' means
    (a) an event or situation which threatens serious damage to human welfare in a place in the United Kingdom,
    (b) an event or situation which threatens serious damage to the environment of a place in the United Kingdom, or
    (c) war, or terrorism, which threatens serious damage to the security of the United Kingdom.
(2) For the purposes of subsection (1)(a) an event or situation threatens damage to human welfare only if it involves, causes or may cause
    (a) loss of human life,
    (b) human illness or injury,
    (c) homelessness,
    (d) damage to property,
    (e) disruption of a supply of money, food, water, energy or fuel,
    (f) disruption of a system of communication,
    (g) disruption of facilities for transport, or
    (h) disruption of services relating to health.
(3) For the purposes of subsection (1)(b) an event or situation threatens damage to the environment only if it involves, causes or may cause
    (a) contamination of land, water or air with biological, chemical or radio-active matter, or
    (b) disruption or destruction of plant life or animal life.
(4) A Minister of the Crown, or, in relation to Scotland, the Scottish Ministers, may by order
    (a) provide that a specified event or situation, or class of event or situation, is to be treated as falling, or as not falling, within any of paragraphs (a) to (c) of subsection (1);
    (b) amend subsection (2) so as to provide that in so far as an event or situation involves or causes disruption of a specified supply, system, facility or service
      (i) it is to be treated as threatening damage to human welfare, or
      (ii) it is no longer to be treated as threatening damage to human welfare.
(5) The event or situation mentioned in subsection (1) may occur or be inside or outside the United Kingdom.

2. Duty to assess, plan and advise

(1) A person or body listed in Part 1 or 2 of Schedule 1 shall
    (a) from time to time assess the risk of an emergency occurring,

(b) from time to time assess the risk of an emergency making it necessary or expedient for the person or body to perform any of his or its functions,

(c) maintain plans for the purpose of ensuring, so far as is reasonably practicable, that if an emergency occurs the person or body is able to continue to perform his or its functions,

(d) maintain plans for the purpose of ensuring that if an emergency occurs or is likely to occur the person or body is able to perform his or its functions so far as necessary or desirable for the purpose of

   (i) preventing the emergency,

   (ii) reducing, controlling or mitigating its effects, or

   (iii) taking other action in connection with it,

(e) consider whether an assessment carried out under paragraph (a) or (b) makes it necessary or expedient for the person or body to add to or modify plans maintained under paragraph (c) or (d),

(f) arrange for the publication of all or part of assessments made and plans maintained under paragraphs (a) to (d) in so far as publication is necessary or desirable for the purpose of

   (i) preventing an emergency,

   (ii) reducing, controlling or mitigating the effects of an emergency, or

   (iii) enabling other action to be taken in connection with an emergency, and

(g) maintain arrangements to warn the public, and to provide information and advice to the public, if an emergency is likely to occur or has occurred.

(2) In relation to a person or body listed in Part 1 or 2 of Schedule 1 a duty in subsection (1) applies in relation to an emergency only if

(a) the emergency would be likely seriously to obstruct the person or body in the performance of his or its functions, or

(b) it is likely that the person or body

   (i) would consider it necessary or desirable to take action to prevent the emergency, to reduce, control or mitigate its effects or otherwise in connection with it, and

   (ii) would be unable to take that action without changing the deployment of resources or acquiring additional resources.

(3) A Minister of the Crown may, in relation to a person or body listed in Part 1 of Schedule 1, make regulations about

(a) the extent of a duty under subsection (1) (subject to subsection (2));

(b) the manner in which a duty under subsection (1) is to be performed.

(4) The Scottish Ministers may, in relation to a person or body listed in Part 2 of Schedule 1, make regulations about

(a) the extent of a duty under subsection (1) (subject to subsection (2));

(b) the manner in which a duty under subsection (1) is to be performed.

(5) Regulations under subsection (3) may, in particular

(a) make provision about the kind of emergency in relation to which a specified person or body is or is not to perform a duty under subsection (1);

(b) permit or require a person or body not to perform a duty under subsection (1) in specified circumstances or in relation to specified matters;

(c) make provision as to the timing of performance of a duty under subsection (1);

(d) require a person or body to consult a specified person or body or class of person or body before or in the course of performing a duty under subsection (1);

(e) permit or require a county council to perform a duty under subsection (1) on behalf of a district council within the area of the county council;

(f) permit, require or prohibit collaboration, to such extent and in such manner as may be specified, by persons or bodies in the performance of a duty under subsection (1);

(g) permit, require or prohibit delegation, to such extent and in such manner as may be specified, of the performance of a duty under subsection (1);

(h) permit or require a person or body listed in Part 1 or 3 of Schedule 1 to co-operate, to such extent and in such manner as may be specified, with a person or body listed in Part 1 of the Schedule in connection with the performance of a duty under subsection (1);

(i) permit or require a person or body listed in Part 1 or 3 of Schedule 1 to provide information, either on request or in other specified circumstances, to a person or body listed in Part 1 of the Schedule in connection with the performance of a duty under subsection (1);

(j) permit or require a person or body to perform (wholly or partly) a duty under subsection (1)(a) or (b) having regard to, or by adopting or relying on, work undertaken by another specified person or body;

(k) permit or require a person or body, in maintaining a plan under subsection (1)(c) or (d), to have regard to the activities of bodies (other than public or local authorities) whose activities are not carried on for profit;

(l) make provision about the extent of, and the degree of detail to be contained in, a plan maintained under subsection (1)(c) or (d);

(m) require a plan to include provision for the carrying out of exercises;

(n) require a plan to include provision for the training of staff or other persons;

(o) permit a person or body to make arrangements with another person or body, as part of planning undertaken under subsection (1)(c) or (d), for the performance of a function on behalf of the first person or body;

(p) confer a function on a Minister of the Crown, on the Scottish Ministers, on the National Assembly for Wales, on a Northern Ireland department or on any other specified person or body (and a function conferred may, in particular, be a power or duty to exercise a discretion);

(q) make provision which has effect despite other provision made by or by virtue of an enactment;

(r) make provision which applies generally or only to a specified person or body or only in specified circumstances;

(s) make different provision for different persons or bodies or for different circumstances.

(6) Subsection (5) shall have effect in relation to subsection (4) as it has effect in relation to subsection (3), but as if

(a) paragraph (e) were omitted,

(b) in paragraphs (h) and (i)

(i) a reference to Part 1 or 3 of Schedule 1 were a reference to Part 2 or 4 of that Schedule, and

(ii) a reference to Part 1 of that Schedule were a reference to Part 2 of that Schedule, and

(c) in paragraph (p) the references to a Minister of the Crown, to the National Assembly for Wales and to a Northern Ireland department were omitted.

3. Section 2: supplemental

(1) A Minister of the Crown may issue guidance to a person or body listed in Part 1 or 3 of Schedule 1 about the matters specified in section 2(3) and (5).

(2) The Scottish Ministers may issue guidance to a person or body listed in Part 2 or 4 of Schedule 1 about the matters specified in section 2(4) and (5) (as applied by section 2(6)).

(3) A person or body listed in any Part of Schedule 1 shall

(a) comply with regulations under section 2(3) or (4), and

(b) have regard to guidance under subsection (1) or (2) above.

(4) A person or body listed in Part 1 or 2 of Schedule 1 may be referred to as a 'Category 1 responder'.

(5) A person or body listed in Part 3 or 4 of Schedule 1 may be referred to as a 'Category 2 responder'.

4. Advice and assistance to the public

(1) A body specified in paragraph 1, 2 or 13 of Schedule 1 shall provide advice and assistance to the public in connection with the making of arrangements for the continuance of commercial activities by the public, or the continuance of the activities of bodies other than public or local authorities whose activities are not carried on for profit, in the event of an emergency.

(2) A Minister of the Crown may, in relation to a body specified in paragraph 1 or 2 of that Schedule, make regulations about
   (a) the extent of the duty under subsection (1);
   (b) the manner in which the duty under subsection (1) is to be performed.

(3) The Scottish Ministers may, in relation to a body specified in paragraph 13 of that Schedule, make regulations about
   (a) the extent of the duty under subsection (1);
   (b) the manner in which the duty under subsection (1) is to be performed.

(4) Regulations under subsection (2) or (3) may, in particular
   (a) permit a body to make a charge for advice or assistance provided on request under subsection (1);
   (b) make provision of a kind permitted to be made by regulations under section 2(5)(a) to (i) and (o) to (s).

(5) Regulations by virtue of subsection (4)(a) must provide that a charge for advice or assistance may not exceed the aggregate of
   (a) the direct costs of providing the advice or assistance, and
   (b) a reasonable share of any costs indirectly related to the provision of the advice or assistance.

(6) A Minister of the Crown may issue guidance to a body specified in paragraph 1 or 2 of that Schedule about the matters specified in subsections (2) and (4).

(7) The Scottish Ministers may issue guidance to a body specified in paragraph 13 of that Schedule about the matters specified in subsections (3) and (4).

(8) A body shall
   (a) comply with regulations under subsection (2) or (3), and
   (b) have regard to guidance under subsection (6) or (7).

5. General measures

(1) A Minister of the Crown may by order require a person or body listed in Part 1 of Schedule 1 to perform a function of that person or body for the purpose of
   (a) preventing the occurrence of an emergency,
   (b) reducing, controlling or mitigating the effects of an emergency, or
   (c) taking other action in connection with an emergency.

(2) The Scottish Ministers may by order require a person or body listed in Part 2 of Schedule 1 to perform a function of that person or body for the purpose of
   (a) preventing the occurrence of an emergency,
   (b) reducing, controlling or mitigating the effects of an emergency, or
   (c) taking other action in connection with an emergency.

(3) A person or body shall comply with an order under this section.

(4) An order under subsection (1) may
   (a) require a person or body to consult a specified person or body or class of person or body;
   (b) permit, require or prohibit collaboration, to such extent and in such manner as may be specified;
   (c) permit, require or prohibit delegation, to such extent and in such manner as may be specified;

(d) permit or require a person or body listed in Part 1 or 3 of Schedule 1 to co-operate, to such extent and in such manner as may be specified, with a person or body listed in Part 1 of the Schedule in connection with a duty under the order;

(e) permit or require a person or body listed in Part 1 or 3 of Schedule 1 to provide information in connection with a duty under the order, whether on request or in other specific circumstances to a person or body listed in Part 1 of the Schedule;

(f) confer a function on a Minister of the Crown, on the Scottish Ministers, on the National Assembly for Wales, on a Northern Ireland department or on any other specified person or body (and a function conferred may, in particular, be a power or duty to exercise a discretion);

(g) make provision which applies generally or only to a specified person or body or only in specified circumstances;

(h) make different provision for different persons or bodies or for different circumstances.

(5) Subsection (4) shall have effect in relation to subsection (2) as it has effect in relation to subsection (1), but as if

(a) in paragraphs (d) and (e)

    (i) a reference to Part 1 or 3 of Schedule 1 were a reference to Part 2 or 4 of that Schedule, and

    (ii) a reference to Part 1 of that Schedule were a reference to Part 2 of that Schedule, and

(b) in paragraph (f) the references to a Minister of the Crown, to the National Assembly for Wales and to a Northern Ireland department were omitted.

6. Disclosure of information

(1) A Minister of the Crown may make regulations requiring or permitting one person or body listed in Part 1 or 3 of Schedule 1 ('the provider') to disclose information on request to another person or body listed in any Part of that Schedule ('the recipient').

(2) The Scottish Ministers may make regulations requiring or permitting one person or body listed in Part 2 or 4 of Schedule 1 ('the provider') to disclose information on request to another person or body listed in any Part of that Schedule ('the recipient').

(3) Regulations under subsection (1) or (2) may be made only in connection with a function of the provider or of the recipient which relates to emergencies.

(4) A Minister of the Crown may issue guidance to a person or body about the performance of functions under regulations made under subsection (1).

(5) The Scottish Ministers may issue guidance to a person or body about the performance of functions under regulations made under subsection (2).

(6) A person or body shall

(a) comply with regulations under subsection (1) or (2), and

(b) have regard to guidance under subsection (4) or (5).

7. Urgency

(1) This section applies where

(a) there is an urgent need to make provision of a kind that could be made by an order under section 5(1) or by regulations under section 6(1), but

(b) there is insufficient time for the order or regulations to be made.

(2) The Minister may by direction make provision of a kind that could be made by an order under section 5(1) or by regulations under section 6(1).

(3) A direction under subsection (2) shall be in writing.

(4) Where a Minister gives a direction under subsection (2)

(a) he may revoke or vary the direction by further direction,

(b) he shall revoke the direction as soon as is reasonably practicable (and he may, if or in so far as he thinks it desirable, re-enact the substance of the direction by way of an order under section 5(1) or by way of regulations under section 6(1)), and

(c) the direction shall cease to have effect at the end of the period of 21 days beginning with the day on which it is given (but without prejudice to the power to give a new direction).

(5) A provision of a direction under subsection (2) shall be treated for all purposes as if it were a provision of an order under section 5(1) or of regulations under section 6(1).

8. Urgency: Scotland

(1) This section applies where
  (a) there is an urgent need to make provision of a kind that could be made by an order under section 5(2) or by regulations under section 6(2), but
  (b) there is insufficient time for the order or regulations to be made.
(2) The Scottish Ministers may by direction make provision of a kind that could be made by an order under section 5(2) or by regulations under section 6(2).
(3) A direction under subsection (2) shall be in writing.
(4) Where the Scottish Ministers give a direction under subsection (2)
  (a) they may revoke or vary the direction by further direction,
  (b) they shall revoke the direction as soon as is reasonably practicable (and they may, if or in so far as they think it desirable, re-enact the substance of the direction by way of an order under section 5(2) or by way of regulations under section 6(2)), and
  (c) the direction shall cease to have effect at the end of the period of 21 days beginning with the day on which it is given (but without prejudice to the power to give a new direction).
(5) A provision of a direction under subsection (2) shall be treated for all purposes as if it were a provision of an order under section 5(2) or of regulations under section 6(2).

9. Monitoring by Government

(1) A Minister of the Crown may require a person or body listed in Part 1 or 3 of Schedule 1
  (a) to provide information about action taken by the person or body for the purpose of complying with a duty under this Part, or
  (b) to explain why the person or body has not taken action for the purpose of complying with a duty under this Part.
(2) The Scottish Ministers may require a person or body listed in Part 2 or 4 of Schedule 1
  (a) to provide information about action taken by the person or body for the purpose of complying with a duty under this Part, or
  (b) to explain why the person or body has not taken action for the purpose of complying with a duty under this Part.
(3) A requirement under subsection (1) or (2) may specify
  (a) a period within which the information or explanation is to be provided;
  (b) the form in which the information or explanation is to be provided.
(4) A person or body shall comply with a requirement under subsection (1) or (2).

10. Enforcement

(1) Any of the following may bring proceedings in the High Court or the Court of Session in respect of a failure by a person or body listed in Part 1 or 3 of Schedule 1 to comply with section 2(1), 3(3), 4(1) or (8), 5(3), 6(6), 9(4) or 15(7)
  (a) a Minister of the Crown,
  (b) a person or body listed in Part 1 of Schedule 1, and
  (c) a person or body listed in Part 3 of Schedule 1.
(2) In proceedings under subsection (1) the High Court or the Court of Session may grant any relief, or make any order, that it thinks appropriate.

11. Enforcement: Scotland

(1) Any of the following may bring proceedings in the Court of Session in respect of a failure by a person or body listed in Part 2 or 4 of Schedule 1 to comply with section 2(1), 3(3), 4(1) or (8), 5(3), 6(6), 9(4) or 15(7)
    (a) the Scottish Ministers,
    (b) a person or body listed in Part 2 of Schedule 1, and
    (c) a person or body listed in Part 4 of Schedule 1.
(2) In proceedings under subsection (1) the Court of Session may grant any remedy, or make any order, that it thinks appropriate.

12. Provision of information

Regulations or an order under this Part may, if addressing the provision or disclosure of information, make provision about

    (a) timing;
    (b) the form in which information is provided;
    (c) the use to which information may be put;
    (d) storage of information;
    (e) disposal of information.

13. Amendment of lists of responders

(1) A Minister of the Crown may by order amend Schedule 1 so as to
    (a) add an entry to Part 1 or 3;
    (b) remove an entry from Part 1 or 3;
    (c) move an entry from Part 1 to Part 3 or vice versa.
(2) The Scottish Ministers may by order amend Schedule 1 so as to
    (a) add an entry to Part 2 or 4;
    (b) remove an entry from Part 2 or 4;
    (c) move an entry from Part 2 to Part 4 or vice versa.
(3) An order under subsection (1) or (2)
    (a) may add, remove or move an entry either generally or only in relation to specified functions of a person or body, and
    (b) may make incidental, transitional or consequential provision (which may include provision amending this Act or another enactment).

14. Scotland: consultation

(1) A Minister of the Crown shall consult the Scottish Ministers before making regulations or an order under this Part in relation to a person or body if or in so far as the person or body exercises functions in relation to Scotland.
(2) The Scottish Ministers shall consult a Minister of the Crown before making regulations or an order under this Part.

15. Scotland: cross-border collaboration

(1) Where a person or body listed in Part 1 of Schedule 1 has a duty under section 2 or 4, the Scottish Ministers may make regulations
    (a) permitting or requiring a person or body listed in Part 2 or 4 of that Schedule to co-operate, to such extent and in such manner as may be specified, with the person or body listed in Part 1 of that Schedule in connection with the performance of the duty;
    (b) permitting or requiring a person or body listed in Part 2 or 4 of that Schedule to provide information, either on request or in other specified circumstances, to the person or body listed in Part 1 of that Schedule in connection with the performance of the duty.

(2) The Scottish Ministers may issue guidance about a matter addressed in regulations under subsection (1).

(3) Where a person or body listed in Part 2 of Schedule 1 has a duty under section 2 or 4, a Minister of the Crown may make regulations

    (a) permitting or requiring a person or body listed in Part 1 or 3 of that Schedule to co-operate, to such extent and in such manner as may be specified, with the person or body listed in Part 2 of that Schedule in connection with the performance of the duty;

    (b) permitting or requiring a person or body listed in Part 1 or 3 of that Schedule to provide information, either on request or in other specified circumstances, to the person or body listed in Part 2 of that Schedule in connection with the performance of the duty.

(4) A Minister of the Crown may issue guidance about a matter addressed in regulations under subsection (3).

(5) If a Minister of the Crown makes an order under section 5(1) imposing a duty on a person or body listed in Part 1 of Schedule 1, the Scottish Ministers may make an order

    (a) permitting or requiring a person or body listed in Part 2 or 4 of that Schedule to co-operate, to such extent and in such manner as may be specified, with the person or body listed in Part 1 of that Schedule in connection with the duty;

    (b) permitting or requiring a person or body listed in Part 2 or 4 of that Schedule to provide information, either on request or in other specified circumstances, to the person or body listed in Part 1 of that Schedule in connection with the duty.

(6) If the Scottish Ministers make an order under section 5(2) imposing a duty on a person or body listed in Part 2 of Schedule 1, a Minister of the Crown may make an order

    (a) permitting or requiring a person or body listed in Part 1 or 3 of that Schedule to co-operate, to such extent and in such manner as may be specified, with the person or body listed in Part 2 of that Schedule in connection with the duty;

    (b) permitting or requiring a person or body listed in Part 1 or 3 of that Schedule to provide information, either on request or in other specified circumstances, to the person or body listed in Part 2 of that Schedule in connection with the duty.

(7) A person or body shall

    (a) comply with regulations or an order under this section, and

    (b) have regard to guidance under this section.

(8) In this Act, except where the contrary intention appears

    (a) a reference to an order under section 5(1) includes a reference to an order under sub-section (6) above, and

    (b) a reference to an order under section 5(2) includes a reference to an order under sub-section (5) above.

16. National Assembly for Wales

(1) A Minister of the Crown shall consult the National Assembly for Wales before

    (a) making regulations under section 2(3), 4(2) or 6(1) which relate wholly or partly to Wales,

    (b) issuing guidance under section 3(1), 4(6) or 6(4) which relates wholly or partly to Wales,

    (c) giving an order under section 5(1) which relates wholly or partly to Wales,

    (d) giving a direction under section 7(2) which makes provision relating wholly or partly to Wales of a kind that could be made by regulations under section 6(1),

    (e) giving a direction under section 7(2) which makes provision relating wholly or partly to Wales of a kind that could be made by an order under section 5(1),

    (f) bringing proceedings under section 10 in respect of a failure by a person or body where the failure relates wholly or partly to Wales, or

    (g) making an order under section 13(1) in respect of a person or body with, or in so far as the person or body has, functions in relation to Wales.

(2) A Minister of the Crown may not without the consent of the National Assembly for Wales take action of a kind specified in subsection (3) that relates wholly or partly to a person or body specified in subsection (4).

(3) The actions referred to in subsection (2) are

    (a) making regulations under section 2(3), 4(2) or 6(1),

    (b) making an order under section 5(1),

    (c) issuing guidance under section 3(1), 4(6) or 6(4),

    (d) giving a direction under section 7,

    (e) bringing proceedings under section 10, and

    (f) making an order under section 13.

(4) The persons and bodies referred to in subsection (2) are

    (a) a council specified in paragraph 2 of Schedule 1, and

    (b) a person or body specified in paragraph 4, 5, 8, 9, 10, 11 or 21 of that Schedule, if and in so far as the person or body has functions in relation to Wales.

17. Regulations and orders

(1) Regulations and orders under this Part shall be made by statutory instrument.

(2) An order under section 1(4), 5(1) or 13(1) may not be made by a Minister of the Crown unless a draft has been laid before and approved by resolution of each House of Parliament.

(3) An order under section 1(4), 5(2) or 13(2) may not be made by the Scottish Ministers unless a draft has been laid before and approved by resolution of the Scottish Parliament.

(4) Regulations made by a Minister of the Crown under this Part shall be subject to annulment in pursuance of a resolution of either House of Parliament.

(5) Regulations made by the Scottish Ministers under this Part shall be subject to annulment in pursuance of a resolution of the Scottish Parliament.

(6) Regulations or an order under this Part

    (a) may make provision which applies generally or only in specified circumstances or for a specified purpose,

    (b) may make different provision for different circumstances or purposes, and

    (c) may make incidental, consequential or transitional provision.

18. Interpretation, &c

(1) In this Part

'enactment' includes

    (a) an Act of the Scottish Parliament,

    (b) Northern Ireland legislation, and

    (c) an instrument made under an Act of the Scottish Parliament or under Northern Ireland legislation (as well as an instrument made under an Act),

'function' means any power or duty whether conferred by virtue of an enactment or otherwise,

'terrorism' has the meaning given by section 1 of the Terrorism Act 2000 (c 11), and

'war' includes armed conflict.

(2) In this Part a reference to the United Kingdom includes a reference to the territorial sea of the United Kingdom.

(3) Except in a case of contradiction, nothing in or done under this Part shall impliedly repeal or revoke a provision of or made under another enactment.

Part II
Emergency Powers

19. Meaning of 'emergency'

(1) In this Part 'emergency' means

    (a) an event or situation which threatens serious damage to human welfare in the United Kingdom or in a Part or region,

(b) an event or situation which threatens serious damage to the environment of the United Kingdom or of a Part or region, or

(c) war, or terrorism, which threatens serious damage to the security of the United Kingdom.

(2) For the purposes of subsection (1)(a) an event or situation threatens damage to human welfare only if it involves, causes or may cause

    (a) loss of human life,

    (b) human illness or injury,

    (c) homelessness,

    (d) damage to property,

    (e) disruption of a supply of money, food, water, energy or fuel,

    (f) disruption of a system of communication,

    (g) disruption of facilities for transport, or

    (h) disruption of services relating to health.

(3) For the purposes of subsection (1)(b) an event or situation threatens damage to the environment only if it involves, causes or may cause

    (a) contamination of land, water or air with biological, chemical or radio-active matter, or

    (b) disruption or destruction of plant life or animal life.

(4) The Secretary of State may by order amend subsection (2) so as to provide that in so far as an event or situation involves or causes disruption of a specified supply, system, facility or service

    (a) it is to be treated as threatening damage to human welfare, or

    (b) it is no longer to be treated as threatening damage to human welfare.

(5) An order under subsection (4)

    (a) may make consequential amendment of this Part, and

    (b) may not be made unless a draft has been laid before, and approved by resolution of, each House of Parliament.

(6) The event or situation mentioned in subsection (1) may occur or be inside or outside the United Kingdom.

20. Power to make emergency regulations

(1) Her Majesty may by Order in Council make emergency regulations if satisfied that the conditions in section 21 are satisfied.

(2) A senior Minister of the Crown may make emergency regulations if satisfied

    (a) that the conditions in section 21 are satisfied, and

    (b) that it would not be possible, without serious delay, to arrange for an Order in Council under subsection (1).

(3) In this Part 'senior Minister of the Crown' means

    (a) the First Lord of the Treasury (the Prime Minister),

    (b) any of Her Majesty's Principal Secretaries of State, and

    (c) the Commissioners of Her Majesty's Treasury.

(4) In this Part 'serious delay' means a delay that might

    (a) cause serious damage, or

    (b) seriously obstruct the prevention, control or mitigation of serious damage.

(5) Regulations under this section must be prefaced by a statement by the person making the regulations

    (a) specifying the nature of the emergency in respect of which the regulations are made, and

    (b) declaring that the person making the regulations

        (i) is satisfied that the conditions in section 21 are met,

        (ii) is satisfied that the regulations contain only provision which is appropriate for the purpose of preventing, controlling or mitigating an aspect or effect of the emergency in respect of which the regulations are made,

(iii) is satisfied that the effect of the regulations is in due proportion to that aspect or effect of the emergency,

(iv) is satisfied that the regulations are compatible with the Convention rights (within the meaning of section 1 of the Human Rights Act 1998 (c 42)), and

(v) in the case of regulations made under subsection (2), is satisfied as to the matter specified in subsection (2)(b).

21. Conditions for making emergency regulations

(1) This section specifies the conditions mentioned in section 20.

(2) The first condition is that an emergency has occurred, is occurring or is about to occur.

(3) The second condition is that it is necessary to make provision for the purpose of preventing, controlling or mitigating an aspect or effect of the emergency.

(4) The third condition is that the need for provision referred to in subsection (3) is urgent.

(5) For the purpose of subsection (3) provision which is the same as an enactment ('the existing legislation') is necessary if, in particular

(a) the existing legislation cannot be relied upon without the risk of serious delay,

(b) it is not possible without the risk of serious delay to ascertain whether the existing legislation can be relied upon, or

(c) the existing legislation might be insufficiently effective.

(6) For the purpose of subsection (3) provision which could be made under an enactment other than section 20 ('the existing legislation') is necessary if, in particular

(a) the provision cannot be made under the existing legislation without the risk of serious delay,

(b) it is not possible without the risk of serious delay to ascertain whether the provision can be made under the existing legislation, or

(c) the provision might be insufficiently effective if made under the existing legislation.

22. Scope of emergency regulations

(1) Emergency regulations may make any provision which the person making the regulations is satisfied is appropriate for the purpose of preventing, controlling or mitigating an aspect or effect of the emergency in respect of which the regulations are made.

(2) In particular, emergency regulations may make any provision which the person making the regulations is satisfied is appropriate for the purpose of

(a) protecting human life, health or safety,

(b) treating human illness or injury,

(c) protecting or restoring property,

(d) protecting or restoring a supply of money, food, water, energy or fuel,

(e) protecting or restoring a system of communication,

(f) protecting or restoring facilities for transport,

(g) protecting or restoring the provision of services relating to health,

(h) protecting or restoring the activities of banks or other financial institutions,

(i) preventing, containing or reducing the contamination of land, water or air,

(j) preventing, reducing or mitigating the effects of disruption or destruction of plant life or animal life,

(k) protecting or restoring activities of Parliament, of the Scottish Parliament, of the Northern Ireland Assembly or of the National Assembly for Wales, or

(l) protecting or restoring the performance of public functions.

(3) Emergency regulations may make provision of any kind that could be made by Act of Parliament or by the exercise of the Royal Prerogative; in particular, regulations may

(a) confer a function on a Minister of the Crown, on the Scottish Ministers, on the National Assembly for Wales, on a Northern Ireland department, on a coordinator appointed under section 24 or on any other specified person (and a function conferred may, in particular, be

(i) a power, or duty, to exercise a discretion;

(ii) a power to give directions or orders, whether written or oral);

(b) provide for or enable the requisition or confiscation of property (with or without compensation);

(c) provide for or enable the destruction of property, animal life or plant life (with or without compensation);

(d) prohibit, or enable the prohibition of, movement to or from a specified place;

(e) require, or enable the requirement of, movement to or from a specified place;

(f) prohibit, or enable the prohibition of, assemblies of specified kinds, at specified places or at specified times;

(g) prohibit, or enable the prohibition of, travel at specified times;

(h) prohibit, or enable the prohibition of, other specified activities;

(i) create an offence of

(i) failing to comply with a provision of the regulations;

(ii) failing to comply with a direction or order given or made under the regulations;

(iii) obstructing a person in the performance of a function under or by virtue of the regulations;

(j) disapply or modify an enactment or a provision made under or by virtue of an enactment;

(k) require a person or body to act in performance of a function (whether the function is conferred by the regulations or otherwise and whether or not the regulations also make provision for remuneration or compensation);

(l) enable the Defence Council to authorise the deployment of Her Majesty's armed forces;

(m) make provision (which may include conferring powers in relation to property) for facilitating any deployment of Her Majesty's armed forces;

(n) confer jurisdiction on a court or tribunal (which may include a tribunal established by the regulations);

(o) make provision which has effect in relation to, or to anything done in

(i) an area of the territorial sea,

(ii) an area within British fishery limits, or

(iii) an area of the continental shelf;

(p) make provision which applies generally or only in specified circumstances or for a specified purpose;

(q) make different provision for different circumstances or purposes.

(4) In subsection (3) 'specified' means specified by, or to be specified in accordance with, the regulations.

(5) A person making emergency regulations must have regard to the importance of ensuring that Parliament, the High Court and the Court of Session are able to conduct proceedings in connection with

(a) the regulations, or

(b) action taken under the regulations.

23. Limitations of emergency regulations

(1) Emergency regulations may make provision only if and in so far as the person making the regulations is satisfied

(a) that the provision is appropriate for the purpose of preventing, controlling or mitigating an aspect or effect of the emergency in respect of which the regulations are made, and

(b) that the effect of the provision is in due proportion to that aspect or effect of the emergency.

(2) Emergency regulations must specify the Parts of the United Kingdom or regions in relation to which the regulations have effect.

(3) Emergency regulations may not

(a) require a person, or enable a person to be required, to provide military service, or

(b) prohibit or enable the prohibition of participation in, or any activity in connection with, a strike or other industrial action.

(4) Emergency regulations may not
  (a) create an offence other than one of the kind described in section 22(3)(i),
  (b) create an offence other than one which is triable only before a magistrates' court or, in Scotland, before a sheriff under summary procedure,
  (c) create an offence which is punishable
    (i) with imprisonment for a period exceeding three months, or
    (ii) with a fine exceeding level 5 on the standard scale, or
  (d) alter procedure in relation to criminal proceedings.
(5) Emergency regulations may not amend

  (a) this Part of this Act, or
  (b) the Human Rights Act 1998 (c 42).

24. Regional and Emergency Coordinators

(1) Emergency regulations must require a senior Minister of the Crown to appoint
  (a) for each Part of the United Kingdom, other than England, in relation to which the regulations have effect, a person to be known as the Emergency Coordinator for that Part, and
  (b) for each region in relation to which the regulations have effect, a person to be known as the Regional Nominated Coordinator for that region.
(2) Provision made in accordance with subsection (1) may, in particular, include provision about the coordinator's
  (a) terms of appointment,
  (b) conditions of service (including remuneration), and
  (c) functions.
(3) The principal purpose of the appointment shall be to facilitate coordination of activities under the emergency regulations (whether only in the Part or region for which the appointment is made or partly there and partly elsewhere).
(4) In exercising his functions a coordinator shall
  (a) comply with a direction of a senior Minister of the Crown, and
  (b) have regard to guidance issued by a senior Minister of the Crown.
(5) A coordinator shall not be regarded as the servant or agent of the Crown or as enjoying any status, immunity or privilege of the Crown.

25. Establishment of tribunal

(1) Emergency regulations which establish a tribunal may not be made unless a senior Minister of the Crown has consulted the Council on Tribunals.
(2) But
  (a) a senior Minister of the Crown may disapply subsection (1) if necessary by reason of urgency,
  (b) subsection (1) shall not apply where the Council on Tribunals have consented to the establishment of the Tribunal, and
  (c) a failure to satisfy subsection (1) shall not affect the validity of regulations.
(3) Where the Council on Tribunals are consulted by a senior Minister of the Crown under subsection (1)
  (a) the Council shall make a report to the Minister, and
  (b) the Minister shall not make the emergency regulations to which the consultation relates before receiving the Council's report.
(4) But
  (a) a senior Minister of the Crown may disapply subsection (3)(b) if necessary by reason of urgency, and
  (b) a failure to comply with subsection (3)(b) shall not affect the validity of regulations.

(5) Where a senior Minister of the Crown receives a report under subsection (3)(a) he shall lay before Parliament as soon as is reasonably practicable after the making of the regulations to which the report relates

(a) a copy of the report,

(b) a statement of the extent to which the regulations give effect to any recommendations in the report, and

(c) an explanation for any departure from recommendations in the report.

(6) Where a senior Minister of the Crown makes emergency regulations without consulting the Council on Tribunals (in reliance on subsection (2)(a))

(a) he shall consult the Council about the regulations as soon as reasonably practicable after they are made,

(b) the Council shall make a report to the Minister, and

(c) subsection (5) shall apply (with any necessary modifications).

26. Duration

(1) Emergency regulations shall lapse

(a) at the end of the period of 30 days beginning with the date on which they are made, or

(b) at such earlier time as may be specified in the regulations.

(2) Subsection (1)

(a) shall not prevent the making of new regulations, and

(b) shall not affect anything done by virtue of the regulations before they lapse.

27. Parliamentary scrutiny

(1) Where emergency regulations are made

(a) a senior Minister of the Crown shall as soon as is reasonably practicable lay the regulations before Parliament, and

(b) the regulations shall lapse at the end of the period of seven days beginning with the date of laying unless during that period each House of Parliament passes a resolution approving them.

(2) If each House of Parliament passes a resolution that emergency regulations shall cease to have effect, the regulations shall cease to have effect

(a) at such time, after the passing of the resolutions, as may be specified in them, or

(b) if no time is specified in the resolutions, at the beginning of the day after that on which the resolutions are passed (or, if they are passed on different days, at the beginning of the day after that on which the second resolution is passed).

(3) If each House of Parliament passes a resolution that emergency regulations shall have effect with a specified amendment, the regulations shall have effect as amended, with effect from

(a) such time, after the passing of the resolutions, as may be specified in them, or

(b) if no time is specified in the resolutions, the beginning of the day after that on which the resolutions are passed (or, if they are passed on different days, the beginning of the day after that on which the second resolution is passed).

(4) Nothing in this section

(a) shall prevent the making of new regulations, or

(b) shall affect anything done by virtue of regulations before they lapse, cease to have effect or are amended under this section.

28. Parliamentary scrutiny: prorogation and adjournment

(1) If when emergency regulations are made under section 20 Parliament stands prorogued to a day after the end of the period of five days beginning with the date on which the regulations are made, Her Majesty shall by proclamation under the Meeting of Parliament Act 1797 (c 127) require Parliament to meet on a specified day within that period.

(2) If when emergency regulations are made under section 20 the House of Commons stands adjourned to a day after the end of the period of five days beginning with the date on which the regulations are made, the Speaker shall arrange for the House to meet on a day during that period.

(3) If when emergency regulations are made under section 20 the House of Lords stands adjourned to a day after the end of the period of five days beginning with the date on which the regulations are made, the Lord Chancellor shall arrange for the House to meet on a day during that period.

(4) In subsections (2) and (3) a reference to the Lord Chancellor or the Speaker includes a reference to a person authorised by Standing Orders of the House of Lords or of the House of Commons to act in place of the Lord Chancellor or the Speaker in respect of the recall of the House during adjournment.

29. Consultation with devolved administrations

(1) Emergency regulations which relate wholly or partly to Scotland may not be made unless a senior Minister of the Crown has consulted the Scottish Ministers.

(2) Emergency regulations which relate wholly or partly to Northern Ireland may not be made unless a senior Minister of the Crown has consulted the First Minister and deputy First Minister.

(3) Emergency regulations which relate wholly or partly to Wales may not be made unless a senior Minister of the Crown has consulted the National Assembly for Wales.

(4) But

(a) a senior Minister of the Crown may disapply a requirement to consult if he thinks it necessary by reason of urgency, and

(b) a failure to satisfy a requirement to consult shall not affect the validity of regulations.

30. Procedure

(1) Emergency regulations shall be made by statutory instrument (whether or not made by Order in Council).

(2) Emergency regulations shall be treated for the purposes of the Human Rights Act 1998 (c 42) as subordinate legislation and not primary legislation (whether or not they amend primary legislation).

31. Interpretation

(1) In this Part

'British fishery limits' has the meaning given by the Fishery Limits Act 1976 (c 86),

'the continental shelf' means any area designated by Order in Council under section 1(7) of the Continental Shelf Act 1964 (c 29),

'emergency' has the meaning given by section 19,

'enactment' includes

(a) an Act of the Scottish Parliament,

(b) Northern Ireland legislation, and

(c) an instrument made under an Act of the Scottish Parliament or under Northern Ireland legislation (as well as an instrument made under an Act),

'function' means any power or duty whether conferred by virtue of an enactment or otherwise,

'Part' in relation to the United Kingdom has the meaning given by subsection (2),

'public functions' means

(a) functions conferred or imposed by or by virtue of an enactment,

(b) functions of Ministers of the Crown (or their departments),

(c) functions of persons holding office under the Crown,

(d) functions of the Scottish Ministers,

(e) functions of the Northern Ireland Ministers or of the Northern Ireland departments, and

(f) functions of the National Assembly for Wales,

'region' has the meaning given by subsection (2),

'senior Minister of the Crown' has the meaning given by section 20(3),

'serious delay' has the meaning given by section 20(4),

'territorial sea' means the territorial sea adjacent to, or to any Part of, the United Kingdom, construed in accordance with section 1 of the Territorial Sea Act 1987 (c 49),

'terrorism' has the meaning given by section 1 of the Terrorism Act 2000 (c 11), and

'war' includes armed conflict.

(2) In this Part

    (a) 'Part' in relation to the United Kingdom means

        (i) England,

        (ii) Northern Ireland,

        (iii) Scotland, and

        (iv) Wales,

    (b) 'region' means a region for the purposes of the Regional Development Agencies Act 1998 (c 45), and

    (c) a reference to a Part or region of the United Kingdom includes a reference to

        (i) any part of the territorial sea that is adjacent to that Part or region,

        (ii) any part of the area within British fishery limits that is adjacent to the Part or region, and

        (iii) any part of the continental shelf that is adjacent to the Part or region.

(3) The following shall have effect for the purpose of subsection (2)

    (a) an Order in Council under section 126(2) of the Scotland Act 1998 (c 46) (apportionment of sea areas),

    (b) an Order in Council under section 98(8) of the Northern Ireland Act 1998 (c 47) (apportionment of sea areas), and

    (c) an order under section 155(2) of the Government of Wales Act 1998 (c 38) (apportionment of sea areas);

but only if or in so far as it is expressed to apply for general or residual purposes of any of those Acts or for the purposes of this section.

PART III

GENERAL

32. Minor and consequential amendments and repeals

(1) Schedule 2 (minor and consequential amendments and repeals) shall have effect.

(2) The enactments listed in Schedule 3 are hereby repealed or revoked to the extent specified.

33. Money

There shall be paid out of money provided by Parliament

    (a) any expenditure incurred by a Minister of the Crown in connection with this Act, and

    (b) any increase attributable to this Act in the sums payable under any other enactment out of money provided by Parliament.

34. Commencement

(1) The preceding provisions of this Act shall come into force in accordance with provision made by a Minister of the Crown by order.

(2) But the following provisions of this Act shall come into force in accordance with provision made by the Scottish Ministers by order

    (a) section 1(4) in so far as it relates to the Scottish Ministers,

    (b) sections 2(4) and (6), 3(2), 4(3) and (7), 5(2) and (5), 6(2) and (5), 8, 9(2), 11, 13(2), 14(2), 17(3) and (5), and

(c) a provision of section 2, 3, 4, 5, 6, 9, 13 or 17 in so far as it relates to a provision specified in paragraph (b) above.

(3) An order under subsection (1) or (2)

    (a) may make provision generally or for specific purposes only,

    (b) may make different provision for different purposes,

    (c) may make incidental, consequential or transitional provision, and

    (d) shall be made by statutory instrument.

35. Extent

(1) This Act extends to

    (a) England and Wales,

    (b) Scotland, and

    (c) Northern Ireland.

(2) But where this Act amends or repeals an enactment or a provision of an enactment, the amendment or repeal has the same extent as the enactment or provision.

36. Short title

This Act may be cited as the Civil Contingencies Act 2004.

SCHEDULE 1

CATEGORY 1 AND 2 RESPONDERS

PART 1

CATEGORY 1 RESPONDERS: GENERAL

Local authorities

1. In relation to England

    (a) a county council,

    (b) a district council,

    (c) a London borough council,

    (d) the Common Council of the City of London, and

    (e) the Council of the Isles of Scilly.

2. In relation to Wales

    (a) a county council, and

    (b) a county borough council.

Emergency services

3. (1) A chief officer of police within the meaning of section 101(1) of the Police Act 1996 (c 16).

(2) The Chief Constable of the Police Service of Northern Ireland.

(3) The Chief Constable of the British Transport Police Force.

4. A fire and rescue authority within the meaning of section 1 of the Fire and Rescue Services Act 2004 (c 21).

Health

5. A National Health Service trust established under section 5 of the National Health Service and Community Care Act 1990 (c 19) if, and in so far as, it has the function of providing

    (a) ambulance services,

    (b) hospital accommodation and services in relation to accidents and emergencies, or

    (c) services in relation to public health in Wales.

6. An NHS foundation trust (within the meaning of section 1 of the Health and Social Care (Community Health and Standards) Act 2003 (c 43)) if, and in so far as, it has the function of providing hospital accommodation and services in relation to accidents and emergencies.

7. A Primary Care Trust established under section 16A of the National Health Service Act 1977 (c 49).

8. Local Health Board established under section 16BA of the National Health Service Act 1977.

9. (1) The Health Protection Agency established by section 1 of the Health Protection Agency Act 2004 in so far as its functions relate to Great Britain.

(2) Until its dissolution consequent upon the coming into force of section 1 of that Act, the Special Health Authority established under section 11 of the National Health Service Act 1977 and known as the Health Protection Agency.

10. A port health authority constituted under section 2(4) of the Public Health (Control of Disease) Act 1984 (c 22).

Miscellaneous

11. The Environment Agency.

12. The Secretary of State, in so far as his functions include responding to maritime and coastal emergencies (excluding the investigation of accidents).

PART 2
CATEGORY 1 RESPONDERS: SCOTLAND

Local authorities

13. A council constituted under section 2 of the Local Government etc (Scotland) Act 1994 (c 39).

Emergency services

14. A chief constable of a police force maintained under the Police (Scotland) Act 1967 (c 77).

15. (1) A fire authority.

(2) In sub-paragraph (1) 'fire authority' means
   (a) a council constituted under section 2 of the Local Government etc (Scotland) Act 1994, or
   (b) a joint board constituted under an administration scheme made by virtue of the Local Government (Scotland) Act 1973 (c 65) or section 36 of the Fire Services Act 1947 (c 41).

16. The Scottish Ambulance Service Board.

Health

17. A Health Board constituted under section 2 of the National Health Service (Scotland) Act 1978 (c 29).

Miscellaneous

18. The Scottish Environment Protection Agency.

PART 3
CATEGORY 2 RESPONDERS: GENERAL

Utilities

19. (1) A person holding a licence of a kind specified in sub-paragraph (2) and granted under section 6 of the Electricity Act 1989 (c 29).

(2) Those licences are
   (a) a transmission licence,
   (b) a distribution licence, and

(c) an interconnector licence.

(3) Expressions used in this paragraph and in the Electricity Act 1989 shall have the same meaning in this paragraph as in that Act.

20. (1) A person holding a licence of a kind specified in sub-paragraph (2).

(2) Those licences are

    (a) a licence under section 7 of the Gas Act 1986 (c 44), and

    (b) a licence under section 7ZA of that Act.

21. A water undertaker or sewerage undertaker appointed under section 6 of the Water Industry Act 1991 (c 56).

22. (1) A person who provides a public electronic communications network which makes telephone services available (whether for spoken communication or for the transmission of data).

(2) In sub-paragraph (1)

    (a) the reference to provision of a network shall be construed in accordance with section 32(4)(a) and (b) of the Communications Act 2003 (c 21), and

    (b) 'public electronic communications network' shall have the meaning given by sections 32(1) and 151(1) of that Act.

Transport

23. A person who holds a licence under section 8 of the Railways Act 1993 (c 43) (operation of railway assets) in so far as the licence relates to activity in Great Britain.

24. A person who provides services in connection with railways in Great Britain

    (a) without holding a licence under section 8 of that Act, but

    (b) in reliance on Council Directive 95/18/EC on the licensing of railway undertakings.

25. (1) Transport for London.

(2) London Underground Limited (being a subsidiary of Transport for London).

26. An airport operator, within the meaning of section 82(1) of the Airports Act 1986 (c 31), in Great Britain.

27. A harbour authority, within the meaning of section 46(1) of the Aviation and Maritime Security Act 1990 (c 31), in Great Britain.

28. The Secretary of State, in so far as his functions relate to matters for which he is responsible by virtue of section 1 of the Highways Act 1980 (c 66) (highway authorities).

Health and safety

29. The Health and Safety Executive.

Health

29A. A Strategic Health Authority established under section 8 of the National Health Service Act 1977.

PART 4
CATEGORY 2 RESPONDERS: SCOTLAND

Utilities

30. (1) A person holding a licence of a kind specified in sub-paragraph (2) and granted under the Electricity Act 1989 (c 29), in so far as the activity under the licence is undertaken in Scotland.

(2) Those licences are

    (a) a transmission licence,

    (b) a distribution licence, and

    (c) an interconnector licence.

(3) Expressions used in this paragraph and in the Electricity Act 1989 shall have the same meaning in this paragraph as in that Act.

31. (1) A person holding a licence of a kind specified in sub-paragraph (2), in so far as the activity under the licence is undertaken in Scotland.

(2) Those licences are

(a) a licence under section 7 of the Gas Act 1986 (c 44), and

(b) a licence under section 7ZA of that Act.

32. Scottish Water, established by section 20 of, and Schedule 3 to, the Water Industry (Scotland) Act 2002 (asp 3).

33. (1) A person who provides a public electronic communications network which makes telephone services available (whether for spoken communication or for the transmission of data) in so far as the services are made available in Scotland.

(2) In sub-paragraph (1)

(a) the reference to provision of a network shall be construed in accordance with section 32(4)(a) and (b) of the Communications Act 2003 (c 21), and

(b) 'public electronic communications network' shall have the meaning given by sections 32(1) and 151(1) of that Act.

Transport

34. A person who holds a licence to operate railway assets under section 8 of the Railways Act 1993 (c 43) in so far as such operation takes place in Scotland.

35. A person who provides services in connection with railways

(a) without holding a licence under section 8 of that Act, but

(b) in reliance on Council Directive 95//18/EC on the licensing of railway undertakings, in so far as such services are provided in Scotland.

36. An airport operator within the meaning of section 82(1) of the Airports Act 1986 (c 31) in so far as it has responsibility for the management of an airport in Scotland.

37. A harbour authority, within the meaning of section 46(1) of the Aviation and Maritime Security Act 1990 (c 31) in so far as it has functions in relation to improving, maintaining and managing a harbour in Scotland.

Health

38. The Common Services Agency established by section 10 of the National Health Service (Scotland) Act 1978 (c 29).

SCHEDULE 2
MINOR AND CONSEQUENTIAL AMENDMENTS AND REPEALS

SECTION 32

PART 1
AMENDMENTS AND REPEALS CONSEQUENTIAL ON PART 1

Civil Defence Act 1939 (c 31)

1. The Civil Defence Act 1939 shall cease to have effect.

Civil Defence Act (Northern Ireland) 1939 (c 15 (NI))

2. The Civil Defence Act (Northern Ireland) 1939 shall cease to have effect.

Civil Defence Act 1948 (c 5)

3. The Civil Defence Act 1948 shall cease to have effect.

Civil Defence Act (Northern Ireland) 1950 (c 11 (NI))

4. The Civil Defence Act (Northern Ireland) 1950 shall cease to have effect.

Defence Contracts Act 1958 (c 38)

5. In section 6(1) of the Defence Contracts Act 1958 (interpretation, &c), in the definition of 'defence materials' omit paragraph (b).

Public Expenditure and Receipts Act 1968 (c 14)

6. Section 4 of the Public Expenditure and Receipts Act 1968 (compensation to civil defence employees for loss of employment, &c) shall cease to have effect.

Local Government Act 1972 (c 70)

7. In section 138 of the Local Government Act 1972 (emergency powers)
   (a) subsection (1A) shall cease to have effect, and
   (b) in subsection (3) for 'subsections (1) and (1A) above' substitute 'subsection (1) above'.

Civil Protection in Peacetime Act 1986 (c 22)

8. The Civil Protection in Peacetime Act 1986 shall cease to have effect.

Road Traffic Act 1988 (c 52)

9. In section 65A(5) of the Road Traffic Act 1988 (light passenger vehicles and motor cycles not to be sold without EC certificate of conformity) omit paragraph (c).

Metropolitan County Fire and Rescue Authorities

10. (1) The bodies established by section 26 of the Local Government Act 1985 (c 51) and known as metropolitan county fire and civil defence authorities shall be known instead as metropolitan county fire and rescue authorities.
(2) So far as necessary or appropriate in consequence of sub-paragraph (1), a reference in an enactment, instrument, agreement or other document to a metropolitan county fire and civil defence authority shall be treated as a reference to a metropolitan county fire and rescue authority.
(3) In the following provisions for '(fire services, civil defence and transport)' substitute '(fire and rescue services and transport)'
   (a) paragraph 29 of Schedule 1A to the Race Relations Act 1976 (c 74),
   (b) sections 21(1)(i), 39(1)(g), 67(3)(k) and 152(2)(i) of the Local Government and Housing Act 1989 (c 42),
   (c) section 1(10)(d) of the Local Government (Overseas Assistance) Act 1993 (c 25),
   (d) paragraph 19 of Schedule 1 to the Freedom of Information Act 2000 (c 36), and
   (e) sections 23(1)(k) and 33(1)(j) of the Local Government Act 2003 (c 26).

PART 2

AMENDMENTS AND REPEALS CONSEQUENTIAL ON PART 2

Emergency Powers Act 1920 (c 55)

11. The Emergency Powers Act 1920 shall cease to have effect.

Emergency Powers Act (Northern Ireland) 1926 (c 8)

12. The Emergency Powers Act (Northern Ireland) 1926 shall cease to have effect.

Northern Ireland Act 1998 (c 47)

13. In paragraph 14 of Schedule 3 to the Northern Ireland Act 1998 (reserved matters) for 'the Emergency Powers Act (Northern Ireland) 1926' substitute 'Part 2 of the Civil Contingencies Act 2004'.

PART 3
MINOR AMENDMENTS

Energy Act 1976 (c 76)

14. After sections 1 to 4 of the Energy Act 1976 (powers to control production and supply of fuel, &c) insert

'**5. Sections 1 to 4: territorial application**
(1) A power under sections 1 to 4 may be exercised in relation to anything which is wholly or partly situated in, or to activity wholly or partly in
   (a) the United Kingdom,
   (b) the territorial sea of the United Kingdom, or
   (c) an area designated under the Continental Shelf Act 1964 (c 29).
(2) Subsection (1) is without prejudice to section 2(2)(b).'

Highways Act 1980 (c 66)

15. (1) At the end of section 90H(2) of the Highways Act 1980 (traffic calming works regulations) add

'(d) provide that, in such cases or circumstances as the regulations may specify, works may be constructed or removed only with the consent of a police officer of such class as the regulations may specify.'

(2) In section 329(1) of that Act (interpretation) for the definition of 'traffic calming works' substitute

'"traffic calming works", in relation to a highway, means works affecting the movement of vehicular or other traffic for the purpose of
(a) promoting safety (including avoiding or reducing, or reducing the likelihood of, danger connected with terrorism within the meaning of section 1 of the Terrorism Act 2000 (c 11)), or
(b) preserving or improving the environment through which the highway runs;'.

Road Traffic Regulation Act 1984 (c 27)

16. (1) The Road Traffic Regulation Act 1984 shall be amended as follows.
(2) In Part 2 (traffic regulation: special cases) after section 22B insert

'**22C. Terrorism**
(1) An order may be made under section 1(1)(a) for the purpose of avoiding or reducing, or reducing the likelihood of, danger connected with terrorism (for which purpose the reference to persons or other traffic using the road shall be treated as including a reference to persons or property on or near the road).
(2) An order may be made under section 1(1)(b) for the purpose of preventing or reducing damage connected with terrorism.
(3) An order under section 6 made for a purpose mentioned in section 1(1)(a) or (b) may be made for that purpose as qualified by subsection (1) or (2) above.
(4) An order may be made under section 14(1)(b) for a purpose relating to danger or damage connected with terrorism.

(5) A notice may be issued under section 14(2)(b) for a purpose relating to danger or damage connected with terrorism.

(6) In this section 'terrorism' has the meaning given by section 1 of the Terrorism Act 2000 (c 11).

(7) [Repealed by the Scotland Act 1998 (Transfer of Functions to the Scottish Ministers etc.) Order 2005 SI no.849 (S.2)]

(8) In Wales an order made, or notice issued, by virtue of this section may be made or issued only with the consent of the Secretary of State if the traffic authority is the National Assembly for Wales.

22D. Section 22C: supplemental

(1) An order may be made by virtue of section 22C only on the recommendation of the chief officer of police for the area to which the order relates.

(2) The following shall not apply in relation to an order made by virtue of section 22C
(a) section 3,
(b) section 6(5),
(c) the words in section 14(4) from 'but' to the end,
(d) section 121B, and
(e) paragraph 13(1)(a) of Schedule 9.

(3) Sections 92 and 94 shall apply in relation to an order under section 14 made by virtue of section 22C as they apply in relation to an order under section 1 or 6.

(4) An order made by virtue of section 22C, or an authorisation or requirement by virtue of subsection (3) above, may authorise the undertaking of works for the purpose of, or for a purpose ancillary to, another provision of the order, authorisation or requirement.

(5) An order made by virtue of section 22C may
(a) enable a constable to direct that a provision of the order shall (to such extent as the constable may specify) be commenced, suspended or revived;
(b) confer a discretion on a constable;
(c) make provision conferring a power on a constable in relation to the placing of structures or signs (which may, in particular, apply a provision of this Act with or without modifications).'

(3) In section 67 (traffic signs: emergencies &c) after subsection (1) insert

'(1A) In subsection (1)
(a) "extraordinary circumstances" includes terrorism or the prospect of terrorism within the meaning of section 1 of the Terrorism Act 2000 (c 11), and
(b) the reference to 7 days shall, in the application of the subsection in connection with terrorism or the prospect of terrorism, be taken as a reference to 28 days;
but this subsection does not apply to a power under subsection (1) in so far as exercisable by a traffic officer by virtue of section 7 of the Traffic Management Act 2004 (c 18).'

(4) In Schedule 9 (reserve powers of Secretary of State, Scottish Ministers and National Assembly for Wales)
(a) in paragraph 1, after 'sections 1, 6, 9,' insert '14 (in so far as the power under that section is exercisable by virtue of section 22C),', and
(b) after paragraph 12 insert

'12A
Article 2 of the Scotland Act 1998 (Transfer of Functions to the Scottish Ministers etc) Order 1999 (SI 1999/1750) shall not apply to a provision of this Schedule in so far as it relates to the exercise of a power under this Act by virtue of section 22C.

12B

A power conferred upon the Secretary of State by this Schedule shall, in so far as it relates to the exercise of a power under this Act by virtue of section 22C, be exercisable in relation to Wales by the National Assembly for Wales with the consent of the Secretary of State.'.

Roads (Scotland) Act 1984 (c 66)

17. (1) After section 39B of the Roads (Scotland) Act 1984 (traffic calming works regulations) insert

'39BA. Prescribing of works for anti-terrorism purposes

(1) (2) [Repealed by the Scotland Act 1998 (Transfer of Functions to the Scottish Ministers etc.) Order 2005 SI no.849 (S.2)]

(3) Regulations under section 39B of this Act may, if they are made for the purpose of, or in connection with, avoiding or reducing, or reducing the likelihood of, danger connected with terrorism within the meaning of section 1 of the Terrorism Act 2000, provide that, in such circumstances as the regulations may specify, works may be constructed or removed only with the consent of a police officer of such class as the regulations may specify.'

(2) In section 40 of that Act (interpretation of sections 36 to 39C) for the definition of 'traffic calming works' substitute

' "traffic calming works", in relation to a road, means works affecting the movement of vehicular or other traffic for the purpose of

(a) promoting safety (including avoiding or reducing, or reducing the likelihood of, danger connected with terrorism within the meaning of section 1 of the Terrorism Act 2000 (c 11)), or

(b) preserving or improving the environment through which the road runs.'

SCHEDULE 3
REPEALS AND REVOCATIONS
SECTION 32

Short title and chapter	Repeal or revocation
The Emergency Powers Act 1920 (c 55).	The whole Act.
The Emergency Powers Act (Northern Ireland) 1926 (c 8).	The whole Act.
The Air-Raid Precautions Act (Northern Ireland) 1938 (c 26 (NI)).	The whole Act.
The Civil Defence Act 1939 (c 31).	The whole Act.
The Civil Defence Act (Northern Ireland) 1939 (c 15 (NI)).	The whole Act.
The Civil Defence Act 1948 (c 5).	The whole Act.
The Civil Defence Act (Northern Ireland) 1950 (c 11 (NI)).	The whole Act.

(*Cont.*)

Short title and chapter	Repeal or revocation
Short title and chapter	Repeal or revocation
The Criminal Justice Act (Northern Ireland) 1953 (c 14 (NI)).	In Schedule 2, the entry relating to the Civil Defence Act (Northern Ireland) 1950.
The Civil Defence (Armed Forces) Act 1954 (c 66).	The whole Act.
The Defence Contracts Act 1958 (c 38).	In section 6(1), in the definition of 'defence materials', paragraph (b).
The Town and Country Planning (Scotland) Act 1959 (c 70).	In paragraph 2 of Schedule 4, the entry relating to the Civil Defence Act 1948.
The Emergency Powers Act 1964 (c 38).	Section 1.
The Lands Tribunal and Compensation Act (Northern Ireland) 1964 (c 29 (NI)).	In Schedule 1, the entry relating to the Civil Defence Act (Northern Ireland) 1939.
The Emergency Powers (Amendment) Act (Northern Ireland) 1964 (c 34 (NI)).	The whole Act.
The Police (Scotland) Act 1967 (c 77).	In Schedule 4, the entry relating to the Civil Defence Act 1948.
The Public Expenditure and Receipts Act 1968 (c 14).	Section 4.
The Land Charges Act 1972 (c 61).	In Schedule 2, paragraph 1(f).
The Local Government Act 1972 (c 70).	Section 138(1A).
The Drainage (Northern Ireland) Order 1973 (SI 1973/69 (NI 1)).	In Schedule 8, paragraphs 3 and 4.
The Statute Law (Repeals) Act 1976 (c 16).	In Schedule 2, in Part II, the entry relating to the Civil Defence Act 1939.
The Road Traffic (Northern Ireland) Order 1981 (SI 1981/154 (NI 1)).	Article 31G(5)(c).
The Civil Aviation Act 1982 (c 16).	In Schedule 2, paragraph 2.
The Criminal Justice Act 1982 (c 48).	Section 41.
The Police and Criminal Evidence Act 1984 (c 60).	In Schedule 2, the entry relating to section 2 of the Emergency Powers Act 1920.
The Fines and Penalties (Northern Ireland) Order 1984 (SI 1984/703 (NI 3)).	Article 12.
The Civil Protection in Peacetime Act 1986 (c 22).	The whole Act.
The Road Traffic Act 1988 (c 52).	Section 65A(5)(c).
The Water Act 1989 (c 15).	In Schedule 25, paragraph 1(4).
The Electricity Act 1989 (c 29).	In Schedule 16, paragraph 1(3) and paragraph 4.
The Police and Criminal Evidence (Northern Ireland) Order 1989 (SI 1989/1341 (NI 12)).	In Schedule 2, the entry relating to the Emergency Powers Act (Northern Ireland) 1926.
The Local Government Finance Act 1992 (c 14).	In Schedule 13, paragraph 6.
The Local Government etc (Scotland) Act 1994 (c 39).	In Schedule 13, paragraph 24.

(*Cont.*)

Short title and chapter	Repeal or revocation
The Gas Act 1995 (c 45).	In Schedule 4, paragraph 2(5).
The Police Act 1997 (c 50).	In Schedule 9, paragraphs 2 and 17.
The Greater London Authority Act 1999 (c 29).	Section 330.
The Transport Act 2000 (c 38).	In Schedule 5, paragraph 3.
The Civil Defence (Grant) Act 2002 (c 5).	The whole Act.

SELECT BIBLIOGRAPHY

(Frequently cited or key texts and including footnote shortforms)

Anderson, CV (ed), *The Federal Emergency Management Agency (Fema)* (Nova Science Publishers, New York, 2003)

Anderson, I, *Foot and Mouth Disease 2001: The Lessons to be Learned Inquiry* (2001–02 HC 888)

Association of Chief Police Officers, *Local Emergency Centres: Good Practice Guide* (ACPO, London, 2004)

—— and Cabinet Office, *Humanitarian Assistance In Emergencies—Guidance on Establishing Family Assistance Centres* (http://www. ukresilience.info/contingencies/facacpoguidance.pdf, 2005)

Audit Commission, *Worth the Risk* (London, 2001)

Babington, A, *Military Intervention in Britain* (Routledge, London, 1990)

Beck, U, *Risk Society* (Sage, London, 1992)

——, Giddens, A, and Lash, S, *Reflexive Modernization* (Polity Press, Cambridge, 1994)

Bingham, T, 'Personal Freedom and the Dilemma of Democracies' (2003) 52 *International & Comparative Law Quarterly* 841

Bonner, D, *Emergency Powers in Peacetime* (Sweet & Maxwell, London, 1985)

British Standards Institution, *Publicly Available Specification 56: Guide to Business Continuity Management* (http://www.bsi-global.com/Fire/Business/pas56.xalter, 2003)

BSE Inquiry, *Report* (1999–00 HC 887) and Government Response to the report on the BSE inquiry (Cm 5263, London, 2001)

Burkhart, FD, Media, *Emergency Warnings and Citizen Response* (Westview Press, Boulder, 1991)

Cabinet Office, *Emergency Planning Review: The Future of Emergency Planning in England and Wales: A Discussion Document* (London, 2001)

——, *Emergency Planning Review: The Future of Emergency Planning in England and Wales: Results of the Consultation* (London, 2002)

'Cabinet Office, *The Draft Civil Contingencies Bill*'—Cabinet Office, *The Draft Civil Contingencies Bill: Consultation Document* (Cm 5843-I, London, 2003)

'Cabinet Office, *The Draft Civil Contingencies Bill: Explanatory Notes etc.*'—Cabinet Office, *The Draft Civil Contingencies Bill: Explanatory Notes, Regulatory Impact Assessment (Local Responders) and Regulatory Impact Assessment (Emergency Powers)* (Cm 5843, London, 2003)

——, *Dealing with Disaster* (revised 3rd ed, London, 2003)

'Cabinet Office, *The Government's Response to the Report of the Joint Committee*'—Cabinet Office, *The Government's Response to the Report of the Joint Committee on the Draft Civil Contingencies Bill (Incorporating the Government's Response to the Report of the House of Commons Defence Committee on the Draft Civil Contingencies Bill)* (Cm 6078, London, 2004)

——, *Civil Contingencies Act 2004: Consultation on the Draft Regulations and Guidance* (Cm 6401, London, 2004)

——, *Handling a Crisis: Lead Government Departments' Areas of Responsibility* (http://www.ukresilience.info/handling.htm, 2004)

'Cabinet Office, *Central Government Arrangements*'—Cabinet Office, *Central Government Arrangements for Responding to an Emergency* (http://www.ukresilience.info/contingencies/conops.pdf, 2005)

'Cabinet Office, *Emergency Preparedness: Guidance on Part 1*'—Cabinet Office, *Emergency Preparedness: Guidance on Part 1 of the Civil Contingencies Act 2004, Its Associated Regulations and Non-statutory Arrangements* (http://www.ukresilience.info/ccact/emergprepfinal.pdf, 2005)

'Cabinet Office, *Emergency Response and Recovery*'—Cabinet Office, *Emergency Response and Recovery* (http://www.ukresilience.info/ccact/emergresponse.pdf, 2005)

——, *Civil Contingencies Act 2004: A Short Guide* (http://www. ukresilience.info/ccact/3octshortguide.pdf, 2005)

Campbell, D, *War Plan UK* (Burnett Books Ltd, London, 1982)

Central Emergency Planning Unit, *Northern Ireland Standards in Civil Protection* (http://cepu.nics.gov.uk/pubs.htm, 2001)

——, *Civil Protection in Northern Ireland: The Implications of the Civil Contingencies Bill* (http://cepu.nics.gov.uk/consult.pdf, Belfast, 2003)

——, *A Guide to Emergency Planning Arrangements in Northern Ireland* (http://cepu.nics.gov.uk/pubs/emerplanarrange.pdf, 2004)

——, *A Guide to Evacuations in Northern Ireland* (http://cepu.nics.gov.uk/pubs.htm, 2004)

——, *A Guide to Plan Preparation* (http://cepu.nics. gov.uk/pubs.htm, 2004)

——, *A Guide to Risk Assessment* (http://cepu.nics.gov.uk/pubs.htm, 2004)

——, *Business Continuity Management in the Public Sector* (http://cepu.nics.gov.uk/pubs.htm, 2004)

Civil Contingencies Bill (2003–04 HC 14) (2003–04 HL 128) and Explanatory Memorandum

'Civil Contingencies Secretariat, Full Regulatory Impact'—Civil Contingencies Secretariat, Full Regulatory Impact Assessment for Part 1 of the Civil Contingencies Bill (Cabinet Office, London, 2004)

'Civil Contingencies Secretariat, *The Government's Response*'—Civil Contingencies Secretariat, *The Government's Response to the Public Consultation on the Draft Civil Contingencies Bill* (Cabinet Office, London, 2004)

'Civil Contingencies Secretariat, *The Lead Government Department*'—Civil Contingencies Secretariat, *The Lead Government Department and its Role* (http://www.ukresilience.info/lgds.pdf, 2004)

Cmar, T, 'Office of Homeland Security' (2003) 39 *Harvard Journal on Legislation* 455

Council of Europe, *Guidelines on Human Rights and the Fight Against Terrorism* (3rd ed, Strasbourg, 2005)

Creighton, WB, 'Emergency Legislation and Industrial Disputes in Northern Ireland' in Wood, JC, *Encyclopaedia of Northern Ireland Labour Law and Practice* (Labour Relations Agency, Belfast, 1983)

Department for Transport, *Local Government and the Regions, Bellwin Scheme of Emergency Financial Assistance to Local Authorities* (London, 2001)

Department of Health Emergency Preparedness Division, *NHS Emergency Planning Guidance 2005* (Department of Health, http://www.dh.gov.uk/assetRoot/04/12/12/36/04121236.pdf, London, 2005)

Department for Homeland Security, *National Strategy For Homeland Security* (http://www.dhs.gov/interweb/assetlibrary/nat_strat_hls.pdf, Washington, 2002)

——, *National Response Plan* (http://www.dhs.gov/dhspublic/interapp/editorial/editorial_0566.xml, Washington, 2004)

Dewar, M, *Defence of the Nation* (Arms and Armour Press, London, 1989)

Disasters Working Party, *Disasters: Planning for a Caring Response* (HMSO, London, 1991)

EADRCC, *NATO's Role in Disaster Assistance* (http://www.nato.int/eadrcc/mcda-e.pdf, Brussels, 2001)

Enarson, E and Morrow, BH (eds), *The Gendered Terrain of Disaster* (Praeger, Westport, 1998)

European Commission, *Vade-mecum of Civil Protection in the European Union* (http://europa.eu.int/comm/environment/civil/pdfdocs/vademec.pdf, Luxembourg, 1999)

Evelegh, R, *Peacekeeping in a Democratic Society* (Hurst, London, 1978)

Ewing, KD and Gearty, CA, *The Struggle for Civil Liberties: Political Freedom and the Rule of Law in Britain 1914–1945* (Oxford University Press, Oxford, 2000)

Executive Session on Domestic Preparedness, *Beyond the Beltway: Focusing on Hometown Security* (John F Kennedy School of Government, Harvard University, 2002)

Federal Emergency Management Agency, *Responding to Incidents of National Consequence: Recommendations for America's Fire and Emergency Services Based on the Events of September 11, 2001, and Other Similar Incidents* (Washington, 2002)

Fennell, D, *Investigation into the Kings Cross Underground Fire* (Cm 499, HMSO, London, 1988)

Franklin, J, (ed), *The Politics of Risk Society* (Polity Press, Cambridge, 1998)

Furedi, F, *Culture of Fear: Risk-taking and the Morality of Low Expectation* (rev ed, Continuum, London, 2002)

Ghosh, TK, Prelas, MA, Viswanath, DS, Loyalka, SK (eds), *Science and Technology of Terrorism and Counterterrorism* (Marcel Dekker Inc, New York, 2002)

Greer, SC, 'Military Intervention in Civil Disturbances' [1983] *PL* 573

Gregory, F, 'The Civil Contingencies Act 2004 and its Place in the Domestic Management of Terrorism in the UK' (2004) 10.01 *Jane's Intelligence Review* 18

Gunningham, N and Grabosky, P, *Smarter Regulation* (Clarendon Press, Oxford, 1998)

Harrison, S, *Disasters and the Media* (Macmillan, Basingstoke, 1999)

Health and Safety Executive, *Disasters Working Party, Disasters: Planning for a Caring Response* (TSO, Norwich, 1991)

——, *Arrangements for Responding to Nuclear Emergencies* (HSE Books, Sudbury, 1994)

——, *Guide to the Pipelines Safety Regulations 1996* (Stationery Office, London, 1997)

——, *Emergency Planning for Major Accidents—Control of Major Accident Hazards Regulations* (HSE Books, Sudbury, 1999)

——, *Guide to the Control of Major Accident Hazards Regulations 1999* (HSE Books, Sudbury, 1999)

——, *Guide to the Radiation (Emergency Preparedness and Public Information Regulations 2001* (HSE Books, Sudbury, 2001)

——, *Hatfield Derailment Investigation* (http://www.hse.gov.uk/railways/hatfield.htm, London, 2002)

Hennessy, P, *The Secret State* (Allen Lane, London, 2002)

Hilliard, L, 'Local Government, Civil Defence and Emergency Planning (1986) 49 *Modern Law Review* 476

HM Treasury, *The Financial System and Major Operation Disruption* (Cm 5751, London, 2003)

Home Office and Cabinet Office, Guidance on Dealing with Fatalities in Emergencies (http://www.ukresilience.info/contingencies/pubs/fatalities.pdf, London, 2004)

——, *The Needs of Faith Communities in Major Emergencies: Some Guidelines* (http://www.ukresilience.info/contingencies/pubs/faith-communities.pdf, 2005)

Home Office, *Report of the Disasters and Inquest Working Group* (HMSO, London, 1997)

——, *Exercise Planners Guide* (http://www.ukresilience.info/contingencies/business/exercise_planners_guide.htm, London, 1998)

——, *The Release of Chemical, Biological, Radiological or Nuclear (CBRN) Substances or Material: Guidance for Local Authorities* (http://www.ukresilience.info/cbrn/cbrn_guidance.pdf, London, 2003)

——, *The Decontamination of People Exposed to Chemical, Biological, Radiological or Nuclear (CBRN) Substances or Material Strategic National Guidance* (2nd ed, http://www.ukresilience.info/cbrn/peoplecbrn.pdf, London, 2004)

Hood, C, and Jones, DKC, (eds), *Accident and Design* (UCL Press, London, 1996)

——, Rothstein, H, and Baldwin, R, *The Government of Risk: Understanding Risk Regulation Regimes* (Oxford University Press, Oxford, 2001)

House of Commons Defence Committee, *Defence and Security in the UK* (2001–02 HC 518) and Government Response (2001–02 HC 1230)

——, *Draft Civil Contingencies Bill* (2002–03 HC 557)

House of Commons Library, *Civil Contingencies Bill* (Research Paper 04/07, House of Commons, London)

House of Commons Science and Technology Committee, *The Scientific Response to Terrorism* (2002–03 HC 415-I) and Government Reply (Cm 6108, London, 2004)

House of Lords Delegated Powers and Regulatory Reform Committee, *Twenty-Fifth Report* (2003–04 HL 144)

——, *Thirtieth Report* (2003–04 HL 175)

Jasanoff, S, *Designs on Nature: Science and Democracy in Europe and the United States* (Princeton University Press, Princeton, 2005)

Jeffery, K, and Hennessy, P, *States of Emergency: British Governments and Strikebreaking since 1919* (Routledge & Kegan Paul, London, 1983)

Joint Committee on Human Rights, *Scrutiny of Bills and Draft Bills; Further Progress Report* (2002–03 HC 1005, HL 149)

'Joint Committee on the Draft Civil Contingencies Bill'—Joint Committee on the Draft Civil Contingencies Bill, *Draft Civil Contingencies Bill* (2002–03 HC 1074, HL 184)

Lakha, R, and Moore, T, *Tolley's Handbook of Disaster & Emergency Management: Principles & Practice* (2nd ed, Butterworths, Croydon, 2004)

Lash, S, Szersynski, B, and Wynne, B, (eds), *Risk, Environment and Modernity* (Sage, London, 1996)

Lasker, RD, *Redefining Readiness: Terrorism Planning Through the Eyes of the Public* (Center for the Advancement of Collaborative Strategies in Health, New York Academy of Medicine, New York, 2004)

Laurie, P, *Beneath the City Streets* (Granada, London, 1983)

Lee, HP, *Emergency Powers* (Law Book Co, Sydney, 1984)

Levy, BS, and Sidel, VW, *Terrorism and Public Health* (Oxford University Press, New York, 2003)

Lewis, J, 'Risk Vulnerability and Survival' (1987) 13(4) *Local Government Studies* 75

Local Government and Public Services Committee, *Report on the Civil Contingencies Bill* (National Assembly for Wales, Cardiff, 2004)

London Emergency Services Liaison Panel, *Major Incident Procedure Manual* (6th ed, http://www.leslp.gov.uk/LESLP_Man.pdf, 2004)

Lord Cullen, *Public Inquiry into the Piper Alpha Disaster* (Cm 1310, London, 1990)

——, *Public Inquiry into the Shootings at Dunblane Primary School on 13 March 1996* (Cm 3386, HMSO, London, 1996) and Government Response (Cm 3392, HMSO, London, 1996)

——, *The Ladbroke Grove Rail Inquiry* (HSE Books, Norwich, 2001)

Lord Justice Clarke, *Thames Safety Inquiry: Final Report* (Cm 4558, London, 2000)

——, *Public Inquiry into the Identification of Victims Following Major Transport Accidents* (Cm 5012, HMSO, London, 2001)

Lord Justice Popplewell, *Inquiry into Crowd Safety at Sports Grounds* (Cmnd 9585, HMSO, London, 1985 and Cmnd 9710, HMSO, London, 1986)

Lord Justice Taylor, *The Hillsborough Stadium Disaster, 15 April 1989* (Cm 962, HMSO, London, 1989)

McDonald Commission, *Report of the Commission of Inquiry Concerning Certain Activities of the Royal Canadian Mounted Police, Freedom and Security* (2nd Report, Ottowa, 1981)

Maritime & Coastguard Agency, *Search and Rescue Framework for the United Kingdom of Great Britain and Northern Ireland* (Southampton, 2002)

Media Emergency Forum Joint Working Party Report, *9/11: Implications for Communications* (http://www.ukresilience.info/mef/mefreport.pdf, 2002)

Mileti, DS (ed), *Disasters by Design* (Joseph Henry Press, Washington DC, 1999)

Ministry of Defence, *The Strategic Defence Review: A New Chapter* (Cm 5566, London, 2002)

——, *Delivering Security in a Changing World* (Cm 6043, London, 2003)

'Ministry of Defence, *Operations in the UK*'—Ministry of Defence, *Operations in the UK: The Defence Contribution to Resilience* (http://www.ukresilience.info/contingencies/defencecontrib.pdf, London, 2005)

Morris, GS, 'The Emergency Powers Act 1920' (1979) *Public Law* 317

——, 'The Police and Industrial Emergencies' (1980) 9 *Industrial Law Journal* 1

——, 'The Regulation of Industrial Action in Essential Services' (1983) 12 *Industrial Law Journal* 69

——, *Strikes in Essential Services* (Mansell, London, 1986)

New Zealand Law Commission Report No 12, *First Report on Emergencies*, Wellington, New Zealand, 1990; Report No 22, *Final Report on Emergencies*, 1991

Nuclear Emergency Planning Liaison Group, *Civil Nuclear Guidance* (Department of Trade and Industry, London, 2004)

O'Brien, TH, *Civil Defence* (HMSO, London, 1955)

O'Malley, P, *Risk, Uncertainty and Government* (Glasshouse, London, 2004)

Office of the Deputy Prime Minister, *Our Fire and Rescue Services* (Cm 5808, London, 2003)

——, *The Financial Management of Local Disasters* (http://www.odpm.gov.uk/embedded_object.asp?id = 1137068, London, 2004)

——, *Guidance on Development of a Site Clearance Capability in England and Wales* (http://www.odpm.gov.uk/embedded_object.asp?id = 1123763, London, 2005)

Peak, S, *Troops in Strikes* (Cobden Trust, London, 1984)

Perrow, C, *Normal Accidents: Living with High Risk Technologies* (Princeton University Press, Princeton, 1999)

Platt, RH, (ed) *Disaster and Democracy* (Island Press, Washington DC, 1999)

Posner, RA, *Catastrophe: Risk and Response* (Oxford University Press, New York, 2004)

Quarantelli, EL (ed), *What Is a Disaster?* (Routledge, London, 1998)

Robertson, B, 'The Defence Act 1990 and Military Assistance to the Civil Power' (1991) 14 *New Zealand Universities Law Review* 254

Rona, G, 'International Law Under Fire' (2003) 27 *Fletcher Forum of World Affairs* 55

Rowe, P, *Defence: The Legal Implications* (Brassey's Defence Publishers, London, 1987)

Rowe, PJ, and Whelan, CJ (eds), *Military Intervention in Democratic Societies* (Croom Helm, London, 1985)

Royal Society, *Risk: Analysis, Perception, Management* (Royal Society, London, 1992)

——, *Making the UK Safer: Detecting and Decontaminating Chemical and Biological Agents* (Royal Society, London, 2004)

Sagan, S, *The Limits of Safety: Organizations, Accidents, and Nuclear Weapons* (Princeton University Press, Princeton, 1993)

Schneider, SK, *Flirting with Disaster* (ME Sharpe, Armonk, 1995)

Scottish Executive Health Department, *National Health Service in Scotland Manual of Guidance: Responding to Emergencies* (http://www.show.scot.nhs.uk/emergencyplanning/guidance.htm, Edinburgh, 2004)

Scottish Executive, *Dealing with Disasters Together* (http://www.scotland.gov.uk/library5/government/dealdisasters.pdf, Edinburgh, 2003)

——, *Civil Contingencies Bill: Scottish Consultation Report* (Justice Department, Edinburgh, 2004)

——, *Review of the Bellwin Scheme—A Consultation Paper* (http://www.scotland.gov.uk/Resource/Doc/69582/0017087.pdf, 2005)

Scottish Office, *Guidance on Preparing for Emergencies* (http://www.scotland.gov.uk/consultations/justice/crgcca-03.asp, 2005)

Sedgwick, M, 'Al-Qaeda and the Nature of Religious Terrorism' (2004) 16 *Terrorism and Political Violence* 795

Shearer, A, *Survivors and the Media* (John Libbey, London, 1991)

Sheen, B, *mv Herald of Free Enterprise: Report of Court No. 8074 Formal Investigation* (Department of Transport, London, 1987)

Simpson, AWB, *In the Highest Degree Odious* (Clarendon Press, Oxford, 1992)

Task Force on Major Operational Disruption in the Financial System, *Report* (http://www.fsc.gov.uk/upload/public/attachments/5/tfreportwholereport.pdf, 2003)

Thessin, J, 'Department of Homeland Security' (2003) 40 *Harvard Journal on Legislation* 513

Thurlow, R, *The Secret State* (Blackwell, Oxford, 1994)

Topping, I, 'Emergency Planning Law and Practice in Northern Ireland' (1988) 39 *Northern Ireland Legal Quarterly* 336

Townshend, C, *Making the Peace* (Oxford University Press, Oxford, 1993)

Uff, J, *The Southall Rail Accident Inquiry Report* (HSE Books, Sudbury, 2000)

Walker, C, 'Constitutional Governance and Special Powers against Terrorism' (1997) 35 *Columbia Journal of Transnational Law* 1

——, and Reid, K, 'Military Aid in Civil Emergencies: Lessons from New Zealand' (1998) 27 *Anglo-American Law Review* 133

——, *A Guide to the Anti-terrorism Legislation* (Oxford University Press, Oxford, 2002)

——, 'Biological Attack, Terrorism and the Law' (2004) 17 *Journal of Terrorism and Political Violence* 175

——, 'Terrorism and Criminal Justice' [2004] *Criminal Law Review* 311

Whelan, CJ, 'Military Intervention in Industrial Disputes' (1979) 8 *Industrial Law Journal* 222

——, 'Political Violence and Commercial Risk' (2004) 56 *Current Legal Problems* 531

Wood-Heath, M, and Annis, M, *The Role of Non-Governmental Organisations' Volunteers in Civil Protection in European Member States & European Economic Area Countries* (British Red Cross, London, 2002)

——, *Working Together to Support Individuals in an Emergency or Disaster* (British Red Cross, London, 2004)

LEGISLATIVE MATERIALS

Parliamentary debates

Legislative stage	Date	Hansard reference
House of Commons		
First Reading	7 January 2004	Vol 416, col 259
Second Reading	19 January 2004	Vol 416, col 1096–184
Committee	27/29 January 2004	Standing Committee F
	3/5/10 February 2004	
Report and Third Reading	24 May 2004	Vol 421, col 1322–411
Consideration of Lords'	17 November 2004	Vol 426, col 1360–416
amendments	18 November 2004	Vol 426, col 1508–18
House of Lords		
First Reading	25 May 2004	Vol 661, col 1200
Second Reading	5 July 2004	Vol 663, col 603–54
Committee	15 September 2004	Vol 664, col 1206–41, col 1259–84
	14 October 2004	Vol 665, col 392–425, col 440–516,
	21 October 2004	col 716–64
		Vol 665, col 929–59, col 989–1054
Report	9 November 2004	Vol 666, col 755–824, 843–80
Third Reading	16 November 2004	Vol 666, col 1319–64
Consideration of	18 November 2004	Vol 666, col 1609–58
Commons' amendments		
Royal Assent	18 November 2004	Vol 666, col 1659

Secondary legislation

Civil Contingencies Act 2004 (Amendment of List of Responders) Order 2005 SI No 2043

Civil Contingencies Act 2004 (Commencement No 1) Order 2004 SI No 3281

Civil Contingencies Act 2004 (Commencement No 2) Order 2005 SI No 772

Civil Contingencies Act 2004 (Commencement No 3) Order 2005 SI No 2040

Civil Contingencies Act 2004 (Commencement) (Scotland) Order 2005 SI No 493 (C 26)

Civil Contingencies Act 2004 (Contingency Planning) (Scotland) Regulations 2005 SSI No 494

Civil Contingencies Act 2004 (Contingency Planning) Regulations 2005 SI No 2042

Scotland Act 1998 (Transfer of Functions to the Scottish Ministers etc) Order 2005 SI No 849 (S 2)

WEB PAGES

Organization etc	URL (add http://)
Air Accidents Investigation Branch	www.aaib.dft.gov.uk
Air Quality Archive	www.airquality.co.uk
Ambulance Service Association	www.asancep.org.uk
Animal Management in Disasters	www.animaldisasters.com/
Association of British Insurers	www.abi.org.uk
Association of Insurance & Risk Managers	www.airmic.com
Audit Commission	www.audit-commission.gov.uk/ emergencyplanning/index.asp
BASICS	www.basics.org.uk
BBC Connecting in a Crisis	www.bbc.co.uk/connectinginacrisis/ index.shtml
Bellwin Scheme	www.local.odpm.gov.uk/finance/ bellwin.htm
Benfield Hazard Research Centre	www.benfieldhrc.org
British Civil Defence	www.britishcivildefence.org
British Damage Management Association	www.bdma.org.uk
Business Continuity Institute	www.thebci.org
Cabinet Office	www.cabinet-office.gov.uk
Center for Disease Control and Prevention	www.cdc.gov
Central Emergency Planning Unit (CEPU)	cepu.nics.gov.uk
Central Sponsor for Information Assurance & Resilience	www.cabinetoffice.gov.uk/csia
Centre for Catastrophe Preparedness and Response, New York University	www.nyu.edu/ccpr
Centre for Research on the Epidemiology of Disasters	www.cred.be
Civil Contingencies Scotland	www.civilcontingenciesscotland.gov.uk
Civil Contingencies Secretariat	www.ukresilience.gov.uk
Community Documentation Centre on Industrial Risk	mahbsrv.jrc.it/Activities- WhatIsCDCIR.html
Continuity Central	continuitycentral.com/
Continuity Forum	www.continuityforum.org
Control of Major Accident Hazards	www.hse.gov.uk/pubns/comahind.htm
Crisis Research Centre	www.cot.nl
Disaster & Emergency Management on the Internet	www.keele.ac.uk/depts/por/disaster.htm
Disaster Action	www.disasteraction.org.uk
Disaster Central	www.disaster-central.com
Disaster Preparedness and Emergency Response Association	www.disasters.org
Disaster Research Center	www.udel.edu/DRC
Disaster Resource	www.disaster-resource.com

Disaster Survivor Support	www.Egroups.com/group/ DisasterSurvivorSupport
Drinking Water Inspectorate	www.dwi.gov.uk/index.htm
Earthquake Engineering Research Institute	www.eeri.org
Earthquake Hazards Program	neic.usgs.gov
Emergency Information Infrastructure Partnership	www.emforum.org/home.htm
Emergency Management Australia	www.ema.gov.au
Emergency Planning College	www.epcollege.gov.uk
Emergency Planning Society	www.emergplansoc.org.uk
Environment Agency	www.environment-agency.gov.uk
Euro-Atlantic Disaster Response Coordination Centre	www.nato.int/eadrcc/home.htm
EUR-OPA	www.coe.int/T/DG4/MajorHazards/ default_en.asp
European Civil Protection Authorities	europa.eu.int/comm/environment/ civil/prote/cp10_en.htm
European Commission, Civil Protection	europa.eu.int/comm/environment/ civil/index.htm
European Crisis Management Academy	www.ecm-academy.nl
European Union Solidarity Fund	europa.eu.int/comm/regional_policy/ funds/solidar/solid_en.htm
Federal Emergency Management Agency	www.fema.gov
Federal Emergency Management Agency Learning Resource Center	www.lrc.fema.gov
Federation Nationale de Protection Civile (France)	www.protection-civile.org
Fire Service College	www.fireservicecollege.ac.uk
Flood Hazard Research Centre	www.fhrc.mdx.ac.uk
Food Standards Agency	www.food.gov.uk
Gender and Disaster Network	online.northumbria.ac.uk/geography_ research/gdn
Government News Network	www.gnn.gov.uk
Hazchem	www.hazchem.freeuk.com
Health & Safety Executive	www.hse.gov.uk
Health Protection Agency	www.hpa.org.uk
Institute of Civil Defence and Disaster Studies	www.icdds.org
Institute of Emergency Management	www.iem.org
Institute of Risk Managers	www.theirm.org
Institution of Occupational Safety & Health	www.iosh.co.uk
Interdepartmental Liaison Group on Risk Assessment (UK-ILGRA)	www.hse.gov.uk/aboutus/meetings/ ilgra/index.htm
International Civil Defence Organisation	www.icdo.org
International Federation of Red Cross and Red Crescent Societies, World Disaster Reports	www.ifrc.org/publicat/wdr
International Journal of Mass Emergencies and Disasters	www.usc.edu/schools/sppd/ijmed

Internet Journal of Rescue and Disaster Medicine	www.ispub.com/journals/ijrdm.htm
Journal of Homeland Security	www.homelandsecurity.org/newjournal/default.asp
Journal of Homeland Security and Emergency Management	www.bepress.com/jhsem/
London Emergency Services Liaison Panel	www.leslp.gov.uk
London Fire and Emergency Planning Authority	www.london-fire.gov.uk/default.asp
London Prepared	www.londonprepared.gov.uk
London Resilience Forum	www.londonprepared.gov.uk/resilienceteam
Major Accident Hazards Bureau (European Commission)	mahbsrv.jrc.it
Major airline disasters	dnausers.d-n-a.net/dnetGOjg/Disasters.htm
Marine Accident Investigation Branch	www.maib.gov.uk
Maritime and Coastguard Agency	www.mcga.gov.uk
Medecins Sans Frontiéres	http://www.msf.org
Meteorological Office	www.meto.gov.uk
Ministry of Defence	www.mod.uk
Mountain Rescue Council	www.bluedome.co.uk/assoc/mrc/mrc.htm
National Chemical Emergency Centre	www.mountain.rescue.org.uk
National Council for Civil Protection	www.britishcivildefence.org/NCP/ncp.html
National Forum for Risk Management in the Public Sector	www.alarm-uk.com
National Infrastructure Security Coordination Centre	www.niscc.gov.uk/niscc/index-en.html
National Search & Rescue Dog Association	www.nsarda.org.uk
Natural Hazards Center	www.colorado.edu/hazards
Office of the Deputy Prime Minister	www.odpm.gov.uk
Office of the Gas and Electricity Markets	www.ofgem.gov.uk/ofgem/index.jsp
Preparing for Emergencies	www.pfe.gov.uk
Public Entity Risk Institute	www.riskinstitute.org/test.php?pid = pubs&tid = 1129
Public Safety and Emergency Preparedness Canada	www.psepc-sppcc.gc.ca/index-en.asp
Rail Accident Investigation Branch	www.raib.gov.uk
Ready Business Campaign (US DHS)	www.ready.gov/business
Red Cross	www.redcross.org.uk
Relief Web	www.reliefweb.int
Risk Support Team (HM Treasury)	www.hm-treasury.gov.uk/documents/public_spending_and_services/risk/pss_risk_index.cfm

Royal United Services Institute	www.rusi.org
Scottish Emergency Co-ordinating Committee	www.civilcontingenciesscotland.gov.uk
Scottish Environment Protection Agency	www.sepa.org.uk
Society of Industrial Emergency Services Officers	www.sieso.org.uk
St Andrew's Ambulance Association	www.firstaid.org.uk
St John Ambulance	www.sja.org.uk
State Veterinary Service	www.svs.gov.uk
Survive: The Business Continuity Group	www.survive.com
Task Force on Potentially Hazardous Near Earth Objects	www.nearearthobjects.co.uk
Technical Rescue Magazine	www.t-rescue.com
Tornado & Storm Research Organisation	www.torro.org.uk
UK Climate Impacts Programme	www.ukcip.org.uk
UK Financial Sector Continuity	www.fsc.gov.uk
UK Resilience	www.ukresilience.info
UN Central Emergency Response Fund	ochaonline.un.org
UN Environment Protection Awareness and Preparedness for Emergencies On a Local Level	www.uneptie.org/pc/apell/disasters/lists/technological.html
UN International Strategy for Disaster Reduction	www.unisdr.org
(UN) ReliefWeb	www.reliefweb.int
United Kingdom Airlines Emergency Planning Group	www.ukaepg.org/aims.htm
United Kingdom Flight Safety Committee	www.ukfsc.co.uk
US Red Cross	www.prepare.org
World Institute for Disaster Risk Management	www.drmonline.net

INDEX

355

Lightning Source UK Ltd.
Milton Keynes UK

171402UK00001B/1/P